'Americans don't win wars anymore. Recent events, to include the ugly end in Afghanistan and the mess in Iraq, confirm this sad reality. But it doesn't have to be this way. Combat veteran and scholar Ben Zweibelson has thought deeply about the subject. In *Understanding the Military Design Movement*, he diagnoses what went wrong, how it got so mixed up, and most importantly, what to do about it. It's past time for Americans to figure out how to win when we are forced to fight. Ben Zweibelson shows us the way.'

Lieutenant General (ret.) Daniel P. Bolger, PhD,
U.S. Army, Retired, USA

'Military Design can be tricky to comprehend. The design movement has just started to march, there has been a missing piece of the design puzzle in the form of a written theoretical backdrop. With this impactful book full of important insights, Dr. Zweibelson as the pathfinder he is, guides and provides academics, researchers, educators, practitioners and students of Design, a vast, broad and highly relevant theoretical foundation in Design which will help us appreciate and navigate in this infinite and complex environment we are surrounded by. This is a book both for the present as well as for the future.'

Lieutenant General Michael Claesson, Commander Joint Forces Command, *Swedish Armed Forces, Sweden*

'War and commerce traveled parallel paths in the 20th century. In this important book, Dr. Zweibelson provides the historical and academic underpinnings of human-centered design—pioneered and scaled most effectively in the commercial sector—and describes in detail how this tool can now be used in military contexts. It's a foundational text; essential reading for everyone working to defend against the adaptable threat posture of our modern world.'

Phil Gilbert, *Retired Head of Design, IBM Corporation*

'Truly a landmark treatise from a master craftsman and practitioner! Dr. Zweibelson not only describes design across time and space but makes a bold and necessary call for more. He pierces the fog and friction of design, capturing its essence while describing its varied growth and application. In laying out this vast landscape he cogently illuminates what is still missing and the deadly temptation to hold on to the stale industrial tools of yesterday. Ultimately, he drives home that our tomorrows depend on what we do today ... to get better at being "comfortable with being uncomfortable," to change how we think and learn, and to bring about that which did not yet exist in order to bring about a better peace.'

Major General Brook J. Leonard, *Chief of Staff, United State Space Command, USA*

'Through this book, Dr. Ben Zweibelson, captures the various roads and obstacles to adopting a Design Thinking Mindset able to tackle the demands of a volatile, uncertain, complex and ambiguous security environment. Ben, a practitioner grounded by operational and field experience enhanced by Academic rigour and foresight, delivers a design thinking journey for any member of the profession of Arms or Academics. Design thinking now forms the basis for modern military thought and our ability to adapt or perish. Those who do not wish to be left behind would gain from reading this book and reflect on simple, complicated, complex and chaos problem solving tools.'

Major General Simon Bernard, *Director-General Military Personnel – (Strategic and Canadian Armed Forces, Ambassador to the Archipelago of Design)*

'From its origins in the Israel Defense Forces, military design thinking has developed over the past two decades into a vibrant global intellectual movement that has gained an ever-growing cohort of adherents through its promise to disrupt and overhaul inflexible habits of thought within the armed forces and create the conditions for the perpetual innovation necessary in a tumultuous and dangerous world. As its leading proponent and intellectual contributor, Ben Zweibelson has with *Understanding the Military Design Movement* delivered a comprehensive and essential account of its history, fundamental tenets, and future prospects. An indispensable reference work for all military designers and readers interested in creativity in war.'

Antoine Bousquet, *Associate Professor, Swedish Defence University, Sweden*

'The modern military industrial command and control complex has been characterized as a total system of networked communications, material technologies and service personnel whose internal coherence is designed to occupy a designated space (a market, a nation state, a war zone, a region, the stratosphere) utterly, and completely. Ben Zweibelson's book offers a sustained and provocative inquiry into the reasons why such a strategy and design is flawed. Yet by breaking down common epistemological assumptions (power need not come from control, nor control from power) Zweibelson goes further in his analysis. He offers a compelling account of why it is nearly all attempts at imposing a pre-planned design from above, no matter how well intentioned, will fail. Zweibelson provides his readers with alternative design strategies, ones that no longer rely on the cybernetic myth of perfect and transparent system feedback, ones whose effectiveness and force rests with qualities of agility, sensitivity, invisibility and even humility. In the wake of recent military events, it seems to be imperative to attend to Zweibelson's arguments with due seriousness: not because they offer neat and complete solutions, but precisely because they refuse them.'

Robin Holt, *Copenhagen Business School, Denmark*

'As new challenges such as threats from non-state transnational actors or the complexities of pan-domain command and control continue to emerge and evolve, we must remain ever-resolved to continue critical self-reflection of how we think and make decisions, and the search for new ways. Dr. Ben Zweibelson's *Understanding the Military Design Movement* does exactly that by capturing his insights uncovered through years of obsessive review and fearless consideration of what design theories, models and methods can bring to military operations. This is a must read for every military professional, defense academic and conflict armchair quarterback that knows there must be a better way.'

Brigadier General Kevin Whale, *Deputy Commanding General –*
Transformation, at Space Operations Command, US Space Force, USA

'This is a fascinating new work for all studying the art of making war. In a refreshing approach, *Understanding the Military Design Movement* focuses on cognition and explores important issues related to "how" to think about making war, rather than didactically advocating "what" to think. This is a work of considerable sophistication and nuance and which, perhaps surprisingly for a work in this genre, grips the reader. The author's passion leaps out and brings a sense of excitement that ensnares. This book will be a primary text for newcomers to military design but equally be a work that experts will want to return to, each time finding fresh insights and visions of what should be. The book offers much for military professionals, defence policymakers, thinktanks, staff colleges, academics and all concerned with understanding how to think about applying military power in the modern world.'

Peter Layton, *Griffith Asia Institute, Griffith University, Australia*
and the Royal United Services Institute, UK

'Ben Zweibelson provides a bold attempt at creating something that is needed but does not yet exist. That is, a bold attempt at giving justice to the emerging field of design for defence and security purposes across selected NATO members and partners. Zweibelson excels in tentatively charting the uncharted: he brings forth a philosophy of knowledge to better locate and position this field in contrast to traditional planning and commercial design. If only for this reason, this book will become a seminal reference. Yet this book is so much more. It is a foundation manifesto giving credibility to design for defence and security purposes as a full-fledged field of study and practice.'

Philippe Beaulieu-Brossard, *Canadian Forces College and*
Co-Executive President of the Archipelago of Design, Canada

'Teaching Design and applying it in security contexts, to those few who held both positions, is a recurring cycle of triumphs and failures. Despite appearing to be a common practice within whole of government organizations recently, the path of design in defense was paved by individuals who, regardless of institutional sponsoring, invested their totality in the intellectual, political and organizational advancement of the field. Picking upon where the pioneers of military design left off, there is no other person more identified with the spread of security design globally than Dr. Zweibelson, and for good reasons. A prolific scholar, passionate educator, capable operator and skillful diplomat, Ben Zweibelson has positioned himself favorably in writing *Understanding the Military Design Movement* to capture the essence of the security design movement, both theoretically and practically, through a comparable evaluation of and reflection on the changing generations of military design; and, the transcontinental military cultures that adopted it. An instrumental study that will undoubtedly solidify that field once and for all.'

Ofra Graicer, *Co-Founder and Instructor, the Israeli Defense Force's Generals' Course*

'Ben Zweibelson is among the leading theorists of this important design field. He also works as a practitioner of military design, first as a field grade officer in the United States Army, now as a senior civilian in the United States Space Force. If you think about design in today's complex and complicated world, *Understanding the Military Design Movement: War, Change, and Innovation* will challenge you to think about design in significant and useful ways.'

Ken Friedman, *Chair Professor of Design Innovation Studies, College of Design and Innovation, Tongji University, China*

'Unless we know where have been, it is impossible to chart an effective war forward. With *Understanding the Military Design Movement*, Ben Zweibelson does exactly that. The acknowledged leader of the global military design movement, Ben leads us through the military design journey from where we were to where we are today and in doing so provides us a vision of the way forward. His exhaustive examination of "why design" and "how to design" should be read and understood by all leaders, public and private sector, who desire to take their organization to the next level.'

Colonel (ret.) James K. Greer, PhD, *Former Director of the U.S. Army School of Advanced Military Studies (2001–2003), Assistant Professor at SAMS 2019–present*

'Ben Zweibelson tells an important story about a concept that continues to influence military doctrine AND actual operations. This is a superb book.'

Colonel (ret) Kevin C.M. Benson, PhD, *Former Director of the U.S. Army School of Advanced Military Studies (2003–2007)*

'The world seems to be an increasingly a confusing place, where the usual tools and mindsets do not seem to work any longer. Those who seek new ways of understanding complexity, and experiment with unconventional methods will most likely thrive. In Ben Zweibelson's must-read book, he does not just explain how and why design theory can help us in finding the way in the contemporary complex environment, but also provides a glimpse of what the future holds. His book is not only an inspiration to military planners but a useful guide to all leaders, academics, researchers, and practitioners of Design who are looking for the new normal.'

'Veteran and security design expert, Dr. Ben Zweibelson has written a must-read analysis on how and why design theory constitutes the evolution of modern military thinking. This deeply researched book provides a compelling narrative about the challenges of applying postmodern design theory to the western military world, while also offering an invaluable account on the military design movement's origins and implementation across the countries which spearheaded this concept. Zweibelson's unparalleled experience as a security design student and team leader clearly comes out in this superb book, which is going to inspire future military leaders and planners.'

Understanding the Military Design Movement

This book explains the history and development of the military design movement, featuring case studies from key modern militaries.

Written by a practitioner, the work shows how modern militaries think and arrange actions in time and space for security affairs, and why designers are disrupting, challenging, and reconceptualizing everything previously upheld as sacred on the battlefield. It is the first book to thoroughly explain what military design is, where it came from, and how it works at deep, philosophically grounded levels, and why it is potentially the most controversial development in generations of war fighters. The work explains the tangled origins of commercial design and that of designing modern warfare, the rise of various design movements, and how today's military forces largely hold to a Newtonian stylization built upon mimicry of natural science infused with earlier medieval and religious inspirations. Why does our species conceptualize war as such, and how do military institutions erect barriers that become so powerful that efforts to design further innovation require entirely novel constructs outside the orthodoxy? The book explains design stories from the Israel Defense Force, the US Army, the US Marine Corps, the Canadian Armed Forces, and the Australian Defence Force for the first time, and includes the theory, doctrine, organizational culture, and key actors involved. Ultimately, this book is about how small communities of practice are challenging the foundations of modern defence thinking.

This book will be of much interest to students of military and strategic studies, defence studies, and security studies, as well as design educators and military professionals.

Ben Zweibelson is a retired U.S. Army Infantry Officer with over 21 years' service including multiple combat tours in Iraq and Afghanistan. He is the director for the U.S. Space Command's Strategic Innovation Group, and previously educated design, innovation and strategic change for the U.S. Special Operations Command. He has a PhD from Lancaster University (UK) and lectures worldwide to defense organizations and militaries.

Routledge Studies in Conflict, Security and Technology

Series Editors: Mark Lacy, *Lancaster University*, Dan Prince, *Lancaster University*, and Sean Lawson, *University of Utah*

The *Routledge Studies in Conflict, Technology and Security* series aims to publish challenging studies that map the terrain of technology and security from a range of disciplinary perspectives, offering critical perspectives on the issues that concern publics, business and policymakers in a time of rapid and disruptive technological change.

Information Warfare in the Age of Cyber Conflict
Edited by Christopher Whyte, A. Trevor Thrall, and Brian M. Mazanec

Emerging Security Technologies and EU Governance
Actors, practices and processes
Edited by Antonio Calcara, Raluca Csernatoni and Chantal Lavallée

Cyber-Security Education
Principles and policies
Edited by Greg Austin

Emerging Technologies and International Security
Machines, the state and war
Edited by Reuben Steff, Joe Burton and Simona R. Soare

Militarising Artificial Intelligence
Theory, technology and regulation
Nik Hynek and Anzhelika Solovyeva

Understanding the Military Design Movement
War, change and innovation
Ben Zweibelson

For more information about this series, please visit: www.routledge.com/ Routledge-Studies-in-Conflict-Security-and-Technology/book-series/CST

Understanding the Military Design Movement

War, Change and Innovation

Ben Zweibelson

Routledge
Taylor & Francis Group

LONDON AND NEW YORK

First published 2023
by Routledge
4 Park Square, Milton Park, Abingdon, Oxon OX14 4RN

and by Routledge
605 Third Avenue, New York, NY 10158

Routledge is an imprint of the Taylor & Francis Group, an informa business

© 2023 Ben Zweibelson

The right of Ben Zweibelson to be identified as author of this work has been asserted in accordance with sections 77 and 78 of the Copyright, Designs and Patents Act 1988.

Trademark notice: Product or corporate names may be trademarks or registered trademarks, and are used only for identification and explanation without intent to infringe.

British Library Cataloguing-in-Publication Data
A catalogue record for this book is available from the British Library

ISBN: 978-1-032-48178-4 (hbk)
ISBN: 978-1-032-48179-1 (pbk)
ISBN: 978-1-003-38776-3 (ebk)

DOI: 10.4324/9781003387763

Typeset in Bembo
by Newgen Publishing UK

For Ethan, Jack and Luke.

May your generation escape the cognitive traps of past ones.

Contents

Preface

This book is a result of how wickedly complex life is for most all of us. Nothing seems to go according to plan. If that were the case, I would be penning a book about my adventures as an astronaut/firefighter/artist/soldier, which was my original plan around ten years of age. My father, Irving Zweibelson, was a World War II veteran, fine artist, and university professor. In the end, the apple would not fall far from the tree. As a child, I developed some unique gifts in fine art, mostly drawing and such. Family, friends, and teachers strongly urged me to invest entirely in the arts, to the point that by my early twenties in a fine arts program at university, I was burned out. Due to family and financial issues, I also needed to find a way to pay for college. My roommate Garrett took me to the Somers Town Fair in Connecticut, and I met a U.S. Army National Guard recruiter. The rest, they say, is history. Well, not without critical support from the Kritzman family and the Hourihan family, whom I am forever indebted to, and proud to be part of their extended families.

I ended up in the infantry, a highly regimented, no-nonsense sort of entity that did not take kindly to the creative, free-spirited, often sarcastic private I was at the time. For over a decade as enlisted and later an infantry officer, I struggled in that I enjoyed the thrills, adventure, and comradery of the Army Infantry world, but my ever-present interest in challenging things, reconfiguring them, and generating imaginative ideas outside of the norms got me in lots of trouble. Several times in my career, I faced the firing squad in terms of bad evaluations, frustrated commanders, and worse. At the half-way point of my officer career, I got the unusual opportunity to attend the U.S. Army School of Advanced Military Studies (SAMS) at Fort Leavenworth, Kansas. This was just at the tail end of the great military design initial experience, a few months after Shimon Naveh was banished from the campus, and the directors changed over. I learned about military design in the classroom and through my SAMS advisor, Alice Butler-Smith, and it opened conceptual doors I had been struggling to pry open using fine arts tools previously. Quickly, Butler-Smith, and also Alex Ryan (both SAMS professors) introduced me to Naveh, and piled me up with forbidden design books gathering dust in the bowels of the Leavenworth library, all stricken from reading lists in the institutional reaction to the early and wild military design days there. The heretical, counterculture, radical design was intoxicating then, and now unavoidably wired into how I understand reality.

I first published on design while at SAMS, and it caused Chris Paparone, then the Dean of the U.S. Army Logistics University, to contact me and introduce me to Lieutenant Colonel Grant Martin, a SAMS graduate, design thinker, and Special Forces officer. The three of us developed a powerful friendship centered on military design that has us engaging every week, often daily, since 2010. Paparone exposed me to sociology, piling on thousands of more articles and books to combine with Naveh, Ryan, and Butler-Smith's demands that I read postmodern philosophy, complexity theory, systems thinking, and military strategy. These important intellectuals set me off on the adventure in military design that resulted in this book. "Pap" and Grant remain my most trusted design journeymen, perpetually the "Three Musketeers" of military design that hopefully never give up the fight.

After graduating SAMS, I took to writing and research as I continued my military career, which continued to land me either with accolades by some leaders, or in the penalty box with stern finger wags by others. Eventually, I took the academic route, retiring and seeking a doctoral program focused on design philosophy. Along the way, I met and befriended a wide array of international and Department of Defense military designers, educators, theorists, and the occasional crackpots. They all would contribute to this book, with many screening multiple versions of chapters, continuously patient with me. I am deeply indebted to them in ways this foreword is insufficient to ever convey.

Commercial designers tend to wrinkle their noses when they encounter military designers, and often are the fiercest skeptics that militaries need any sort of design outside of what is readily available in industry and academia. Phil Gilbert, former head of design for IBM, was an amazing leader and design sparring partner for me over the years, along with various designers such as Adam Cutler within IBM. Ken Friedman, editor of the *She Ji* design journal, is another deep design thinker without whom the first chapter would not have been possible. Commercial designers such as Donna Dupont, Harold Nelson, Cara Wrigley, Candice Luck, and Nathan Schwagler greatly inspired me throughout the last decade plus. Their influences ring through across this book, including many arguments that likely remain unresolved! I relish further engagements, if only for the next book.

Postmodern philosophers interested in military design are the rarest of a rare breed, and I am fortunate to have met and learned from some of the best. Shimon Naveh and Ofra Graicer are by far the best military postmodern thinkers and educators on this planet, and their teachings and friendship flow through every page of this book. Philippe Beaulieu-Brossard and Philippe Dufort are inspirational civilian postmodern scholars who discovered this strange military community of practice, and now chair the largest non-profit, international community of design educators, theorists, and practitioners.

Their zeal and intellect also runs through these pages, and is the glue binding this strange community of practice together. Special operators Ken "the Redman" Poole, Kenny Cobb, Todd Landis, Brett Bourne, Jim Wetzel, Bogdan Gieniewski, Manny Diemer, and so many more across the U.S. Special Operations Command helped create the perfect design studio for this research to develop. Anders Sookermany, Karena Kyne, Mathieu Primeau, Paul Mitchell, Aaron "Jacko" Jackson, Jim Greer, Jeffrey van der Veer, Travis Homiak, Murray Simons, Jason "TOGA" Trew, Aubrey Poe, Major General Simon Bernard, Major General Brook "Tank" Leonard, Tim Day, Brigadier General Imre Porkolab, Lieutenant General Michael Claesson, and so many others contribute to this broad military design community, and to my understanding of this military design movement. There are dozens more to list, and I deeply apologize on any omissions here.

There are a select few deserving of special mention. George "Hal" McNair, my former director at the Joint Special Operations University for seven years, was a champion, defender, and visionary unlike any other. This book would also not be possible without his leadership and trust. Mark Lacy, my doctoral advisor at Lancaster University, mentored and helped me through this demanding journey. My wife, Jill, demonstrated amazing patience, kindness, and encouragement throughout not just this book, but everything life throws at us.

Ben Zweibelson, Colorado Springs, December 1st, 2022

An Introduction to the Military Design Movement

For most of humanity, "design" is a powerful yet perplexing, even paradoxical word. We associate it with a vast range of methods, disciplines, specialized fields, as well as enterprises to create, improve, and develop the world for some sort of advantage or improvement that did not exist *prior to the design activity*. Humans have designed toward individual and group improvement well before recorded history, with the inventions of cave dwelling, control of fire, and the earliest tools as good examples of design. Slightly more challenging yet deeply significant is the design of language, art, religion, politics, and other social constructions that deeply shape reality.

Early design endeavors emerge from telltale clues excavated from long-lost encampments, with anthropologists and historians attempting to fill in the many gaps. Unknown human designers created the first burial sites and rituals, as well as original cave art demonstrating a uniquely human ability to represent the world conceptually and communicate those abstractions to others beyond the original limits of time and space. With every act of innovation, a human mind was designing the world anew; often early designs would occur over and over to gradually nudge societies forward. In each act of design, people would challenge established shared beliefs and group knowledge regarding "how the world is and functions" to usher in yet another new and transformative way to express human activity and existence. Some designs are small and subtle, while others radically alter entire societies and belief systems. They all address novelty, experimentation, risk, and curiosity.

We return to a simple yet profound question: *what is design?* When humans create change in both the naturally complex world and that of a socially constructed world maintained atop the natural one, they are designing. Designers operate in a "domain of the thinker thinking about complexity"[1] so that they are intentionally bringing novelty into the world through design activities. This is where design brings forth a "newness" that offers emergent and distinct opportunities, experiences, or advantages that were unimagined, unrealized, and unexplored *before the design*. This differs design activities from routinized planning actions that extend yesterday's known into today's relevance. Planning is also essential for humanity, yet it addresses the preservation of yesterday's best knowledge and practices into today … whereas design

DOI: 10.4324/9781003387763-1

pulls tomorrow's unimagined opportunities into today's voids that are unfulfilled by yesterday's ideas and content.

When people talk about design today, they usually mean the formalized and modern industrialized, often technologically enhanced design practice. This designing employs methods, models, as well as language and metaphoric devices therein to produce some transformation of the future toward a stakeholder's novel advantage. One may design toward commerce, whether industrial design, architectural design, or perhaps human–centered design to create new user experiences or new products for the modern consumer. There are countless books already written on myriad commercial design methods, techniques, and deliverables. Many design theorists have already pondered these different methods and design communities, including the overlap of industrial design endeavors with the increasing design appetite for state-sponsored instruments of warfare. Yet designing tools of war is not the same as designing within war itself, as this book will explain. War is unfortunately a powerful and enduring social activity of humanity, where earlier premodern designs up through modern industrial designs have directly contributed, developed, and profited. This brings us to the focus of this book and a new design movement that differs with the dominant community of designers oriented toward commercial applications.

While the broader term "design" acts upon transforming the future toward a stakeholder's novel advantage in some context or setting, in security affairs since the late 1990s there is an emerging branch of designing differently. Here, military designers consider the application of organized violence in some capacity to *transform future security contexts into novel advantage for the design benefactors.*[2] Alongside the iterations of various design activities, the specific history of human warfare also features designs where military leaders, theorists, and practitioners enact theories to corresponding models to produce novel methods of warfare. The first chapter frames these tensions, and explains how designing for commerce and designing toward warfare are overlapping, but also in tension, and within the last decades, increasingly in some hybrid interplay that enables what each individual is incapable of accomplishing.

Particularly since the Industrial Revolution, commercial designers build increasingly sophisticated war tools for defense organizations. At the same time, security professionals within defense contexts seek to wield many of these designed products or new tools so that one exercises the application of organized violence in an advantage that transforms the future toward desired goals. Thus, there is designing toward warfare as well *as designing in war.* This is also where not only the many disciplines and fields of designers get confused, but also where modern military designers confuse their own institutions on what "military design" might be, or how to integrate it with strategic plans down to tactical actions. The military design movement is new, distinctive, and requires a full investigation so that defense organizations, their national policymakers, and also commercial enterprises and design academia grasp these distinctions and differences.

This book will address, clarify, and answer several contemporary problems within the military community on formally recognizing military design as a movement. While "design" writ large is framed in this introduction and in the firstchapter, a treatment of how modern militaries think and act is also required so that the last three decades of military design development can fit within the entirety of human war history. Thus, the first and second chapters form the foundation for why military design sprung into existence in the mid-1990s in the Israeli Defense Force, and subsequently how and why those ideas spread to American, Canadian, Australian, and select other communities and locations through the last three decades.

Commercial design is entirely suitable for the design requirements and needs of any military within the channels of commerce. An army will always need a newer tank, a fleet of advanced drones, or a superior rifle. Militaries need a host of user experience improvements, and the defense industry has an insatiable appetite for new products and developments related to warfare. Yet designing toward commerce is insufficient to address how security organizations must design toward transforming war, as well as how they conceptualize conflict and engage in war itself. Designing security transformation will rarely be some linear accumulation of the sum of new designed war tools and products, nor will any commercial designer be expected to anticipate the dynamic, complex security settings where a security force must imagine and improvise. The military designer must do this with people, tools, and conditions of dangerous (even extreme, existential) violence where transformative outcomes are designed to achieve some anticipated future security advantage. In the chaos of war, a designer is not reliant upon pre-established plans or war tools designed for them alone; the design of war itself is also necessary and demanding different, often emergent concepts and experimental methods. Change changes change, and in complex security contexts, a military unable to design is destined to drift aimlessly through the currents in a reactive orientation.

Thus, military design requires definition and explanation. It should not be framed as just another military methodology or model that employs language and heuristic aids to validate theories and act upon complex reality … we must go deeper. It is at the intersection of designing for commerce (where most design disciplines, fields, and communities of nuanced practice occupy), and the increasingly formal and active community of defense designers where they are addressing new areas of design thought, opportunity, as well as risk in security contexts. To design is human, and humans unfortunately wage war as one violent and destructive consequence of social, political, ideological, cultural, and economic interactions among populations, states, and groups. To design in warfare is also decidedly a human enterprise. I do not condone war in this respect, nor do I admire it. For us to appreciate how humans design broadly, we cannot avoid the difficult and perhaps unpleasant aspect of considering designing in war.

The rise of modern design methods and praxis will be addressed first in this book, so that the intricate overlaps, interplay, and tensions between commercial design and military design are understood and mapped through key developmental periods over the last few centuries. The first chapter creates the foundation for understanding why commercial design alone cannot continue uninterrupted into direct military design requirements in most applications outside of specific technical or service requirements. The second chapter will also showcase how designing in commerce and designing in warfare have a long and storied history of mutual interdependence, inspiration, as well as many shared attributes and behaviors that need only be illuminated with some analysis and history. Commercial designers are far more tangled into the same fabric of complex human society that includes war than many might realize. This answers the question of "why is there a military design, when we already have established commercial design?"

Through this foundation of commercial and security design analysis, we will then examine how and why the modern military and state security apparatuses used informal, amateurish design toward warfare up through the Feudal Age and into the European Renaissance in the second chapter. Western societies would enable the modernization of militaries through a range of important and interrelated developments including the Age of Enlightenment, the rise of natural sciences, the Industrial Revolution, and the political, social, and economic upheavals of the last several centuries. Throughout this volatile, transformative period, military organizations became professionalized, modernized, and scientifically oriented. They became managers of organized violence toward state-on-state engagements through increasingly sophisticated, technological, and eventually existential dimensions of war. Modern warfare also encapsulates any equivalent security expression of violence by non-state entities (done in the same state-on-state orientation and style regardless). This focus is thus western and Euro-centric, suggesting more research is necessary beyond these areas. Militaries designed informally prior to the 1990s, but why a formal design community formed at this point is essential for understanding the surge of military design theory, experimentation, and attempts at doctrine.

This book frames this military transformation away from a premodern military form and function to that of a modern, professionalized force. Modern defense organizations would design toward new ways of warfare in highly technological and managerial models, complete with technical and natural science-inspired metaphoric devices. These narratives and belief systems would subsequently shape all military doctrine, language, methods, and practices through present day. As state-on-state warfare expanded into global, increasingly devastating world wars, militaries would remain tightly wedded to conventional, Newtonian styled models, planning methods and indoctrinated belief systems of military institutions. Modern societies designed a specific and well-defined *way for war to occur*, yet these state-imposed rules for order and rules of warfare would offer only fleeting order and stability in the aftermath of the first atomic weapons employed against adversaries in war. Design of war would continue, requiring new ways of thinking, and thinking about that thinking.

The first military design in a formal sense originates from the Israeli Defense Forces and is the focus of the third chapter. In the mid-1990s, frustrated military intellectuals were granted resources and time to conceptualize a new, different, and entirely disruptive logic for war that was antagonistic in its design to what the institution already possessed. This radical design movement began small, and encountered fierce institutional resistance in Israel, yet still spread to other nations. Chapters four and five detail this leap from Israel to the American Army, and also the U.S. Marine Corps. Both organizations would develop their own versions of a military design methodology, and the path of these forms and their functions would deviate strongly from the original Israeli ideas. Chapter six provides the second part of the Israeli story, occurring a decade after these first transitions into international militaries, and denotes a maturation of the original military design ideas over time and through significant experimentation. The seventh chapter provides insights into the Canadian, Australian, and, to a lesser degree, the broad international interest in military design over the last decade. Across the globe today, allies, partners, and adversaries appear poised to generate their own design interpretations within their military forces. How and why this is transpiring forms the purpose of this book, and the Conclusion attempts to forecast where things might move next.

Throughout this still developing and emerging military design movement some thirty years later, many stakeholders and design actors have overlapped from commercial and other design disciplines into warfare applications, with multiple "tribes" of designers often unable to recognize or communicate with others designing upon the same landscape of organized violence. Some security design methodologies draw from select aspects of commercial design theory and models, while others pull heavily or even exclusively from complexity theory and/or systems theory while jettisoning things like postmodern theory. Some militaries insist on what they term an "operational design" as an overarching framework of traditional military campaigns reliant upon established terms and doctrine. This is done so that subordinate and nested planning endeavors can proceed in an integrated and comprehensive fashion by an architected design for defense concerns. Today there seems to be an "archipelago of designers" across various services and military institutions, and whether they can converge as a formal profession remains an unanswered question. Many tensions abound that are unique to *military* designers due to the form, function, and purpose of such practitioners.

With the military design community, there remains significant disagreement. Some design purists seek to extend the original Israeli ideas (steeped in postmodern theory, complexity theory, and other eclectic disciplines) into deconstructing and transforming entire military paradigms and belief systems; these unorthodox security professionals are often decried as "heretics" and "nomads" by critics and opponents defending the military institution. Others call for uniformity, standardization, and a clear order to military design so that innovation works like a checklist or recipe that can be formulated into military doctrine and training. This book seeks to untangle this Gordian knot of designers across the vast landscape of various design communities of practice.

Motive to Change: Why Consider Military Design?

Designing in security applications refers specifically to the military enterprise of generating novel and innovative warfare conceptualizations, prototypes, and novel activities to gain advantage in a complex and emergent context *where established methodologies and practices are no longer sufficient alone.* A military thus needs innovation in the form of a design so that that they can gain a novel advantage to accomplish goals that were unreachable, unimaginable or unrealized prior to the design activity. More confusingly, much of complex warfare is chaotic, emergent, and impossible to even anticipate at the start of a conflict. This suggests that perpetual designing must occur throughout any conflict timeline, including well before any sort of competition or deterrence fails. This is so one might navigate uncharted spaces to discover unimagined opportunities that were hidden at the beginning of defense action.

Horst Rittel, one of the earliest design theorists and covered in the next chapter, posited that designers are "need fulfillers". Yet these needs will range from designing fulfillment of existing needs to an unimagined, unexpected, or unrealized innovation that designers must bring into reality for the most deadly, dangerous, and chaotic of human affairs. They must generate fulfillment of an emergent, transforming need that is beyond the horizon of the organization's expectations, and thus must convince a skeptical stakeholder of the new design value as it is emerging. Thus, we have fulfillment of needs that are clear, historically validated, and wanted, but also fulfillment of new needs that are unlike the known ones. New needs may seem foreign, alien, or unrelated to what we feel we want instead. This is difficult for designers to envision, and perhaps doubly difficult to execute and then convey to an organization that may resist the change. Although there is nothing like a crisis to open minds to radical, even heretical ideas, we should not make it a habit of sending militaries into harms' way equipped with only institutionalized ways of thinking and acting. Militaries need to be able to design not just to survive, but to transform the future in ways that cannot be understood or recognized using yesterday's constructs alone. This is most readily apparent in the shifting design of what "victory" is through organized violence.

Militaries today appear to be in a greater need of design applied to security contexts, if only based on the significant increase in military design literature, academic discourse, doctrinal manifestations, as well as educational and case study examples to be presented in the third through sixth chapters. Yet within security forces and across the civilian and commercial design world few theorists have been able to explain design *outside of their own school, field, or institution*. There remains a fierce debate on what "design" means, whether designing in warfare is indeed distinct from other designing, and within militaries whether one or multiple ways to design is useful or not. Design theorist Buchanan provides a useful summary of why design is such a challenging term to define across a range of practitioners as well as diverse design contexts:

> Despite efforts to discover the foundations of design thinking in the fine arts, the natural sciences, or most recently, the social sciences, design eludes reduction and remains a surprisingly flexible activity. No single definition of design, or branches of professionalized practice such as industrial or graphic design, adequately covers the diversity of ideas and methods gathered together under the label.[3]

As stated earlier, design is ultimately *concerned with change*; the need to prepare for today's demands while appreciating that the tools one used yesterday may be not only useless tomorrow, but also irrationally favored to prevent the discovery of new ones.[4] Krippendorff even proposes a formal design philosophy that is epistemologically informed of various contextual frames in which design is practiced, yet within his proposed meta-framework for human-centered design discourse he does not consider design in security contexts.[5] He does leave the door open by promoting a dynamic design philosophy that continues to critically reflect and transform itself through continued discourse. Thus, design methods and models might change depending on client needs, and a defense client with specific security needs might spawn design innovation of an entirely different orientation to that of commercial considerations. This also positions the Conclusion for exploring where this military design community of practice might travel to next.

Another tension between security design as a manifestation of creative thinking on war and that of convergent, established military planning practices (established strategy making and operational decision-making) is the difference between efficiency and effectiveness. Convergent, systematic processes feature a preference toward incremental gain in efficiencies so that clearly defined goals and objectives as originally crafted can be reached faster and with lower strain on resources and organizational interest. Modern militaries fixate upon this in how they make strategies, form plans, and organize to act, as the second chapter will demonstrate. Systematic logic functions with inputs linked to clear outputs, and where linear-causal relationships work mathematically, even mechanically to sequence discrete and reducible activities across time and space to lead toward overarching objectives and goals. However, the act of becoming more efficient does not correlate directly with becoming *more effective* in complex systems; effectiveness frequently requires innovation, imagination, and a willingness to break away from the very practices that an organization might be attempting to improve efficiency with. This creates significant paradox for militaries concerning whether to design or plan.

This is the first of several core tensions between how militaries plan security activities drawing from a historic and analytic orientation (repeating yesterday's successes), and how security forces seek design praxis through invoking a systemic and imaginative, forward-facing perspective (imagining a tomorrow unlike anything understood today). In commercial contexts the rivalry or range of choices between various design methodologies, schools of thought, disciplines, and even trademarked design models is largely understood through commercial competition or specialization of a design. Yet in security contexts, militaries do not select one design methodology over another for those reasons. Instead, as the second chapter will outline, militaries prefer one design methodology over others for service cultural reasons, institutional or doctrinal adherence, organizational and also paradigmatic compliance purposes. Military organizations manifest as centralized hierarchies driven by self-interests and in their zeal to create their own design methodologies to indoctrinate throughout their members, they produce yet another serious design tension. These tensions come to light in the Israeli, American, Canadian, European, and Australian armed forces in this book. Subsequent chapters present how security organizations place identity and exclusivity of concepts, language, and warfare models in a more significant and institutionally self-referential way than commercial design counterparts.

Why might a security force seek to create a particular way to do design in war, and how might these methods and models present dissimilar as well as overlapping (potentially counterproductive as well) consequences in modern, complex warfare? Military forces frequently adhere to a single and usually service-created mode for designing in warfare that is not often tailored to a particular design field, focus, or outcome as commercial design communities do. Military design is expected to address all design needs, in all military design contexts, with complete compatibility to other military decision-making methods. Additionally, in numerous cases the chosen design methodology attempts to accommodate instead of disrupting most contemporary military planning methods based on culture, identity, and belief systems upheld by a military force. They frequently draw from the same institutionally structured language, models, as well as epistemological and ontological underpinnings as well. Unlike commercial design methodologies and models where difference and paradox may help highlight why one type of design community might be better suited to address certain challenges, many security organizations create and use design methods that reinforce institutional norms at the expense of disrupting, dismantling, or challenging them.

More radical forms of military design thus face greater adversity in acceptance and use due to this core tension triggering both confusion and institutional defense of one over the other. Militaries tend to seek increased stability, uniformity, reliability, and continuity as these are cherished qualities in the chaos and uncertainty of the modern battlefield. Designs that threaten institutionally sacred ideas, values or tenets on war appear to generate adversaries within the military itself on whether design is useful or not. Some vocal critics within military communities even reject both the word "design" and any hint of a formal design methodology outright when considering future activities within security contexts.[6] They operate cognitively from an efficiency-based orientation that demands better iterations of the contemporary planning or strategy-making; the argument frequently used is that planning is "proven to work" while any design action that ushers in experimental ideas, new language, or unfamiliar war applications are "unproven" and inherently less likely to succeed.

War is changing, and modern societies have political and military leadership that struggle with how armed forces ought to adapt and change in order to sustain some level of relevance and competence. When nations employ military forces in the application of organized violence, the "playing field" of warfare is expected to be properly managed, rationalized, and contained to ensure necessary perceptions of control, management, risk, and causality in accordance with internationally agreed upon state-level "rules for warfare". Warfare in this century is supposed to look and feel like the last few centuries of war, despite technological and social developments. These analytic-oriented, hierarchical structured militaries of the industrialized, democratic West continue to apply a rational decision-making frame toward ideally similar opponents, particularly in the establishment and maintenance of state war plans, contingencies, and modern constructs such as "great power competition" among top western adversaries. Warfare designs continue to promote concepts, beliefs, and patterns held in high regard due to historical qualities, relation to military identities, as well as some ritualization into unassailable stances on war itself. There are proposed "rules of war" that all competitors are expected to follow, otherwise risk is framed in the starkest negative tone. Yet such security designs grow increasingly inappropriate and insufficient for confronting non-hierarchical, networked, and divergent threats that are increasingly familiar in twenty-first century competition. This is precisely where the more disruptive, radical forms of military design offer the most chance for transformation as well as the strongest reason for institutional resistance.

Few contemporary adversaries whether Westphalian nation-state or non-state actor are willing to play by these war rules[7]; the game has changed regardless of how much technological, analytical, or political capital is invested to retain the older system for organized violence built upon rather mechanistic, linear, and systematic logics of planning. This military design movement uses various forms and combinations of design praxis to disrupt, deconstruct, and transform thought and action in the application of organized violence. Postmodern theorist Michel Foucault, who would likely find any military utilization of postmodern ideas as an unexpected and perhaps unwelcome development, still upheld the supreme importance of thinking differently and across multiple conceptual frames at what is an ever-changing, complex and ultimately human designed reality. He declared that "discourse in general, and scientific discourse in particular, is so complex a reality that we not only can, but should approach it at different levels and with different methods".[8] Any attempt to construct a science of warfare thus demands a design critique and approach that transcends any singular scientific paradigm. This theme of disruption, paradox, and transformation of the modern military using a design approach that differs from commercial enterprise will be explored in all of the chapters, building upon foundations established in the first and second chapter treatments.

Whether this is better understood within a reframing of organized violence because of society entering postmodernity, or that some postmodern theory is now needed to design in war differently by disrupting, deconstructing, and radically altering the modern war frame is left both to the reader and the ongoing internal debate across the diverse design communities. In these areas of still unanswered or underdeveloped security design praxis, many research opportunities abound. Postmodern critique is controversial, yet the cliché of "all is fair in love and war" still must stretch to address how we conceptualize what war is, and what we posit it is not … and why this is. This book will detail the extensive investment many military designers have made into postmodern ideas, as well as robust resistance and rejection by other military designers, broader institutions, and military academia. The design movement continues to develop, from Sydney to Toronto, within the American Army and in North Atlantic Treaty Organization (NATO), and in the coming decade, likely this movement will gain the attention of allies and adversaries alike in how it is disrupting the modern war paradigm.

Design remains a universal construct created within human minds, and enacted upon a complex reality. *Homo sapiens* use design to make sense of reality, and extend the already complex natural world into a human-generated, socially constructed reality beyond those physical limits. Within this second order of existence, societies design and maintain war. War, a human design, was thrust upon a world that prior to the rise of humanity had no such phenomena of organized violence paired to concepts of politics, identity, culture, and belief systems. Since premodern design of original war, the human species have invested greater thought, technologies, resources, and collective purpose into war so that today it continues to manifest and require further designs, whether for peaceful or destructive motives. Militaries previously designed in a variety of ascientific, and later scientifically inspired ways. Modern militaries now are on the edge of another transformation in the twenty-first century, where advanced technology, new planes of human existence such as cyberspace, quantum space, and space itself now unavoidably become areas of new war contemplation. Military designers represent the custodians of innovation, heretics of institutionalism, and the visionaries of what is otherwise unimaginable or unrealized tomorrow. They may carry the keys to future safety, or bleakly, the ultimate horrors of future wars not yet waged. This book explains who they are, why they now take a formalized, professional form, and what function they provide that did not exist previously for defense forces.

Notes

1 Haridimos Tsoukas and Mary Jo Hatch, "Complex Thinking, Complex Practice: The Case for a Narrative Approach to Organizational Complexity," *Human Relations* 54, no. 8 (August 2001): 980.
2 A design benefactor in defense contexts might be the state, population, particular group, as well as other entities or individuals in a wide range of scale, scope, and context.
3 Richard Buchanan, "Wicked Problems in Design Thinking," *Design Issues* 8, no. 2 (Spring 1992): 5.
4 Karl Weick, "Drop Your Tools: An Allegory for Organizational Studies," *Administrative Science Quarterly* 41 (1996): 301–313; Klaus Krippendorff, "Principles of Design and a Trajectory of Artificiality," *Product Development & Management Association* 28 (2011): 416.
5 Klaus Krippendorff, "Propositions of Human-Centeredness; A Philosophy for Design," in *Doctoral Education in Design: Foundations for the Future: Proceedings of the Conference Held 8–12 July 2000, La Clusaz, France* (Staffordshire (UK): Staffordshire University Press, 2000).
6 Milan Vego, "A Case against Systemic Operational Design," *Joint Forces Quarterly* 53 (quarter 2009): 70–75.
7 Sean McFate, *The New Rules of War*, First Edition (New York: William Morrow, 2019).
8 Michel Foucault, *The Order of Things: An Archaeology of the Human Sciences*, Vintage Books Edition, April 1994 (New York: Vintage Books, 1994), xiv.

1 Designing Commerce, Designing War

Of Chickens, Eggs, and Hand Grenades

Over the last decade, militaries and commercial enterprises discovered how designing toward commerce and that of war are far more intertwined in history, goals, and beliefs than previously assumed.[1] Design is most readily associated with innovation, change, and novelty which are prioritized in how humans engage in changing their world around them for advantage in both areas of focus. War and commerce both require the manufacture of goods, and the design of things as well as new conceptualizations for how to transform reality. These may seem different, but frequently commercial design overlaps directly with design of warfare. Indeed, both are symbiotic, and whether designers are disturbed or enlightened by this is irrelevant for the purposes of this chapter. How design formalized into a modern profession cannot be fully explained without engaging in both sides of this design equation.

This chapter illuminates the real and historical interdependence of war and commerce with design, from practical, prescientific origins in Medieval guilds and artisans to booming consulting firms that today sell slick design brands and weekend "boot camps" that grant design certificates. Over the last century within the tensions that produced today's booming societies and industrialized nations, the modern design profession emerged. This occurred within political and social developments that spurred new design abilities into instruments of military power. This led to the design of new wars on a terrifyingly larger scale than ever before. Commercial designers would rise to their current and well-defined forms through the invisible pull of economic growth and trade that is impossible to separate from national security demands. Industrialization and the horrific capacity for organized violence by humanity is sadly an ever-present element of our species today. Informally yet in the same period of modernization, militaries too would navigate through the dangerous jabberwocky of organized violence for national interests and aspirations. Militaries would design, innovate, reflect, adapt and reframe. Yet along the way, both communities would often lose sight of the other, assume their own design theory, models, or methods were sufficient and uniform for either context, and misunderstand the other in a growing chasm of "us" and "them" concerning design, theory, practice, and institutional identities.

DOI: 10.4324/9781003387763-2

Today, many that identify as a design community member align with a particular tribal methodology such as "I design with Agile" or "we use the IDEO approach" and so on. At the same time inside of militaries, a similar tribal design identity is taking root, but often from different designs than that of commercial inspiration. Designers riding their respective beasts now collide with one in often confusing exchanges of incompatible terminologies, dissimilar methods, as well as overarching goals that do not coincide. There is scant evidence of multidisciplinary design, and even less common are examples of design collaborations between commercial and military groups where a mixing of theory and practice is sought outside of preferred (often singular) design perspectives. Today's multidisciplinary landscape of designers has in many respects isolated themselves into warring tribes on independent islands, so much that the largest military design community uses the metaphor "archipelago" for their title.[2] This book attempts to frame how and why this happened, and this chapter provides the fertile soil from which military and commercial designers grew from, albeit in quite different directions.

A conservative window for when design in some modern, organized and recognized sense started coincides with early industrialization efforts in the seventeenth through eighteenth centuries. Artistry existed well before this, and design specialization first oriented toward products and mass production, whether for tools of commerce or tools of war. Yet it would not be until the twentieth century where we see the emergence of formal schools and design communities of practice complete with unique language, methods, culture, and philosophies on design. This coincides with the rise of modern management theory, the rapid growth of factories, urban sprawl, public education, and many significant social reforms where universal standards and formal regulations stimulated specialization and certification of who is recognized as a designer, and who is not. Designer Victor Papanek framed this with:

> the ultimate job of design is to transform man's environment and tools, and by extension, man himself. Man has always tried to change himself and his surroundings, but only recently have science, technology, and mass production made this more nearly possible.[3]

Although the military profession is often separated from civilian or commercial ones due to their unique roles and contexts for action, this chapter explains the development of parallel and often overlapping design paths for commercial and military professions. One essential epistemological tension does exist between the commercial/civilian design domain and that of security affairs and must be highlighted accordingly. By *epistemological*, this addresses the study of knowledge itself and how/why people interpret reality as they do. Designers must be able to consider epistemological constructs as those form the way to understand how and why one school or community of design thinks and acts as it does, and how other design groups also approach designing in profoundly different ways. Thus, thinking about our thinking becomes a serious goal for designers attempting to consider not just the what and how of designing, but also the *why*.

The word and ideas behind "design" mean all sorts of things to many people, blurred across a vast landscape of disciplines, fields, groups, and communities of practice. Bruce Archer, in attempting a "science of design" primarily for architectural, industrial, and commercial design defined it as follows: "to make the plans and drawings necessary for the construction of" and "to fashion with artistic skill or decorative device".[4] Commercial designer Kees Dorst, in further abstraction, posits design as a process that dates back to our earliest intelligent ancestors and represents the urge "to consider a situation, imagine a better situation, and act to create that improved situation".[5] Buckminster Fuller offers that design can be "a weightless, metaphysical conception or a physical pattern … when we say there is a design, it indicates that an intellect has organized event into discrete and conceptual inter-patterning … the opposite of design is chaos".[6]

There is much divergence in how design integrates with other decision-making endeavors such as planning or management of non-novel, routinized endeavors. Richard Boland and Fred Collopy address the creative risk and managerial strategy of design as different from traditional planning as follows: "the design attitude appreciates that the cost of not conceiving of a better course of action than those that are already being considered is often much higher than making the 'wrong' choice among them".[7] Thus, design involves some aspect of risk, experimentation, and creative ideation beyond traditional or established norms and orthodoxies while planning reinforces such institutionalized behaviors and patterns. Planning orients toward extending the best knowledge and techniques from yesterday into today's challenges, while design in turn attempts to improvise, explore, and innovate toward tomorrow's emergent uncertainties so that novel advantages and abilities are ushered into today.

Thus, designing in the broadest sense is about creating something new that provides some advantage to the user or organization, yet the essential requirements to create this "newness" is paradoxical to routine planning activities and requires entirely dissimilar ways of thought and action.[8] One could even suggest that planning itself is a subroutine of designing, in that all planning actions at some point are first designed, and subsequent changing of the planning method remains a deliberate act of design. Humanity requires both designing and planning. As societies interact commercially, culturally, socially, economically as well as through policy and application of instruments of state power including military forces, the design of commerce and the design of war (including the prevention therein) toward future transformative applications are unavoidably interdependent. People must plan to design, and design to plan. It is iteratively woven into virtually all human activities from the highest pursuits of human achievement to the most terrifying and tragic of organized violence and most everything in between.

An unusual quality of design writ large is that tribalism tends to isolate groups that see design in one way from all others that differ. Frequently, one learns a design methodology and joins that particular community, with incentives and rewards to continue to remain adherent to that design. Few designers branch into a multidisciplinary practice, with fewer still considering design at abstract, philosophical levels. In systems theory, "bounded rationality" is when people attempt to make rational decisions but are relying upon unreliable, oversimplified, or ill-informed knowledge of what ultimately is a complex and unknowable, dynamic system.[9] Many designers advocate their own design worldview unaware of other design theories, models, or communities of practice. Wittingly or unwittingly, they may discount such concepts as irrelevant, unnecessary, or in violation of their own core design values and beliefs. This is noticeable in commercial and military design communities that seem quite unaware of one another.[10] People like to form groups, and within these collectives, they usually form identities and belief systems that frame "us" and "not us" in clear, bounded ways. Designers do this just like any other profession.

In this quagmire of competing belief systems on what design is, how it is practiced, as well as the potential turf wars of single-discipline design advocates against challengers, many design groups still use the same language, techniques, theories, and models. Krippendorff stresses the importance of language as well as this potential incommensurability between various design tribes. To design effectively, designers must be able to "realize the limits of their own language" by exploring beyond their own paradigm, and entertaining alternative language as well as being quite open to "the need to create new language" and retire outdated or insufficient terminology currently maintained by the organization.[11] Indeed, the design community often experiences a "Tower of Babel" in design communication, which spans academia, industry, and also security organizations entering the design mix. Rarely do groups acknowledge design limitations, while numerous insist their particular design approach is better than the rest.

The Rise of a Modern Design

Broadly speaking, the modern interpretation of "design" developed largely within the twentieth century after multiple military organizations in both World Wars struggled with complexity and increasingly difficult challenges with scale, size, and emergent conditions that come with advanced industrialized societies.[12] Earlier eighteenth and nineteenth century industrial design movements were necessary precursors, yet still largely wedded to earlier apprenticeship and ascientific or localized modes of professionalization. Over the last 300 years of rapid industrialization and technological advancement, design expanded from artisan or localized processes into formalized, modern enterprises, with new demands made utilizing technological, social, and political developments. War would follow suit, with the myriad and sophisticated war plans of increasingly technological forces spanning multiple time zones and strategic objectives requiring new design cohesiveness.

World War I is often considered the first nation-state *total war* involving the mobilization of entire societies toward the complete destruction and defeat of other nation-states. Siniša Malešević argued it also was the last war of the Feudal Era. "The medieval warrior legacy remained firmly entrenched such that by the beginning of the twentieth century most European states were still ruled by the landed aristocracy".[13] This compliments how to illustrate the design shift from earlier, pre-industrial forms into a modern construct, complete with commercial and warfare needs. Societies as well as warfare transitioned away from earlier agrarian, smaller-scale, and slower existences toward dynamic modern ones; the design of modern tools of war and the demands of nation-states would create the need of industrialization itself as the first modern design schools.

The rapid industrialization of western societies created a vibrant, socially mobile reality sustained by economic growth, scientific progress, and increasingly powerful bureaucratic militaries. Industrial design methods, schools, and communities of practice sustained these new nation-states, provided them the weapons and capacity to wage vastly more violent wars, and the appetite for winning said wars would enrich industrial design practice further. The end of the nineteenth century would also be the origin for mechanistic doctrine governed by irreducible, law-like relationships within which the first wave of modern designers would think with.[14] Theory and decision-making in commercial enterprise as well as politics and war would assume scientific, hard-science attributes, models, language, and methods. Modern design would begin with formulaic categorization and a preference to emulate natural science methods, models, and metaphoric devices.

This modernization of western societies expanded virtually all aspects of earlier national interests, from economics to health, education, urban migration as well as far more sophisticated security apparatuses and abilities to protect this new prosperity. Giddens, in defining the consequences of modernity, placed this relationship "between military power and industrialism, one main expression of which is the industrialization of war" as a primary dynamic for societal advancement.[15] Designing for industry and designing for industrialized warfare would become implicitly integrated into the rapidly expanding nation-state as premodern societies gave way to the modern Westphalian Order. In terms of all contemporary commercial design, the direct lineage from artisans to sprawling industrial factories directly integrated into society, war, and design praxis in the twentieth century.

On the security affairs side of things, design as a term in pure military contexts has been loosely associated with various military modes of thinking and organizing throughout the twentieth century. The word is used in many ways, depending on what period, service, doctrine, or theorist one examines. Some military scholars today interpret "design" retrospectively so that earlier operational planning and strategy might qualify throughout the twentieth century (or earlier), in that the management of multiple theaters and campaigns in war seen in World War II might therefore be one way to correlate design with military activities. However, this is not a suitable nor comprehensive definition of design for several significant reasons. In most cases of declared design action in military theory and doctrine in the twentieth century, the term is synonymous with the design or construction of military campaigns and major operations toward strategic/political goals and not toward some transformation or novel innovation to gain future advantage in war.

This use of the term in military contexts often implies a hierarchical and integrated form of military planning where the design is the abstract and highest level of strategic thought and political correspondence, with all subsequent planning subordinate to the military design. Doctrinally, this first occurred in the early 1980s where the U.S. Army would critically challenge many long-held beliefs on warfare following their strategic defeat in Vietnam, although they also would whitewash key aspects of military history in order to reframe various narratives on how to win against the Soviet Union.[16] The U.S. Army accomplished this through the revision of foundational military doctrine. The highly influential 1982 edition of the American Army's Field Manual 100-5 first introduced German and Soviet recognition of operational art and the "design" of campaigns and operations.

This interpretation of designing warfare would, as described by Kelly and Brennan, "spread through the Anglophone world like a virus ... [With] minor variations in spelling, the same definitions had appeared in British, Canadian, and Australian doctrine by the early years of the 1990s where they remain relatively unchanged to this day".[17] Thus, for much of recent modern military doctrine and theory, to design is to be associated with the creative organization of well-structured military campaigns arranged in time and space for grand strategy or overarching national goals in war. One is expected to design in war to strategize or plan operationally, so that warfare designs sit atop the entire centralized hierarchy for how all military affairs are conceptualized and managed. Militaries would not seek to formalize beyond this immediate adherence to a singular, mechanistic mode of decision-making in war until decades later when such a design frame proved insufficient.

Modern military thought has since the 1980s paired the design of military campaigns to broad, cohesive war strategies, while only in the late 1990s would a new form of design emerge to challenge not just the conceptualization of campaigns, but virtually everything a military force thinks or does. The association of designing military campaigns requires no further abstraction from the modern military frame concerning warfare, while the new military design as a meta-methodology, philosophy and creative ethos requires far more abstract thinking coupled with critical self-reflection beyond institutionally sanctioned norms and beliefs. This tension between thinking systematically about war (breaking things down, isolating, arranging into discrete, causal relationships orchestrated by universal laws) and that of a systemic framing of complex warfare would by the mid-2000s produce two military design camps in opposition. One would demand abstraction and the deconstruction of all established orders, while the other would insist upon design only through institutionally approved, orderly forms and functions to enable subsequent planning and action.

Militaries, industry, and in some ways, academia have misunderstood, ignored, or simply forgotten many of the linkages, overlaps, as well as divergent paths that different design manifestations undertook through the last two centuries of human enterprise and development. People often do not know how they arrived at some current framework of beliefs, practices, and methods, and are also largely unaware of their interdependence on seemingly unrelated fields and disciplines. Military organizations are not alone in their confusion over how design functions within security affairs and organized violence. As commercial design enterprise and academia have only recently turned focus to military design activities in the last few years, this is also a strong indication that modern commercial designers are unaware of their own historic origins with that of modern security affairs and warfare.

Commercial Assimilation of the Military Industrial Mode: Designing Products

Did modern warfare usher in industrial design and all subsequent manifestations, or did the Industrial Revolution create the ability for nation-states to wage modern warfare expressed in a scale and scope unprecedented in human history? Did one come first, and then the other? Is industrial design the father of modern warfare, or did modern warfare encourage the rise of industrial design? This becomes the chicken and egg argument that likely is impossible to resolve. The best answer is that both occurred together with significant codependence. The influence of military modes of organizing and acting would accelerate the industrialization of war, and subsequently project these successful decision-making and design efforts back into the non-military fields of human enterprise. Malešević supports this notion as follows:

> it is crucial to emphasize that these industrial techniques originated, and were developed first and foremost, for military purposes ... [and] from the second half of the nineteenth century onwards, science, technology, administrative organization and the military power of nation-states became so integrated and interdependent that now it is almost unthinkable that they would operate independently.[18]

This highlights an important area of mutual consideration for designing toward commerce and that of designing toward warfare.

While Malešević was discussing industrialization in the twentieth century writ large, the design methodologies therein are addressed in greater detail by commercial design theorists across six decades of theory, research, and experimentation.[19] Industrial design represented the first modern form of "design" according to Papanek, with subsequent human-centered design (HCD) methods branching from this first form in the 1950s–1960s.[20] However, most of these design theorists throughout the 1950s–2000s never mention the military or any commercial design origin in modern war as relevant to the design story.[21] There appears to be a collective dismissal of any military-commercial origin story, which begs the question of why commercial designers might object to any direct correlation with the designing of war with commerce. Could military clients, as well as the nation-state wielding instruments of military power, be detrimental or perhaps even institutionally unattractive to the ethos, belief systems, or values that comprise a commercial design community of practice? Or perhaps do designers seek to sidestep the violent ways that newly designed war tools/products might be used after handed to the client? This in some regard might be true. Yet this alone cannot explain the commercial detachment from design and war.

Some argue that much of civilian enterprise from academia to many disciplines of industry at some philosophical level harbor this sentiment. Malešević proposed a casual rejection of war and violence as the necessary precursor for nearly all industrialization and modern bureaucratic form and function.[22] Chris Gray reinforced this modern sociological or policy "blind spot" of acknowledging the post-1945 industrial military complex where technology, science, academia, and industry are deeply integrated into creating greater and greater weapons of war for the military client.[23] Organized violence accelerated industrialization for humanity in part due to existential threat of terrifying new weapon designs that could eliminate *Home sapiens* as a species. Failure to modernize toward "total war" national industry for organized violence put a society at the mercy to those willing to do so. In return the Industrialized Age would usher in modern "total war" collective violence unlike anything previously seen. Today, commercial, civilian design communities practice design that directly stems from a commerce–war relationship, yet contemporary design may no longer recognize the linkages. The best starting point for these strange bedfellows occurred in the fertile interwar period between World Wars I and II.

The first modern design educational enterprise started in Germany in 1919 and was called the *Staatliches Bauhaus*, more commonly referred to as "Bauhaus" or "school of building" when translated from German.[24] Originally an eclectic home for combining industrialized crafts with new styles and ideas from the fine arts disciplines, it would blossom into a powerful industrial design movement that included architecture, graphic design, advertising, interior design and above all, an intellectual growth for creative design expression.[25] There are some examples of earlier industrial design education, yet the Bauhaus was arguably the first formally structured and implemented as a modern design school with a recognizable form of knowledge, practice, education and collective vision. Further, their origin as part of the reaction and reflection of how horrid and destructive the first

"Great War" was provides a crucial aspect of the theme offered here. The eventual demise of the Bauhaus school is also significant here historically and in terms of how and why creativity becomes a desired and feared commodity in both commerce and warfare. With the Bauhaus, the overlap and interplay between military institutions and commercial design enterprise is a striking example of this perpetual symbiotic relationship between designing for war and designing for commerce.

This first design school's founding and *raison d être* came in the aftermath of World War I, where German society stood disillusioned and dejected by 1917.[26] The first world-wide expression of "total war" was made possible by an expansion of design thinking in how to produce not just the technologically advanced tools of war, but why employing them in new ways could generate different outcomes than seen in any earlier battlefield. Tragically, total war is totally destructive on a scale and scope that utterly shatters previous understandings of organized violence. Of the many German military officers that survived the war only to return to a broken society, Walter Gropius would in his own intellectual soul-searching develop the vision to establish this radical and disruptive school of commercial design.[27] Gropius wrote about his war experiences and a desire to: "start building [his] life anew;" that in the chaos that was the interwar period Germany, he hoped "that through a new art, a new order could be created. Gropius called for a unification of the arts".[28] He would in reaction to the horrors of World War I lead the establishment of the first modern school of design oriented toward transformation of society away from such chaos and pain.

Not only would war bring about the conditions to create the first modern design school that featured theory, education, experimentation, and a cohesive ethos, but political, social, and military fallout as a reaction to World War I would also later cause the closure of it. Bauhaus occupied the interwar period of 1919–1938 almost exactly, inhabiting a difficult and increasingly tense position in German society as a disruptive, counterculture, and source of innovation as well as controversy in an unstable and chaotic period. The German Nazi regime in 1933 forced the closure of the Bauhaus school due to an association of the school with communist intellectual thoughts as well as the disruptive, critical, and influential influence of the Bauhaus community of practice upon European society.[29] New social Marxists (also termed critical theorists) established themselves in Germany due to direct frustration with the war's outcomes, as well as how Marx's revolution occurred not in industrialized London or Berlin, but the largely agrarian, pre-industrial Russia.[30] Despite many of the Bauhaus designs significantly impacting international (as well as Nazi societal and industrial) products and industrialized effects, the creative energy associated with such innovation troubled Nazi leadership. The Nazis elimination of the Bauhaus would provide a telling example of a modern military bureaucracy recognizing design education as a threat to their power despite a competing need for innovation, creativity, and different ways to think and act. Those that dare think differently and inspire new designs come from the same soil that potentially threatens the stability and dominance of the current establishment itself.

Despite the closure of the Bauhaus in 1933, the radical ideas concerning a modern exploration of design with art, multiple new theories, and the need to scale design education to a larger and more structured format would live beyond the shuttered schoolhouse. This Bauhaus movement would deeply influence subsequent commercial as well as military design developments, especially after 1945. This birth of modern design challenged societal norms, reframe belief systems and conceptual frameworks in a ripple effect spanning well beyond the reach of Nazi oppressors. Totalitarian and oppressive regimes would learn that one might burn many books ... only to have new books appear and replace those censored or obliterated. The movement could not be silenced. Heretics and radical thinking, if stimulating and inspirational, can live on well after someone is burned at the stake or a school shuttered. The Bauhaus movement promoted an "anti-academic arts school" andragogy for fusing avant-gardist strategies with emergent technology and mixed disciplines.[31] Their interest in ideas considered controversial, anti-establishment, or hostile to those holding power would be both dangerous and intoxicating. Design educators and advocates of the Bauhaus school would exit Germany and in the 1930s–1950s open similar schools across Europe and North America. The match was lit, and while an emerging "Cold War" would increase international tensions between the American and Soviet nuclear superpowers, a modern design movement was formalized into an international community of practice.

In 1953, Max Bill, Ingre Aicher-Scholl, and Olt Aicher founded the Ulm School of Design or *Hochschule für Gestaltung* in West Germany, drawing inspiration from the earlier Bauhaus efforts. The Ulm School would only last a little over a decade but would extend the earlier Bauhaus design philosophy forward while moving into deeper theoretical and methodological explorations, liberated from the earlier oppressive Nazi regime and now beyond the grasp of Soviet oppression. Horst Rittel, a professor at the Ulm School, would subsequently emigrate to California to introduce the foundations of HCD thinking to the University of Berkeley and elsewhere. Rittel would help usher in the next major commercial design movement and scale it considerably in American academia through the 1960s and 1970s. His academic design work would, along with a handful of other design pioneers in the late 1950s and 1960s, establish the foundations of modern design theory and practice.

All these commercial design educational developments would link back to the merging of the twentieth century's industrialization, total-war application of organized violence, and the societal reactions to changing cultural, technological, and societal belief systems. Yet the direct linkages of organized violence and the role of creative design thinking would also become decoupled in this same period. Design movements after 1945 were unlike the interwar period movements in that Bauhaus practitioners drew heavily from their wartime experiences, while there is little evidence of post-World War II designers explicitly extending their wartime experiences into their design school identities, purposes, or inspirations. Perhaps in the post-World War II era, despite the rising threat of nuclear annihilation and a new "Cold War" construct, people sought to return to civilization and distance themselves from the horrors of war that had essentially plagued two generations of society through the 1945. Or, after two devastating world wars the surviving societies found international "total warfare" incompatible with modern international markets, shared values, and increased interdependence on other nations and regions for prosperity. Either way, modern design would accelerate and break away from earlier war interdependency. Commercial design would also forget these earlier interdependencies.

During the twentieth century and particularly after World War II, a generation of military professionals as well as those in the war-production industry would continue industrial design practice in expanding fields of design application well beyond the focus of organized violence, although external companies designing war products and experiences would grow into the "military industrial complex" too. Ideas on strategy as well as large-scale planning would migrate from the military over to industry as thousands exited military service after 1945, with an expansion of military ideas surging in the 1950s in commercial settings for the first time.[32] The exodus of wartime draftees back into civilian enterprise would inject military models, methods, and decision-making directly into commercial industry in the 1940s–1960s; concepts such as "strategy" would enter businesses in formal, modern constructs that were matured in the wide-scale complex management of organized violence in World War II. Industry existed before both World Wars and accelerated after each of them, but the exchange of people, ideas, and behavior patterns would loop back and forth between commerce and war as one influenced the goals and desires (and designs) of the other. Design thinking became mainstream in architecture, products and services, advertising, and urban planning in the post-war world.[33] Military veterans schooled in warfare would quickly fill the ranks of industrial, professional, and academic leadership roles, influencing the strategic and organizational directions of entire disciplines and sectors. Here, the design of new products and user experiences set in an industrialized, technology-centered economy would mature and expand the application of design thinking well into the twentieth century.[34]

Perhaps due to changing societal values and contemporary belief systems, militaries were (and often still are) viewed as oppositional to certain company or community values, or contemporary political and cultural shifts make such associations undesirable. There is clear history of state apparatuses taking on the direct censorship of design movements, such as the Nazis shutting down the Bauhaus design schools within Germany in 1933. The Soviet artistic movement known as *Proletkult* emerged like the Bauhaus after World War I and peaked in 1920, only to be terminated by Lenin over fears of the designers and artists being contrary to Soviet ideals. Lenin advocated a highly disciplined elite to guide the socialist revolution, but these elites needed to advance the political theory,[35] not progressively disrupt society or introduce new ideas the way post-World War I art communities sought to revolt toward. Later, the anti-war sentiments of the 1960s in places such as the United States over Vietnam and the specter of nuclear annihilation also placed a wedge between government (and defense) and societal counter-cultural movements, reform struggles and efforts to transform toward alternative values and beliefs. This pattern of designers in tension with political and governmental aspirations continues today. As recent as 2018, Google employees rebelled against company leadership and demanded termination of a Pentagon contract for artificial intelligence work associated with offensive military capabilities.[36] Indeed, this tension of designers seeking disruption and change seems to continue, in perpetual friction with governmental interests that include the need for warfare capabilities and capacities through design.[37]

Militaries became associated with the best and worst of government, particularly as the twentieth century progressed. As the nation state and all instruments of power associated with it were the focus of postmodernism in the 1960s–1980s, postmodern ideas would focus on the tensions between the state and the individual, power, and control, as well as what is real versus illusionary. Postmodernists such as Virilio argued that all aspects of organized society to include industry, cities, and general human progress are a result of war, or at least the preparation of war – however, these positions remain less prominent across various disciplines addressing the intersection of war and societal progress.[38] Some postmodern ideas do play a major role in how design spread in defense starting in the 1990s, which the second and third chapters will elaborate upon. However, postmodern ideas initially would gain influence only in civilian design and in rather limited applications. Design theorists such as Bernard Tschumi, Bruno Munari, and Victor Papanek would include postmodern themes and ideas into design work in architecture, commercial design, and design philosophy.[39] Robert Chia and Robin Holt would be in the minority of postmodernists later interested in defense, strategy, and politics.[40] Postmodernism influenced academia as well as some design communities in the 1970s–1990s, but not make any detectable inroads into military thinking until the mid-1990s where Israeli Defense Forces would experiment in radical ways (to be explained in the third chapter).[41]

Design as an ever-expanding multidiscipline community of practice continued to develop in commercial and academic contexts throughout the mid-twentieth century. Modern design would develop and branch into many different areas of focus, such as architecture, graphic design, advertising, urban planning, and later still into software development and more from the original industrial design trunk. Yet this maturation in the 1950s and forward largely occurred outside of the original military context from which it was born in the violence of state-on-state warfare. While commercial design theories, models, and methods developed well beyond their original applications, the military decision-making methods, doctrine, and overarching belief system would not develop much beyond the core form and function that emerged during the two World Wars. The original design thinking demands of World War I would largely stagnate within military organizations and be the focus of the second chapter. This highly analytic mode of decision-making and planning preferred by modern militaries would move towards a positivist epistemology (breaking things into simpler parts and isolating core laws, rules to apply to reassembled wholes) and the expectation that increased control and prediction would make complex war contexts "solvable". Design for militaries morphed into linear, mechanistic planning methodologies fixated on uniformity, reliability, risk reduction, and predictive performance.

Not everything would split down different paths for security and commercial design interests. Even within civilian design disciplines, theorists such as Bruce Archer and Herbert Simon in the 1960s would promote *a positivist design orientation* to break design problems "into its logical parts, independent of time"[42] denoting a reductionist epistemological proclivity across military and civilian fields alike. If everything worth understanding in war could be broken down into precise, engineered equations, one need only crunch numbers better than the enemy to win any battle or war. In commerce, the only difference between earlier analog advertisers and graphic design efforts and those in the digital age of social media would be the scale, scope, and speed of "meta-data" for reductionist advocates. Such design endeavors promised to render individual behaviors and actions into predicable, rational, and reducible patterns to exploit for profit. Whether selling cigarettes, soda or arranging military strategic options against a nuclear armed competitor, a scientific rationalized calculus for designing manifested across the various fields. One need only be able to combine speed and accuracy with ever-increasing information yields to outpace and defeat competitors and adversaries.

The rise of military managers such as Robert McNamara and a generation of "Whiz Kids" using analytic optimization methodologies applied toward military goals created an industrialized military complex from the 1940s onward that consumed designed materials for war, but itself fixated largely on linear, reductionist planning to use those war tools.[43] Thus, in the 1950s–1990s there are significant patterns and branches of military theoretical development, new doctrine, and novel practice within western militaries, but there is little evidence of any significant deviation from the core ontological and epistemological underpinnings of the same military paradigm used in World War II. Wars in Korea and Vietnam would be expected to unfold like the earlier World Wars, except the newest technology and enhanced warfighter abilities were expected to win contemporary conflicts faster and safer than earlier ones (or deter an adversary through mutually assured destruction). This somewhat stagnant military thinking occurred against a vibrant backdrop of commercial design innovation and expansion into many new disciplines and fields. The surge of modern capitalism invited tremendous creativity, curiosity, and design expansion throughout commerce, academia, and society.

Although design theorists and philosophers might quibble over whether state-directed violence or economic production first triggered modern design in a "chicken or the egg" paradox, the arrival of the modern military frame for war and a separate yet interacting frame for designing commerce ushered in dramatic change for both. In what had previously been a localized, artisan affair for products and user experiences, the industrial design movement launched universal, widespread production with a trade-off in artisan quality for low-cost, mass-produced, and reliable alternatives.[44] Industrial design generated this tension between the artisan strengths of craftsmanship, time, and availability with the industrialized and novel expansion of speed, uniformity, and disposability. For Archer, designers *were different from artists* in that the ideological adherence to "creating art for art's sake" is not the same as designing for the consumer's interests.[45]

From industrialized modes of designing for organized violence as well as economic capitalism, design as a movement rapidly spread through the various disciplines of architecture, art, advertising, public administration, engineering, and other sub-fields though the mid-twentieth century. Further, the mass exodus of experienced military professionals after 1945 and the peacetime downsizing of massive militaries provided an infusion of industrialized professionals across industry. What had previously been applied in a total-war, whole-of-society mobilization to achieve difficult strategic goals (and for some, realizing existential threats previously unimaginable) would now transfer those bureaucratic as well as decision-making activities into pure commercial settings. All of this would occur under a new and terrifying Nuclear Age, where the innovative design of new existential weapons promised both a possible end to traditional warfare as well as a potential end of civilization positioned in the hands of elite political leadership in Washington and Moscow.

Moving from a Natural Science, Industrial Design to Human-Centered Thinking

Industrial design could craft superior products and user experiences, but it also frequently created great things that were misused or made for the wrong reasons. New vehicles would fail to sell despite being designed to function well, new user experiences would also usher in a host of unexpected problems such as tobacco health problems, and architectural designers paired with city planners construct perfectly crafted buildings and neighborhoods only to later tear them down due to not enough people willing to relocate out of slums to live in centrally designed, sterile housing projects.[46] Particularly in areas of societal or social design applications, industrial designers would see analytical sound, practical products that should work as intended, fail because people did not want them, or found such designs ineffective. Designers were able to master the objective, and still somehow miss the subjective that together demonstrates the complexity of human existence.

The rational decision-making of designers seemed in tension with human subjectivity, intuitiveness, and even subconscious human decision-making. Ideas such as "empathy" were not deeply considered early the industrial design movement aside from broad, universal concepts such as a job, house, and two cars in the garage … the contextual and emotional aspects of design were underdeveloped. One might design the perfect office building that was superior to all others and priced for utilization but fail to generate enough occupancy to function. Klaus Krippendorff would address design failure in this regard with the importance of designers needing to make things meaningful to the culture it is being used in. Designers that violate this axiom risk failing; designers that disregard it may just be designing for themselves and only accidently for others.[47] Furthermore, the powerful grip upon modern designers of all stripes for scientific rationalism (natural sciences formed the framework for all other human challenges and endeavors) and a positivist epistemology concerning industrial design would soon be challenged with new ideas and theories on what, how, and why to design for complex societal needs and wants. Modern design would be redesigned, through the expansion of thinking differently through new disciplines, knowledge, and communities of curious practitioners.

Postmodern design, while expanded in the second chapter, requires introduction here concerning design writ large in modernity. The postmodern movement is difficult to bound and even more confusing to identify goals and purposes, as postmodernists themselves often disagree and distinguish themselves apart from one another in perplexing, fluid ways. Yet postmodernists would play important influential roles in the shift of commercial design theory in the 1960s–1980s and later the pivot of military forces toward security design theories in the 1990s–2000s. If modernism represents the fruits of the Industrial Revolution and the ideas of the Age of Enlightenment where reason could frame scientifically sound knowledge and human advancement, postmodernism emerged as the antithesis to that thesis.[48] Postmodern thinking challenges, deconstructs, and destroys the theories, models, and methods of modern society while still maintaining almost a symbiotic relationship with modernism.[49] Often, postmodernism's large umbrella is confusing with artists, writers, political theorists, activists, architects, and even military designers declaring themselves postmodern in vastly different ways and forms. The postmodern landscape is overwhelmed with strange bedfellows.

Yet at this intersection of design for human enterprise, whether toward commerce or organized violence and societal aims, postmodernism is present and significant. Postmodernism requires some sort of modernist world or steppingstone so that it can justify itself and challenge the thesis with antithesis; whether modernism requires postmodernism is another debate entirely.

I offer those postmodern theorists often provide fascinating and paradoxical perspectives, ideas and interpretations of reality that can aid societies in some ways, while perhaps cause great suffering, misery and death in other considerations. The postmodernists of the 1960s–1980s would impact design in architecture, product design, policy and urban planning, philosophy, and design ethics. Later, those same postmodern ideas would creep into militaries through security design theory and provide similar disruption, introspection, and divergent thinking.

Taking a postmodern architectural perspective, Kilduff and Mehra describe the rise of modern city planning and architectural endeavors of the post-World War II societies as a primary stimulant for change. These positivist-oriented epistemological stances of modern city planning attempt to calibrate humans as quantitative, rational components of a large, mechanistic system. People were supposed to be rational actors (if in aggregate), and if each consumer made choices relying on analytical, rationalistic decision-making, then the superior designed products should succeed against sub-optimized or otherwise quantifiably inferior competitors. In the systematic logic underpinning this perspective, known inputs should lead to predictable outputs, in that "A plus B leads to C" in a stable, ordered reality to design within. Yet this did not occur in the real world where paradox, surprise, confusion, and irony abound. While some design theorists discuss the lack of synthesis and empathy by these urban planners, postmodernists dispute the epistemological choices of the modernist stance for artifacts such as new buildings that are built efficiently but produce user dissatisfaction, nonetheless. Kilduff and Mehra explain:

> One of the major reasons for postmodernist architects' disenchantment with the modernist style … was the perceived failure of the "machine-for-living" ethos. The giant housing projects, for example, that disfigure the cities of the world appear, from a postmodernist perspective, to lack any connection to classical ideals of harmony with surroundings … Postmodern architecture, by contrast, seeks to both comment upon, and integrate within, such features of the environment that promote a sense of continuity with the local past.[50]

The rise of commercial design thinking in the twentieth century provided significant development and growth that would influence and later extend into military design endeavors, even if militaries ignored the revolution until far later in its development.[51] First, the industrial designer emerged from artistry to render artifacts into mass production using the technology and values of the Industrial Revolution. The rise of the industrial designer brought with it a growing tension between timeless artisans and what could be described as a growing gap between technologically enhanced functionality and artisan individuality.[52] Particularly since the Industrial Revolution, military dependence on sophisticated technological capabilities to execute increasingly organized and more powerful applications of violence had increased dramatically.[53] Militaries would capitalize on the scientific methodology underpinning such technology, while simultaneously continue to struggle with adaptive strategies and innovative thinking more associated with military art versus a scientific approach to warfare. This tension between classical artisanship and mass production through advancing technology and enhanced social networks reflects not just a significant paradoxical theme for humanity's design for lifestyles as well as organized violence, but would extend into how designers today across industry, academia and security affairs make sense of their own communities of practice, purpose, and design goals.

By the late 1950s, designers employing various industrial design methods within a range of commercial disciplines would begin encountering challenges and paradoxical outcomes to what seemed rather rational and solve-able to the designers. These often related to subjective or wickedly complex design challenges that did not appear "solvable" using objective, analytic processes alone. Dynamic and complex systems featured a *messiness* that prevented any solution from working, and often featuring a *wickedness* that denied previous solution frames from being reapplied to future ones.[54] Often, the best optimized, industrialized solutions would meet all design requirements, yet unexpectedly fail in unforeseen ways when applied to the real world. As some designers sought to map out analytically rational design models and theories to make the design process more objective and structured, while others began to consider the emerging fields of complexity theory, sociology, as well as early postmodern and counter-culture constructs.

Perhaps outside or beyond the cool rationale of analytic reasoning, designers struggled with intuitiveness, subjectivity, contextuality, as well as irrationality if one framed anything that could not be deemed rationale within the existing frame as "irrational". How might one design for a client or consumer that thought and felt differently than they did, and could a designer even imagine what others might consider if they themselves did not critically self-examine their own belief systems, values, and identity? The ability to empathize with others and provide user experiences and design complex projects where multiple stakeholders held paradoxical needs and views would spawn the need for a new way of designing.[55] This development in design emerged amongst the angst, confusion, and dynamic activities of a generational period of protest, unrest, and social change covering the 1950s-1970s in western industrialized societies. This same period featured two more deadly and increasingly unpopular wars, a nuclear arms race between emerging superpowered nations, as well as the social unrest and frustrations of the Civil Rights movement. The fantastic designs of rockets to carry humans to space and the moon were also designed to unleash nuclear annihilation of entire cities. Nuclear power plants could provide essential energy to large populations, yet the same technology could design their extinction.

Human-centric design would rise out of the oversights of an analytically rationalized, quantitatively optimized industrial design movement. Precision, speed, value, and efficiency would still be foundational to design activities, but no longer in exclusion to other design concerns of empathy, context, subjective meaning, and the social construction of reality that rendered it more complex than what natural science inspirations would assume it to be. In the mid-twentieth century, the foundations of this second design wave would be established by design theorists such as Horst Rittel, Victor Papanek, Bruno Munari, Christopher Alexander, and Melvin Webber (among others) moving to challenge the natural science inspired design constructs advocated by theorists such as Bruce Archer and Herbert Simon. The former wanted to increase the awareness of humans within the design process and design context, while the latter sought to marginalize or even eliminate human volatility from what could be a more efficient and stable design logic.

One group offered a powerful critique of the industrial design methods and theoretical underpinnings, and the other sought to defend it as a sound, objective, and highly analytical mode for designing with control. The defenders positioned the human as the consumer, whereas the aggressors placed humans as central to the design activities, *hence "human-centeredness" design praxis*. In the HCD applications, design praxis became an innovative, highly creative, and cross-disciplinary tool that differed with positivist based military analytic decision-making methodologies and industrial design models popularized through the end of World War II.[56]

The different positions of Simon, Archer, Rittel, Webber, Alexander and Papanek demonstrate a maturation of design during the 1950s-1970s. Design applications expanded, particularly into areas that the hard-science, industrialized origins would require new theories, approaches, and design frames. Rittel in the late 1950s pioneered much of the theoretical and educational groundwork for HCD,[57] joining Simon and to a lesser extent Archer on framing the orientation of design as a practice in the 1960s.[58] Although Simon was a pioneer in framing design as a distinct and modern discipline, he would be challenged as well for taking a positivist approach and rendering design into an engineering affair, seeking an objective, analytical, ordered world for design to function within.[59] Krippendorff would reflect upon the first decades of formal design thinking with: "[Simon's] positivism led to what we now recognize as a universalist epistemology which is no longer suitable in information-rich environments such as ours".[60] Rittel similarly criticized a purely positivist approach as insufficient in designing for action in complex reality, where in his design foundations lecture, he noted:

> There is one heuristic argument in favor of taking the world as a world of messages: that the assumption of an independent and objective world in the old sense tends to be, or is, a permanent temptation to be taken as a frame of reference, and its proponents might say that they are closer to it than someone else and argue as if they were nearer to the final truth than others. This is a danger, and therefore it might be more appropriate heuristically to take the other point of view.[61]

Archer promoted a rationalized "science of design" thesis using mathematical equations and geometric graphics to isolate design from artistry, moving closer to Simon's orbit and attempting to render the subjective qualities of design artistry as also reducible to scientific measurements and procedures.[62] This would, with Simon's objective orientation on design theory, help define the early commercial design movement to include methods, models, language, techniques and philosophical perspectives. Design would demonstrate a natural science practicality, yet would increasingly struggle with the subjective, social, and complex dynamics of human needs, desires and beliefs that would not fit so neatly into formulas and outputs.

Regardless, Simon's pioneering work influenced many subsequent design efforts in the 1970s–80s, setting the stage for the next generation of design theorists such as Victor Papanek disrupted and dismantled cherished design beliefs of the commercial practices and capitalistic constructs to advocate alternatives.[63] By the early 2000s, design theorists such as Krippendorff advocated for a design philosophy doctorate, nested in his core argument that "design must continuously redesign its discourse and itself".[64] He would advocate that designers apply their design principles also to themselves, indicating that even the current popular design methodologies across the commercial design world were temporary and needing to be challenged, deconstructed, and replaced with novel, emergent forms of design. The design discipline would, in the 1990s, expand horizons on mature, philosophical levels but also show a heavy investment into trademarked, branded "design methodologies" that created tribes of commercial design consultants.

It is the design philosophical developments that require further examination. Haridimos Tsoukas promoted a dialogical approach to the creation of new knowledge. "Novel combinations create new categories to describe or bring about changes in something familiar … the new concept may have *emergent attributes,* that is, attributes that are different from those of either of the constituent parts".[65] Emergence is important for design consideration in that the legacy or original context that generated the change does not itself possess the ability to explain it. Emergent properties deny most of the analytic, rationalized tools of prediction, control, as well as description until after said emergence occurs. Designers thus should not just think toward the original goal or objective, but inward at the designer's own logic, belief system, biases, and at how being part of a dynamic and complex system one cannot project one design methodology or model upon all possible emergent contexts. Donald Schön, Tsoukas, and Krippendorff emphasized the important of "thinking about thinking" in order to disrupt one's own frame, break through set epistemological choices by realizing and then stepping beyond them.[66] Krippendorff would advocate for an epistemologically informed "philosophy of design" that acknowledges differing design paradigms as well as the perpetual transformative nature of design practice itself.[67] Design required new designs, and the 1990s featured an explosion of new design directions, practices, as well as new mediums such as social media and computerized applications to explore and create within.

The Rise of Design Certifications: Of Tribes, Brands, and Trademarks

The HCD development, starting in the 1970s and rapidly developing in the 1980s–1990s, would encompass what would become multiple types of design applied in different contexts and toward competing (even paradoxical) human endeavors. By shifting away from humans as some monolithic, categorical block of consumers to a subjective and centered, contextual position within design praxis, this second movement of commercial design spread into areas that industrial design had not previously ventured. HCD would broadly address the prominence of human perspective within the design methodology and modeling toward human experiences, meaning, beliefs, and how humanity renders an already complex reality further complex through social construction of ideas upon the real. The first designer to describe design in this sense was Richard Buchanan: "the problem for designers is to conceive and plan what does not yet exist, and this occurs in the context of the indeterminacy of *wicked problems,* before the final result is known".[68] Much of this complexity would not be in a natural science ordered reality, but a decidedly human affair requiring a refocus upon how and why people experience life in unique, complex ways that resist standardized or mechanistic norms.

Archer, while advocating a positivist orientation on design, emphasized the "element of innovation is always present in design" while describing the overall purpose of design activities relates to the planned production of something (product, user experience) or to be fashioned with artistic skill; mere description of existing and known artifacts is not really an act of design for Archer. It must have "at least a modicum of originality".[69] This echoes a later sentiment by military designer Chris Paparone that military planning is a specific and extremely linear form of design, but only in a mechanistic, systematic sense of the term. Later, designers Harold Nelson and Erik Stolterman would promote a similar design definition echoing Buchanan's aforementioned design theme,[70] and military designers Shimon Naveh and Ofra Graicer would overlap with similar phrasing in their overarching approach in Israeli military design applications, to be explained in the third and sixth chapters. Peter Rowe in his architectural design book *Design Thinking* outlined a societal trend starting at the end of the nineteenth century where "a mechanistic type of doctrine can be observed to recur that sought to explain problem-solving behavior through the use of irreducible lawlike relationships deemed to govern mental processes".[71] His ideas on commercial design would also inspire many developments in the 1990s, including the rise of HCD methodology, models, and terminology. Problems in HCD are framed so that the users must be deeply considered, design is done holistically, and purely analytic approaches devoid of synthesis and extensive theoretical develop-ment will be insufficient and potentially hazardous.[72]

Thus, the second development in commercial design would in some ways challenge, disrupt, and redefine the first movement. While industrial design relied heavily upon analytic optimization toward greater efficiencies and developments,[73] HCD would focus on the irrational, subjective, and often nonlinear aspects of humanity that industrial design methods sought simple, often quantitative majorities to streamline.[74] These empathy-based design methodologies first emerged in the early 1960s after early theoretical contributions of the late 1950s and went mainstream across design academia, industry, and practice by the 1980s.[75] HCD helped practitioners appreciate that the strengths of industrial design's ability to optimize, engineer, and accelerate efficiency gains could frequently miss the mark and result in well-designed products failing. That failure was not necessarily due to flaws in the design output although certainly many new products were defective or poorly designed. Rather, designers by the mid-twentieth century began to appreciate that some things could be designed strictly within an indus-trial methodology that did not consider or explore vital systemic tensions or minority perspectives.[76] Krippendorff, in critiquing the early over-engineering emphasis of design theory such as Simon's, saw the reason of design product failures as part of the emphasis on a centralized hierarchical mode of industrialized design:

> Simon came to celebrate hierarchy and the kind of mono-logical ration-ality that is typically pursued in the design of highly functional products. In the human use of artifacts, the application of this techno-logic creates the need of forcing diversity into common frameworks, applying uni-form standards on subordinates by a central authority, a government, a leading industry – including by "ingenious" designers … Simon could not anticipate the trajectory of artificiality we seem to be pursuing. He could not experience that hierarchical systems of some complexity hardly survive in democratic, market-oriented, and user-driven cultures … Design tasks involving teams or stakeholders can no longer be organized hierarchically. Although traditional designers might decry the loss of control that hierarchies provided, chaos, heterarchy, diversity, and dia-logue are the new virtues that design must embrace today.[77]

Analytic optimization and improved efficiencies without subjective, complex, and paradoxical qualities that involve human emotion, curiosity, amusement, and other non-objective constructs were now recognized as not just relevant but essential in many design contexts. Krippendorff would frame this as a purely engineering mindset or discourse attempts to rationalize design challenges so that one can solve problems by optimizing techniques and search for strategic approaches to achieve predetermined ends. This system-atic mode of designing though optimizes in natural science contexts rather well, it struggles in social settings. Krippendorff highlighted this tension as follows: "Science inquires into what is, design into what could be".[78] Rowe also observed that "design is much more than mere problem solving", illustrating a major shift in commercial design orientation through the 1980s toward greater appreciation that many design challenges are emergent, ill-structured, and simply cannot be mapped out sequentially with a client at the start of any project with purely analytical optimization and objectivity.[79] The very qualities that earlier industrial designers attempted to marginalize or eliminate entirely from manufacturing and production processes were instead critical to the notions of creativity, innovation, exploration, and opportunity in human design.

By the 1970s, formal civilian design schools would start touting various methodologies as well as entire disciplines for different design applications. Often, these schools would manifest on campus settings but feature a strong rotation of design theorists, practitioners, and educators from industry to academia and back again. In the 1980s and 1990s, multiple academic programs helped stimulate significant research on different design approaches occurring across various disciplines, fields, and human endeavors. Military design theorist Jackson provides a useful summary of this commercialization of a HCD process where the Hasso Plattner Institute of Design (at Stanford University) popularized a five-stage method templated in the figure below:

> Teaching of design thinking methods have proliferated within higher education institutions since the mid-1990s, accompanied by a revival of its processual aspects. This revival was triggered by Richard Buchanan's influential 1992 article *Wicked Problems in Design Thinking* which broadened the focus of private-sector design from *product to service* design. Buchanan also substantially developed a two-tiered process of problem definition and problem solution that had been advocated by various earlier design thinkers, [popularizing] this approach to the point where it has since become central to the design methodologies taught by most civilian higher education institutions. These methodologies include *participatory design, user-centered design, interaction design, transformation design* and *service design* … to give merely a few examples. While their details differ, each of these design methodologies includes a problem defining (also called problem framing) component and a [problem-solving] component.[80]

The last decade of the twentieth century witnessed a rise in formal, often trademarked HCD methodologies. These tended to rise within collaboration between private design firms and universities seeking new, formalized, and educational manners to teach design. Design moved once again into schools, guilds, and associations defined often by trademarked or copyrighted design methodologies, complete with formal certification or academic degrees for those that graduated as formal designers. Unlike the interwar period when German and Russian schools associated more with social and political movements to challenge art, society, and governments, this new wave of designing appeared nested within economic, proprietary, or similar brand-identity constructs. One would learn to design with empathy, but credentials came with a new sort of validation just as a law degree from an ivy league university implicitly translates to supposed superiority over any state school.

These methodologies would feature new emphasis on sociology, psychology, narrative theory, and design lessons from architecture, advertising, and other related disciplines. In this new, formal methodological structuring, universities and associated companies formulated design thinking into a blend of design education, facilitation, commercial enterprise, and design outreach into government, social, and private sectors. Critics of this compressed format include ethnographers and complexity theorists that reject the notion of a brief workshop in design being useful in accomplishing the deep cultural study for truly complex challenges beyond superficial or topical design requirements. While Stanford University (home of the *"d.school"*), other competing university, and private companies such as IDEO continuously refresh and update how they depict their design methodology, Figure 1.1 takes the broad patterning of these "design ways" to show overlap across myriad design groups and their schools.[81]

While critics often challenge some vulnerabilities of HCD approaches and what some call an overly commercialized context that might misplace symbolic activity for deep change, the basic patterns of most HCD methodologies have for several decades become very popular in industry, academia, and private practice. There are hundreds if not thousands of variations of Hasso Plattner Institute's original "hexagonal" design methodology (adapted by Stanford) across various commercial design disciplines today. The original modes developed are depicted in the below figure and are an iterative sequence commencing with "empathize", then "define", followed by "ideate", and then "prototype" that leads to "test". These concepts are positioned in an order that likely draws extensively from the theoretical works of Rittel, Simon, Papanek, Alexander, Rowe, and other influential designers throughout the 1960s–1980s. IBM's 2015 first version of their "Enterprise Design Thinking Process" is included in Figure 1.1 to illustrate how pervasive the HCD constructs overlap with different versions.[82] Figure 1.1 illustrates common methodological patterns between various HCD styles, brands, or communities of practice.

Nearly all HCD methodologies start with empathy, or in the IBM modification, a near identical combination of understanding and *user-centric* statements. Both demonstrate the importance of human subjectivity and contextual uniqueness in how to design in this way. Understanding empathy includes designers conducting extensive interviews, visits, and other interactions to gain appreciation of individual perspectives, organizational culture and norms, and how alternative perspectives might be overlooked, in tension or possibly marginalized. Thus, the designer needs to frame various tensions concerning key stakeholders including the consumer, the producer, and those

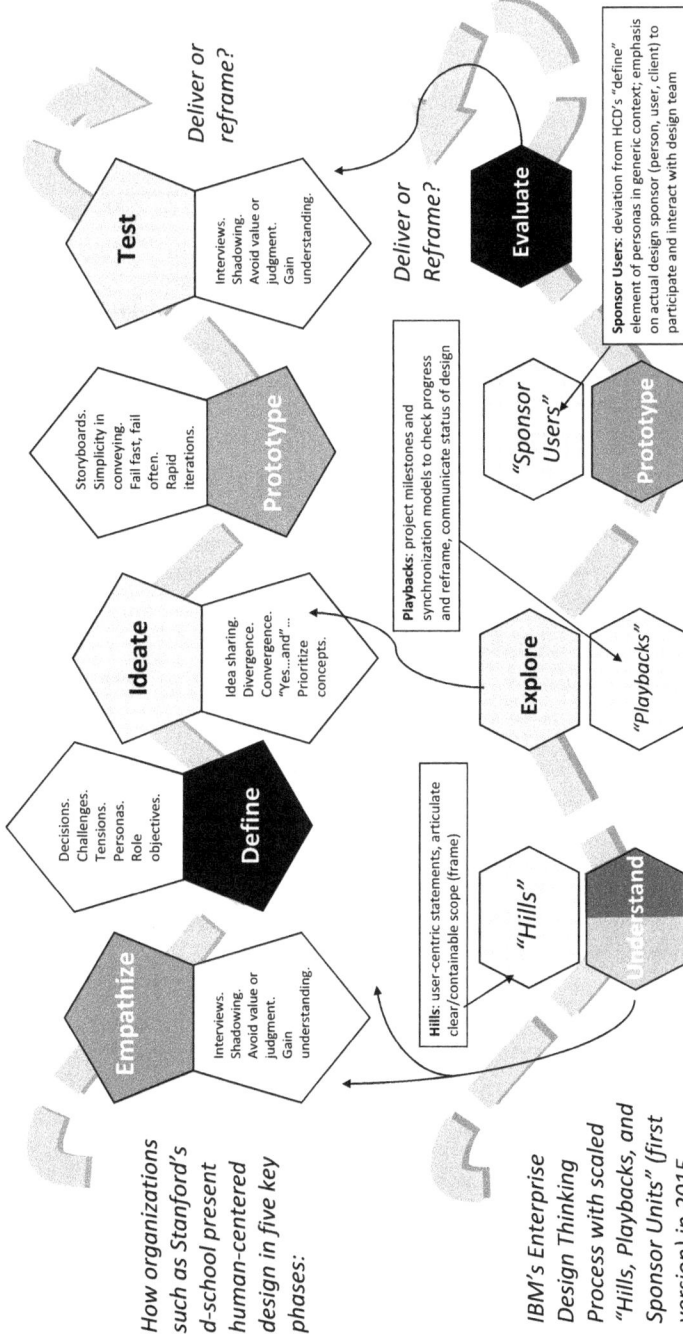

Figure 1.1 Human-Centered Design Methodologies and Patterns.
Source: Author's Creation.

within the system that may influence either of those groups. Stanford's HCD process defines the design challenge next, while IBM positions their definition framing with "understand". Both methods move to an ideation phase where they explore in divergent ways to get to working prototypes. IBM factors in additional project milestones and sponsor engagements into their variation. Both methods test and evaluate their prototypes to eventually bring a final design to production. Figure 1.1 shows the significant overlap and interplay between two popular design processes that fall under the umbrella of "human-centered design".

The methodologies shown above (and countless others) remain quite popular in contemporary design communities, and have in turn inspired many variations, hybrids, and copycats that apply different frameworks yet still draw from similar language, metaphoric devices, as well as models and theories. Yet this methodology as depicted merely shows the *how to* of design, without explaining the *why* and relating some of the activities to the belief systems of the stakeholders, clients as well as the design facilitators themselves. Why is HCD different from earlier industrial design processes? One could line up various methods to try to explain, but this may only remain descriptive in a "what-how" form of discussion. Methodological explanation is insufficient to explain the form and function; philosophical inquiry of the design will illuminate such deeper questions.

An Epistemological Examination of Human–Centered Design

Often, people quickly get flustered if "why" is asked repeatedly. There is a deep desire to solve problems, reach objectives, and validate purpose through action. Philosophical inquiry often rubs against these desires. Yet there is a need to deeply consider how and why designers do what they do in terms of pursuing design activities, performing design methodologies as they choose to, and how these design methods are constructed through cognitive models, associated theories and overarching belief systems shared by different groups of people involved. In contemporary design schools the aforementioned design methodologies and techniques form the foundation for how each design community understands what design is, and how they go about exercising design.

Breaking into how and why design methods themselves are constructed provides a philosophical inquiry into the deeper ontological, epistemological, and systemic frames for how various organizations, schools, and communities think and act as they do. Designers unwittingly bound to a methodology can indeed become astoundingly proficient at performing a particular design way. They may even instruct, train, and operate with mastery. Yet without philosophical inquiry and proper critical reflection, unwitting designers will remain trapped. Witting designers usually understand how and why their methodology forms and functions as it does, where tensions and institutionalized barriers are, and what might lie beyond the design frame in a complex reality full of many other possibilities.

This brings us considering why an epistemological exploration of design methods can enable designers to consider beyond design practices toward a multi-design *praxis* (reflectively practicing design while aware of theory in action). We must move beyond descriptive loops where one only considers what a design method does, and how one might better perform that particular design process. Figure 1.1 only provides a description of the "what-how", while the next illustration takes us to why HCD has the form and function it chooses and not something else. In Figure 1.2, the IBM original HCD methodology shown in Figure 1.1 is expanded to show an epistemological framing of *why* the designers form and function as so. Empathy as a HCD construct might be expressed as either the exploration of multiple social paradigms, or it instead projects a single dominant paradigm upon all other stakeholders. This occurs when a designer frames their beliefs about empathy so that their personal or cultural values are invisibly projected upon all actors and design constructs regardless of difference, potentially resulting in misunderstood or irrelevant empathetic frames. Without being sensitive and open to divergent perspectives and frames than one's own, designers may end up imposing their own worldview and all design options upon every challenge they encounter. Designers aware of this promote a multi-paradigmatic, or self-reflective, or "knowing-in-action" designing approach.[83] Krippendorff provides an excellent summary as follows:

> Discourses construct vastly different realities into which the ideas of a discourse are inscribed and in turn become available for inquiry and elaboration … Different discourses not only construct *incommensurate realities,* their pursuit of different *paradigms* yields different kinds of knowledge: Experiments are not treatments, and neither are technical inventions. Finally, discourses create their own *communities of experts* whose members may not be able to communicate across discourses.[84]

Krippendorff uses the term "discourse" here as the mode in which designers will use language and reflect not just on the language and content therein, but the metaphors above the language that reveal the epistemological and ontological underpinnings that outline a paradigm. As the second chapter will explain, some military designers use postmodern theory as one way to explore the "metarules" of tropes that are conveyed exclusively through metaphors and narrative. Designers critically self-aware of their own paradigm and willing to explore alternative ones gain greater cognitive space for ideation and new definition of the design challenge. Empathy ends up meaning quite different things when considered through either a single-paradigm lens or that of a multi-paradigmatic one, yet this distinction is never clearly addressed in HCD methodology or basic education and practice.

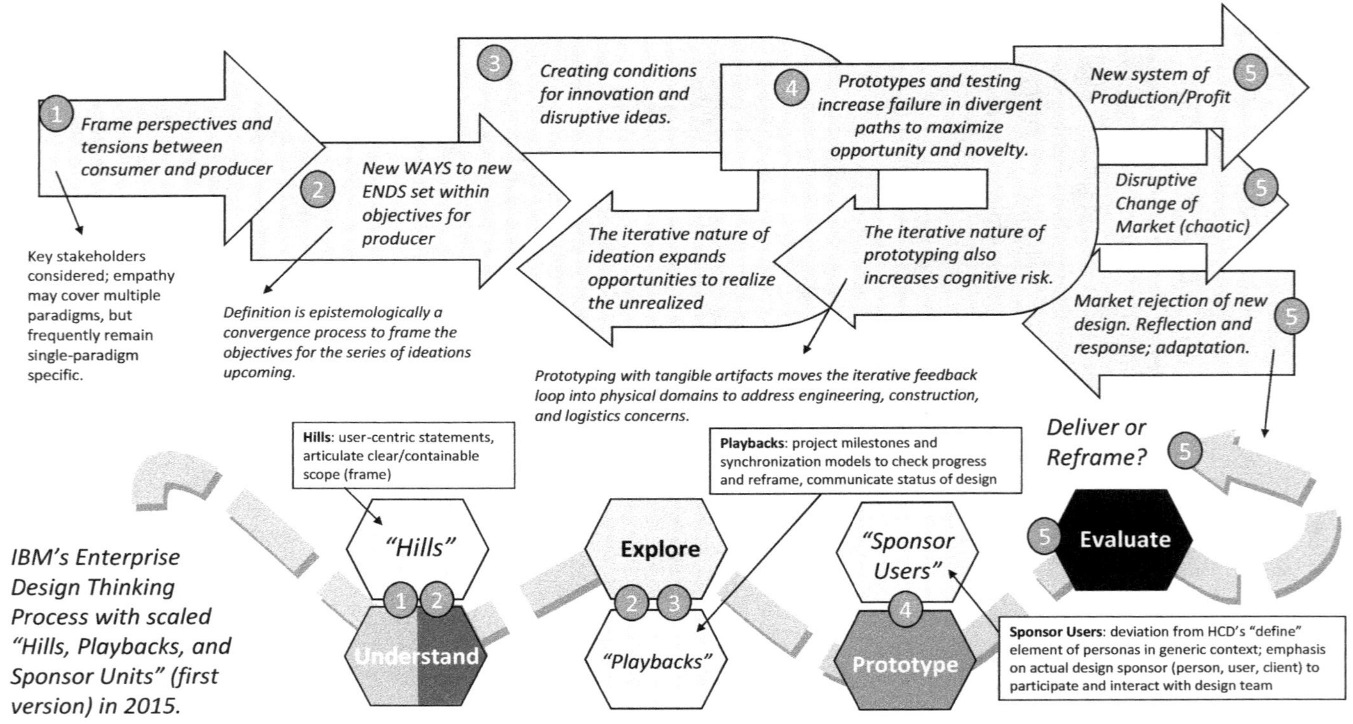

Key stakeholders considered; empathy may cover multiple paradigms, but frequently remain single-paradigm specific.

Frame perspectives and tensions between consumer and producer

Definition is epistemologically a convergence process to frame the objectives for the series of ideations upcoming.

New WAYS to new ENDS set within objectives for producer

Creating conditions for innovation and disruptive ideas.

The iterative nature of ideation expands opportunities to realize the unrealized

Prototypes and testing increase failure in divergent paths to maximize opportunity and novelty.

The iterative nature of prototyping also increases cognitive risk.

Prototyping with tangible artifacts moves the iterative feedback loop into physical domains to address engineering, construction, and logistics concerns.

New system of Production/Profit

Disruptive Change of Market (chaotic)

Market rejection of new design. Reflection and response; adaptation.

Deliver or Reframe?

Hills: user-centric statements, articulate clear/containable scope (frame)

Playbacks: project milestones and synchronization models to check progress and reframe, communicate status of design

Sponsor Users: deviation from HCD's "define" element of personas in generic context; emphasis on actual design sponsor (person, user, client) to participate and interact with design team

"Hills"

"Playbacks"

"Sponsor Users"

Understand

Explore

Prototype

Evaluate

IBM's Enterprise Design Thinking Process with scaled "Hills, Playbacks, and Sponsor Units" (first version) in 2015.

Figure 1.2 Epistemological Frame Using the 2015 Enterprise Design Thinking Example.
Source: Author's Creation.

The empathy step in HCD is not the only area of profound significance for design emphasis. In the subsequent (although iterative) periods of "definition" and "ideation", the epistemological assumptions occurring in HCD are structured to define new ways to accomplish new design outcomes within the design objectives for the producer. These design experiences may be a new product, a new user experience, or a development within the organization so that the organization gains advantage in the future emergent system. The ideation portion of the HCD methodology is, at the epistemological level, attempting to foster conditions for design innovation and disruptive ideas as depicted in Figure 1.2. This can, in the iterative fashion of the design process, increase cognitive risk for the designer, the design group, the organization, and beyond. This is done so that divergent thinking occurs as well as deeply critical reflective practice, so that increasingly wider ranges of different, even radically oppositional, concepts get experimented with for design applicability.

Prototyping and testing, done in the last two portions of the HCD methodology, exist for an iterative process that moves from highly divergent ideation concepts toward tested design innovations that meet some established or emergent concept of success. It is illustrated (in the epistemological mapping for Figure 1.2) with three possible future outcomes of any HCD endeavor. The system continues so that the designer establishes a production model with profit for the producer and the consumer(s), or potentially the design deliverable radically disrupts and transforms the market toward a novel, emergent form. In these outcomes, the design is a systemic disruptor where all other stakeholders and rivals must adapt according to this design innovation or remain in their legacy system behaviors and sustain damage. Lastly, the market may also reject this design output, even if the concept is highly innovative, novel, and needed. Complex systems rarely follow linear patterns, and unexpected outcomes including design rejection need to always be considered. In the case of design failure, the HCD practitioner will re-enter the methodology and conduct reflection to respond, reconsider, ideate further and adapt.

HCD has in the last few decades evolved into highly commercialized, often trademarked and promoted certification courses that designers can attend, whether through a company or through academia where formal degrees can be earned as well. The assimilation of human-centered design methods and models into bureaucratic, linear, and process-oriented activities has drawn the irk of designers to include even founders at IDEO, a highly influential private design company. Little over a decade after the rise of these HCD methodologies illustrated in Figure 1.1, Bruce Nussbaum blogged: "Companies absorbed the process of Design Thinking all too well, turning it into a linear, gated, by-the-book methodology that delivered, at best, incremental change and innovation".[85] He also quotes IDEO's founder, Tim Cook, lecturing at Parsons University on the current state of design:

Design consultancies that promoted Design Thinking were, in effect, hoping that a process trick would produce significant cultural and organizational change. From the beginning, the process of Design Thinking was a scaffolding for the real deliverable: creativity. But in order to appeal to the business culture of process, it was denuded of the mess, the conflict, failure, emotions, and looping circularity that is part and parcel of the creative process. In a few companies, CEOs and managers accepted that mess along with the process and real innovation took place. In most others, it did not. As practitioners of design thinking in consultancies now acknowledge, the success rate for the process was low, very low.[86]

The multitude of HCD methods all might suffer from these critiques, while individual applications clearly may not. The focus here is not to throw out the HCD baby with the bathwater, or suggest that certain methods or schools might have superior design abilities over all others. The critiques are mentioned here briefly so that one might critically examine the methods, models, theories, and language of any other design methodologies they are already familiar with or preparing to explore further. Why do certain groups design as such, and why are they also choosing *not to design* in alternative ways, wittingly or unwittingly? For these commercial design methods in the 2005–2020 period, the debate over moving toward highly bureaucratic, process-oriented, recipe-styled design activities and away from earlier theoretical, artistic, as well as industrialized manifestations seems to be gaining momentum. The critique that commercialization has created a cottage industry of high-priced design consultants wielding various gimmicks and trademarked techniques is a pattern across the more biting of commentaries.[87] Yet many designers seem to appreciate this, and the multidisciplinary design community also appears to be growing.

Human-centered design is explained as an iterative and not necessarily a step-by-step process. Yet it still shows a sequenced process that is linear in causality when considering procedure and iteration flow; the iterations suggest that each process could skip steps, but overall, the experience will span all of the methodological components in some collection of loops. Epistemologically, HCD links a "problem definition" toward creating or discovering some "problem solution" relationship.[88] Although some designers may consider the theories and models that form the HCD methodology for commercial design challenges, most designers employing the hexagonal method might only follow the sequences as they were instructed to do so, repeating the methodology over and over.

This design practice may lack consideration of broader theoretical, andragogic, and epistemological design considerations. Unreflective designers, without thinking inward and upward, may instead focus on process compliance and immediate design objectives. Several of the military design methodologies covered later in this book also share this organizing logic, showing another overlap of designing in commerce and that of warfare. Interestingly, some of the original design theorists in the 1960s–1980s sought to prevent such a thing from happening. Rittel, as one of the visionary design theorists of the HCD movement in the 1960s and a professor of the *Hochschule für Gestaltung Ulm* as well as later a prominent design educator at the University of Berkeley, sought to disrupt this linear model for designing.[89] Designers should not be performing "paint by number" processes oriented toward compliance of the methodology and ignorant of why their designs form and function as such.

Rittel would propose that the "problem definition" sequence is analytic in nature, where designers reverse-engineer their design sequence by first framing a future goal, considering what is preventing that from occurring, and subsequently define the design problem in what Rittel called a rigid manner.[90] Designers then, as Buchanan elaborates along the same theme with, move immediately into specifying all the necessary solution requirements such a configured design normatively should possess.[91] This is rationalization, where one is expected to apply systematic logic (input leads to known output) and analytical optimization, while seeking to prevent intuition or subjectivity from creating user bias or poor understanding. Rittel and others saw that despite the appearance of ordered, systematic designing toward defining the right problem, people tended to either get overwhelmed in highly complex situations or force preferred solutions in irrational ways. Rowe reinforced with: "without a doubt design is to be seen as a normative enterprise … Even if the presence of such normative reasoning is clear, however, its role and character remain vague".[92]

Instead of reinforcing this analytic focus on "problem-solution" constructs, Rittel advocated the concept of "wicked problems" while drawing upon complexity theory to propose nonlinear, emergent, and dynamic design systems. As Buchanan explains,

> the linear model of design thinking is based on *determinate* problems which have definite conditions. The designer's task is to identify those conditions precisely and then calculate a solution. In contrast, the *wicked problems* approach suggests there is a fundamental *indeterminacy* in all but the most trivial design problems.[93]

Rittel would advocate for design theory to appreciate the properties of wicked problems in that they cannot be solved in isolation, and any solution is a one-shot effort that both changes the system and produces design consequences that are non-reversible. These design challenges are unique, systemically linked to larger and more abstract problems across an increasingly complex system, and the designers are fully responsible for their own actions.[94] This translates to design requiring a blend of both analytic, rationalistic thinking as well as intuition, artistry, and a designer's appreciation of complexity rejecting design applications featuring only one or the other in praxis.

Thus, designers for the pursuit of commercial enterprise might critically reflect upon various contemporary design methodologies (in particular, the one they advocate or identify strongest with) and consider whether they adhere to Rittel's philosophical stance on designing in complexity. Or perhaps their community's preferred methodology may instead be a cognitive barrier for them to escape the limitations of "problem definition-problem solution" epistemology.[95] To illustrate this, Figure 1.2 positions the popular HCD methodology with a deeper appreciation of the models, theories, and belief systems operating above the method itself. Rowe critiqued previous commercial design theorists of failing to ever go beyond methodological inquiry of design praxis; that "the kind of theory we need to if we are to explain what is going on when we design must go beyond matters of procedure".[96] Figure 1.2 provides one epistemological framing of how human centered design methods, in this case the 2015 Enterprise Design Thinking process, are exercised in complex reality. This takes the "what/how" of a methodology and explains the "why" at a design philosophical level of inquiry.

Figure 1.2 presents one way to epistemologically frame how most HCD methods function for designers beyond the sequences shown in the methodology alone. This of course might be done with any other design methodology across the commercial enterprise, including others described in this chapter. While these epistemological stances might be debated by design practitioners, few epistemological studies of design exist at all. Figure 1.1 shows the overlaps and interplay between the Stanford design school approach and IBM's hybrid design methodology, but this also extends to many other design schools and tribes. Figure 1.2 attempts to epistemologically frame not just the IBM 2015 method, but *most all human-centered designs in general*. There are exceptions, but philosophically, most HCD communities differ not by design ideas, but by design context. Software design differs from architectural design the way code is not stone. Beyond contextual matters, most HCD acts along the same shared beliefs and processes.

In the first arrow shown in Figure 1.2, designers attempt to frame various perspectives that establish what a consumer might perceive as well as the producer, ideally the design client if a company is attempting to produce a new product or user experience. This combines both the steps of "empathy" and "define" in that epistemologically, a designer that cannot realize the frames, identity, and belief systems of both stakeholders may unintentionally move toward ideation with false or insufficient "empathy" and "define" deliverables. If the design group maintains what sociologists call a "single paradigm" orientation, they may even ignore or mischaracterize any stakeholders employing another social paradigm (termed paradigm incommensurability).[97] Unwitting designers might plow right into projecting their own worldviews upon the consumer, assuming away valuable tensions and alternative perspectives to get to a predetermined design solution.

Papanek provides numerous design examples of this in a range of phys- ical, psychological, cultural, and ethnic contexts where designers prepared a product intended for a single frame. They, trapped in a single-paradigm perspective, would design products that would only function for able- bodied, right-handed people from one lifestyle, education status or culture because the designers had not even considered outside of that frame.[98] Those same designers likely had designed for themselves, if they too had the same characteristics. A military example of this is easily illustrated in American policymakers and strategists expecting the people of Afghanistan to want and need police officers that were literate, controlled at a federalized level through centralized government in Kabul, following a western-based code of law instead of a religious one, and featuring female and male officers rep- resentative of local ethnic ratios.[99] While this law enforcement model clearly matches that of western democratic societies involved in fighting terrorism and insurgency in Afghanistan, it overlooked and rejected traditional Afghan belief systems, values, needs, and wants. Ironically, after the collapse of the Afghanistan government in 2021, a former Taliban bomb cell leader now is the Police Chief of Kabul, enforcing Sharia (not the Afghan republic penal code) law, working with all-male, largely illiterate forces, using cheap and low-technology equipment to maintain order.[100]

The second arrow in Figure 1.2 addresses the ideation phase. Designers work to ideate and think divergently so that the design team replaces outdated or irrelevant "ways" and "ends" within a shared "ends–ways–means" systematic construct so that new inputs and outputs (and new systematic relationships) can be configured. Epistemologically, most ideation occurs in this system- atic rationalization that "old A plus old B led to old C", and divergence from that merely substitutes different "A plus B leads to C" conceptualizations that still maintain the same systematic logic. Natasha Jen, critiquing this linear- causal quality of most HCD thinking, blogged: "Prescriptions create a kind of prison, in terms of how we can think about things and how we work. But a very linear methodology-based way of working completely removes other possibilities".[101] When designers are unaware that they are using processes to attempt divergent thinking only to recycle back to convergent reinforce- ment of the same models adhered to by that company or organization, they are potentially substituting new "ways" and "ends" within a narrow band of ideation.

Katharine Shwab blogged on various critiques of IDEO's model to include acknowledgments by IDEO leadership in 2018. She observed that, "part of the problem is that many people use the design thinking methodology in superficial ways". Shwab goes on to cite IDEO partner Michael Hendrix who called it the "theater of innovation". Shwab paraphrases Hendrix's obser- vation that: "Companies know they need to be more creative and innova- tive, and because they're looking for fast ways to achieve those goals, they cut corners".[102] Figure 1.2 illuminates at the epistemological level that many

HCD processes may appear divergent and innovative with a concoction of symbolic behaviors such as covering walls with sticky notes and writing vision statements, but without reflective practice on why organizations think and act as they do, much of this may never break out of institutionalized groupthink. The design might generate a new product or gimmick, but does it transform anything institutionally? Shwab quotes Hendrix on this problem with:

> We get a lot of the materials that *look* like innovation, or *look* like they make us more creative … [yet in some cases] they end up being a theatrical thing that people can point to and say, "oh we did that".[103]

Designers can follow a methodology rigorously, but when they lack self-reflection, they may end up repeating core institutionalized behaviors that go unchecked, but now carry a new coat of shiny design paint.

In the third numbered arrow in Figure 1.2's epistemological mapping, designers using a HCD methodology use some variation of "prototype" and "test" to create the conditions for design innovation. They encourage divergent thinking and disruptive ideas that may challenge the institutional norms or core beliefs of the legacy system, and the multiple iterations of ideation will over time increase both the range of potential prototypes as well as the cognitive risk for designers and leadership to select particular paths to design toward. They also may remain compliant to deep institutionalized beliefs, meaning the design outputs are "new" but still obedient to core value sets. Whether one is designing with tangible artifacts (actual products, tools, things) or user experiences (subjective, contextual, with abstraction), the prototyping and testing phase of the methodology can produce one of three design outputs in the epistemological framework below.

Upon completion of an HCD effort, designers should generate some design deliverable that extends into a new system of production and profit. The team creates something new that did not exist before but was needed by the consumer and can now be produced by the company in the perpetual design of commerce. HCD produces new things, along with new experiences, and designers emphasize the user experience as a systemic objective for every design output to satisfy in a novel, meaningful way. In some of those deliverables, the new design disrupts or radically changes the market so that many other actors and stakeholders must respond and adapt to the transformation. Instead of just creating a new electric car that competes with the rest of automakers, a company might design something so novel and game-changing that the new car shatters all existing systems and forces widespread adaptation or collapse of competitors. Such a design puts that company in the unique and coveted position of a game-changing market innovator. All competitors must now realize the change, adapt, or die resisting. This is the proverbial "lighting in a jar" and occurs in an emergent, nonlinear, and highly complex way that is difficult to repeat. Not all design unfolds this way, yet often organizations assume that HCD is the magical key to accomplishing this in some way that paradoxically is not supposed to disrupt the existing institutional norms and patterns already in place.

Of Double Diamond Designs: Another Human-Centered Methodology Explained

The British Design Council introduced its own design model, termed the "Double Diamond" methodology in 2005. This is a late-comer to the international commercial design community and arguably based upon earlier HCD sources as well as potentially some creative problem solving methodologies from the 1970s.[104] The Double Diamond depiction of cycles of divergent thinking leads to convergent design thinking and back and forth through different aspects of a design process. The original Double Diamond method was used by the British Design Council from 2005 through 2019, when the Council updated the original model with additional elements depicted below. Although still informally best known as the Double Diamond, the Council retitled their new version as "Framework for Innovation". Despite this HCD method appearing different graphically, it still demonstrates most of the hallmarks of HCD constructs. This additional design process will be deconstructed epistemologically so that, as shown in Figure 1.2, the next two figures (Figures 1.3 and 1.4) extend this broad pattern where virtually all commercial design activity adheres to the same design logic, regardless of trademarks and school certifications.

This Double Diamond methodology, like the HCD methods presented in Figure 1.1, employs a sequential articulated process expressed via a geometric, sequential model. The Double Diamond breaks the design process down into four distinct phases of discover, define, develop, and deliver. The entire process focuses on the commercial user experience and new products, with many sequential overlaps with other HCD ontological postures. Epistemologically, designers using this double diamond method are expected to iteratively cycle through it, with designers traveling back and forth through the methodology and repeating the diamond divergent and convergent ideation experiences. Hence, it creates a pattern of double diamonds as a conceptualized model for designers to orient where they are in the process. While all HCD processes perform iterations of divergent and convergent thinking, the Double Diamond process prioritizes this and configures their graphic to highlight these design activities.

The above representation approximates how the Double Diamond process unfolds. Methodologically, designers iteratively cycle through both diamonds, drawing from design principles and a "methods bank" as they move through discovery into high divergence, with a convergent effort to define what ideas appear valuable. A second development phase takes a similar divergent path, moving back into convergence to deliver some final selected design. Referring back to Figures 1.1 and 1.2, all of these design methodologies state that they are iterative, imply some nonlinearity of the

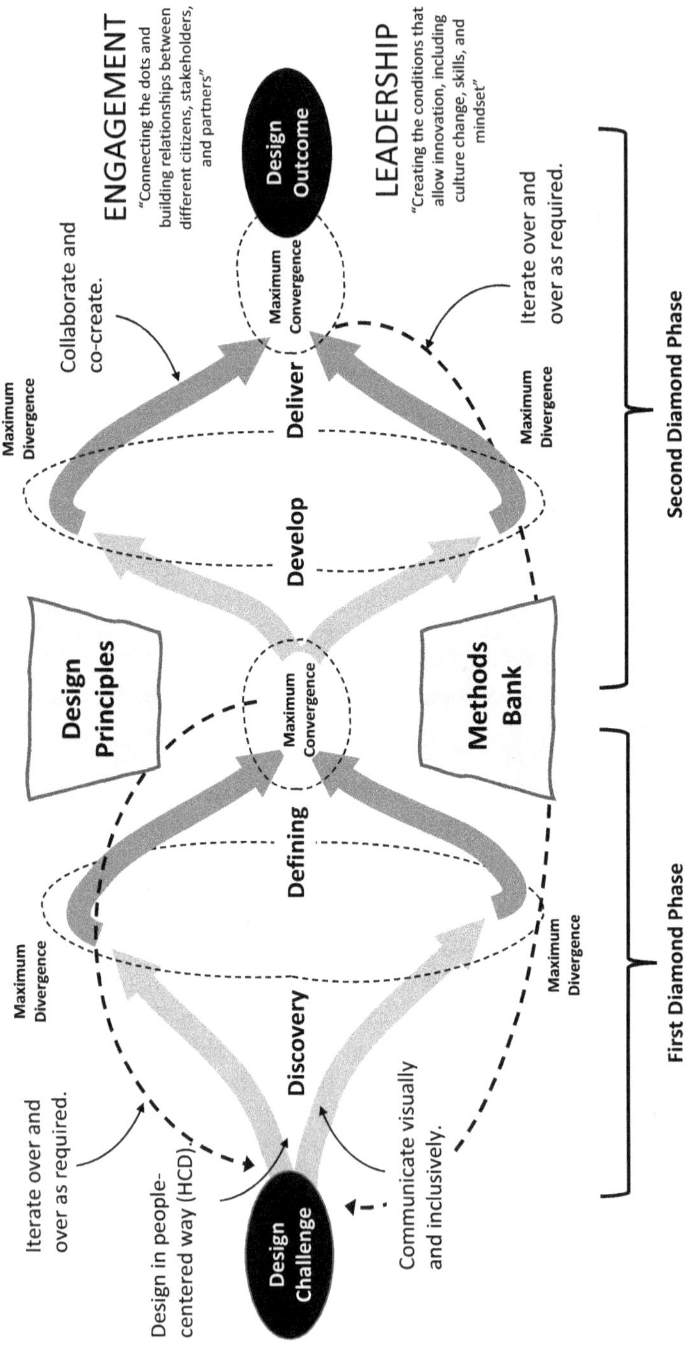

Figure 1.3 Explaining a Divergent/Convergent Diamond Design Methodology.
Source: Author's Creation.

DD process relies upon functionalist analytics such as "market research", researching the user, management information, and building design research groups.

Frame of consumer's user needs coupled with a discovered design "inspiration"

Discovery is an intentionally divergent, iterative process for novelty

Organizational recognition of design novelty accomplished.

Company interprets and "aligns" design novelty into quantifiable solutions.

Solutions (analytically optimized) are tested in a progressive outlook.

The iterative nature of testing implies a stable design sequencing.

System of Production/Profit

Disruptive Change of Market

Market rejection of new design. Reflection and response; adaptation.

DD process finishes with "delivery." This features final testing along with evaluation and feedback loops. A new designed product may compete in the existing market, disrupt that market to some stakeholder's advantage, or fail to perform.

ENGAGEMENT
"Connecting the dots and building relationships between different citizens, stakeholders, and partners"

LEADERSHIP
"Creating the conditions that allow innovation, including culture change, skills, and mindset"

Design Outcome

Design Challenge

Design Principles

Methods Bank

Discovery

Defining

Develop

Deliver

Maximum Convergence

Maximum Divergence

Collaborate and co-create.

Iterate over and over as required.

Design in people-centered way (HCD).

Communicate visually and inclusively.

The "define" stage pre-conditions this discovery stage to generate a range of recognizable business "objectives" that can be defined subsequently.

DD process tests (develops) the design through the alignment of key activities nested with in-house design objectives that lead to a resulting "product" or "service".

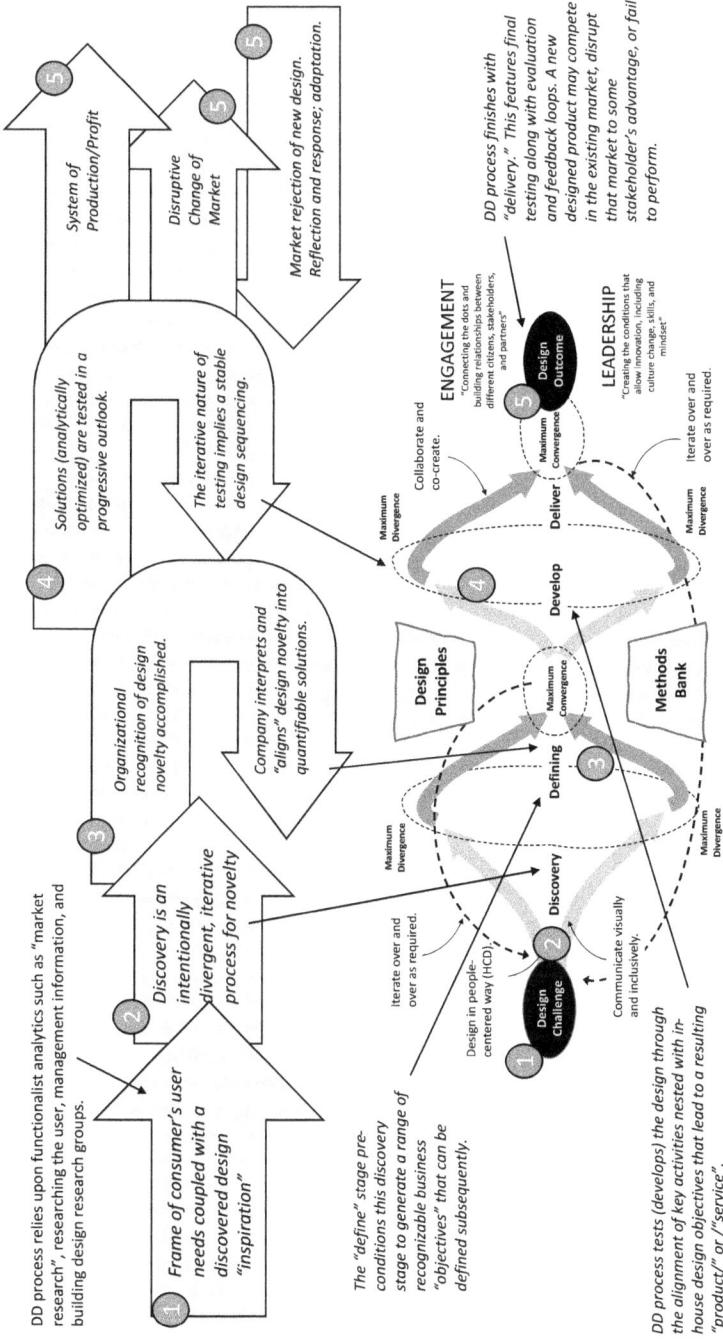

Figure 1.4 Epistemological Framing of Diverge/Converge Design Methods.
Source: Author's Creation.

journey, and show various activities that are more divergent or convergent in the design process. The methods appear quite different, yet under the hood, most HCD methodologies run on the same fuel, use the same parts, and handle on the road in greater homogeneity than likely assumed. Indeed, the divisions between these communities of design practice deal more with their identities and associations (and competition therein) than specifically how they go about conducting design.

In Figure 1.4, the Double Diamond method undergoes the same epistemological framing as done in Figure 1.2. The Double Diamond shares numerous overlaps, with subtle differences being in how the client's needs, interests, and perspective are considered as well as a greater prioritization of distinguishing between convergent and divergent activities. Design methods such as the Stanford version begin with "empathy", which certainly coincides with a "people-centered" declaration of the Double Diamond and a divergent orientation to explore the design challenge. IBM's version closely follows this with "understand" paired with interviews and user-centric statements. How designers using various HCD methods interpret these distinctions and apply their own styles and techniques of course provides even greater variation in how design teams will journey through either method. Taking Figures 1.2 and 1.4 together, these various HCD methodologies differ most significantly in methodological depiction over all else.

The British Design Council frames the consumer's user needs by associating this "design challenge" starting point with some design inspiration either from the client, leadership, or possibly another source. This initial step of defining the design challenge and establishing their discovery phase of divergent design iterations is coupled to deliberate market research, user research, management analysis as well as setting up design research groups. This may also create analytical overemphasis, depending how the design team goes about creating the research, the methods they use, and how they ontologically and epistemologically consider what the research means in the contexts they are operating within. Just as with previous HCD methods considered, this one might project a single paradigm upon all design actors and goals, forcing the research to reach conclusions already envisioned by nonreflective designers in action.

As the designers using the Double Diamond method diverge iteratively to generate a wide range of possible and emergent designs, at some point the facilitators will move the team into a convergent process of "definition" where recognizable business objectives begin to narrow down the designs assessed as useful for subsequent development. This includes an elimination of certain other design proposals. With a design team unaware or unwittingly projecting their social paradigm upon critical alternative ones in operation, this methodology could create self-rationalizing loops of business objectives in the convergent "define" and "deliver" phases that fail to appreciate design emergence, nonlinearity, dynamic system transformation as well as destructive innovation.

In the second pronounced phase of divergent, iterative design, the Double Diamond method aligns design novelty into quantifiable design *solutions*. This suggests analytic optimization or validation within existing business or market beliefs if the designers are not able or willing to frame those earlier within the process. Testing and delivery imply further commercial application, although such efforts could provide policymakers, nongovernmental entities, or perhaps a military organization with design options, provided the designers include necessary integration into the paradigms, belief systems, and institutional theories, models, and methods used by those noncommercial clients. Below, the Double Diamond method is framed epistemologically based upon similar examination as the HCD versions in Figures 1.1 and 1.2.

The Double Diamond method was chosen for this chapter due to the unique graphical configuration it presents, which is different from competing HCD methodologies such as with IDEO, Stanford University, or how IBM depicts "Enterprise Design Thinking". Understandably, each organization must in modern commerce differentiate and legally safeguard intellectual property, and in the ever-present need to sell their way of design over that of competitors, the rise of trademarked "our way to design" communities of practice is realized today in a dizzying array of thousands of different design tribes. Many take specific steps to cater to select markets, such as how IBM incorporated software design methods due to what markets they focus upon, while another company that builds physical artifacts might blend HCD with industrial design concepts. This chapter does not advocate for one over another. Instead, the epistemological framing of *why* HCD forms and functions as it does across these wide arrays of different design groups illuminates the core commercial design logic at play beneath the surface.

HCD methodologies appear quite useful for organizational transformation, creating conditions for innovation, and gaining positions of advantage in emergent future systems. There are tremendous numbers of positive outcomes in commercial design activities, and the synthesis provided in this chapter offers readers considerations at the epistemological level how human-centered design can be employed in productive as well as less appropriate applications. Human-centered design is oriented toward new products, user experiences, and gaining market advantage within competitive dynamics. Defense-oriented organizations also use HCD developed outputs. Militaries need better weapons, vehicles, software, and information management. Security forces also use industrial design outputs. More importantly, these defense organizations working within dynamic conflict environments also require unique design methodologies *dissimilar* from human-centered and industrial design methods. This chapter will conclude with this controversial point on designing toward war and that of commerce. Designing for commerce is not the same as designing toward war.

Design Distinctions: Of Warfare and Markets

There remains the question of whether the design landscape is large enough to accommodate designing toward commercial enterprise and security affairs where each necessitates different and potentially dissimilar methods, models, theories and praxis. Buchanan describes this tension of design form/ function with:

> Industrial design tends to stress what is *possible* in the conception and planning of products; engineering tends to stress what is *necessary* in considering materials, mechanisms, structures, and systems; while marketing tends to stress what is *contingent* in the changing attitudes and preferences of potential users.[105]

The inclusion of empathy and the incorporation of dissimilar stakeholders in commercial design methodologies would aid in introducing complexity theory, new organizational theory, sociology, and a deliberate tilt away from the grip of natural sciences in designing differently for a complex reality.[106]

Meanwhile, the military is largely absent from this, except for isolated and anecdotal interactions with commercial design applications. There are outlier examples of some radical, counter-culture concepts experimented within militaries such as the U.S. Army's temporary and originally classified "First Earth Battalion" construct.[107] This small-scale experimental unit created in the 1970s demonstrated a wide range of blended disciplines to include new-age philosophy, radical perspectives on gender, age, sexual identity, holistic medicine, and the exploration of psychic powers for war, yet this classified and wildly counter-culture entity was only briefly entertained and subsequently disbanded.[108] Led by Special Forces Officer Jim Channon, he would critique the Vietnam War failure with the comedic irony that "we relied on smart bombs instead of smart people".[109] Channon was essentially designing with the military institution, and was disrupting and challenging some of the deepest held beliefs at the time.

Some of the biggest differences concerning military and commercial design deal with the context for each in what they are designing toward. Security organizations seek a monopolization of the application of organized violence where they gain and retain an ever-moving point of advantage, so that adversaries are deterred, put into a disadvantageous situation, or defeated and unable to effectively resist. From local police departments to drug cartels, terrorist groups, organized crime syndicates, or national armed forces must be able to present the ability to accomplish some form of organized violence to shape and guide various behaviors and deter others. Violence here is specifically tied to "organization" in that individual acts of violence are different and not applicable in this context. Malešević makes this distinction in *The Sociology of War and Violence* where individual violence tends to be isolated, whereas organized violence that manifests in war requires significant preparation and organization on behalf of the group of humans attempting to compel the organization to execute said violence.[110]

Monopoly of violence is not necessarily the active use of it. Frequently, nonaction is preferred, particularly in the Nuclear Age where deterrence and "limited wars" are intended to prevent the elimination of all life on earth. Organized violence as a concept must include seemingly innocuous things such as peacekeeping, humanitarian aid, and even peace treaties and disarmament policies, as such activities are possible for nation-states capable of retaining their physical forms along with social constructs through some form of enforcement. This monopolization can be expressed in passive, localized manners, such as the presence of a police car on the side of a highway causing passing drivers to slow down. For warfare, nations wield military instruments of power toward rival nations and also valid non-state actors, creating some gray areas for distinguishing organized violence from different types of violence addressed by law enforcement. Sidestepping the differences between terrorism and local criminal enterprises, the ability to impose organized violence (or the potential therein) are activities designed to change human behavior toward the desires of those in control of said violence.[111] Militaries thus must think, act, and reflect on their thinking and acting in security contexts by addressing their ontological (what is and is not real) and epistemological (how knowledge is constructed) frameworks toward organized violence. This permits them to adjust methodologies and experiment with novel ones as the emergent, dynamic conditions in complex security contexts change.

Commercial forms of design do develop novel products and enhance user experiences that are clearly intended for military use. Commercial designers certainly can work on designs within a military field or topic area, particularly in specialized or sophisticated areas the military profession itself lacks the necessary skills. This can easily be found in military affairs in cyberspace, space, quantum applications, or with artificial intelligence and human–machine teaming for security contexts.

For example, army psychological operations efforts using posters and leaflets to influence an enemy or other population likely will benefit from advertising and graphic design experts, while architectural design experts can be brought in to help determine building vulnerabilities in targeted bombing or other destructive munitions in a military context. The distinction made here is that while commercial designers can develop products and user experiences that are directly intended for use by a military or other entity in the pursuit of the monopoly of violence, the commercial enterprise *does not ever do this directly*. The design mission of the commercial sector ends at the transition from an economic transaction between client and producer. The context in which the item is designed and how and why it is applied are, in security forces two different things.[112]

For instance, industrial designers can design and build a new rifle, a different form of tear gas, or perhaps special humanitarian meal packages.

They might design and produce an infinite variety of war tools as well as conceptual constructs for intended utilization in the application of organized violence. What they cannot and do not address is the application of these tools within a complex security context. The security organization is charged with thinking and acting with the range of designed tools as well as designed ideas. In the example of the new rifle, tear gas, and humanitarian meals, it is entirely up to the security force in that context to design how and why they might use any combination of these tools, as well as deal with the design consequences within a complex and emergent security context. The requirements at the ontological, epistemological levels for security design are distinct and unlike commercial ones. Thus, all associated design methodologies nested in these different ontological and epistemological stances will be unlike commercial design models.

The application of violence is itself not a user experience, nor a novel product, although these effects can be imposed through designed war tools. It is also hardly ever a "solution" – rather it is a managed assemblage of various solutions and experiments that systemically produce change in a complex system … it is nonlinear, emergent, and rarely is there a designable "end state" or finish line to hang a finite "solution" or "goal" upon.[113] This of course conflicts not only with commercial design methods, but also more importantly with legacy system military strategy and operational planning logics. The application of organized violence is dynamic, and not contained within pure economic or perhaps capitalistic confines. This sets up the focus of the next chapter in that the modern military institution readily expects goals and end states through traditional strategic and operational planning efforts. Commercial design, in offering much the same, does not accomplish what the recent military design movement does. Military design challenges these conceptualizations, and offers unorthodox alternatives for complex warfare.

Militaries, despite technological and resource advantage in the west, have encountered significant failures in the last several decades, potentially spanning back to 1945. The success of poorly equipped groups such as the Vietcong versus American forces, the Taliban against Soviet and later American forces, the Kurdish rebellion against the Hussain Regime in Iraq, the Arab Revolt against the Ottoman Empire, and many others show that being larger, wealthier, or more technologically advanced does not promise victory. Security forces considered the favorite at the onset of a conflict have often possessed what appears to be better designed ideas or items of warfare, yet nonetheless they somehow misapplied it or failed to achieve security goals. The U.S. Air Force and the Central Intelligence Agency (CIA) dropped more bombs on Laos, a neutral, poor, and agrarian nation in the Vietnam conflict than the combined tonnage of bombs dropped on Germany and Japan in World War II without achieving remotely similar results.[114] Better designed war technology helps, but certainly does not assure clear advantage or victory.

In another key distinction, militaries work through a monopolistic process of production and employment toward the application of organized violence, unlike most other business contexts where commercial design occurs. Militaries tend to routinely enjoy what a business is most often denied – the ability to completely control all aspects of a commodity, resource, or activity. Militaries come in monopolized form by default, as a national instrument of power would not oblige competition from within the state for the same exercise of power. Within military organizations, there is only one way to produce a commando, a specialized ordinance disposal technician, or a fighter pilot. They must enlist or join that particular service, attend and pass that training, and be assigned in those positions. This process breeds intense specialization as well as institutionally controlled and protected means of pro-duction. This means that security forces do gain exceptional control over the quality, quantity, and application of many aspects within their organiza-tion as state-sanctioned monopolies. This also means that unlike businesses where specialized professionals routinely move from industry to industry, the military monopoly restricts professionalization to rigid and sequential career paths. This can impact flexibility, innovation, and divergent thinking. The exclusiveness of monopolistic control creates conditions for stagnation, ritu-alization of irrelevant or outdated practices, and potentially a rigid inability to critically reflect or change with emergent demands or developments.

A third way in which militaries differ through context is the level of restriction for that organization to apply more of its abilities, resources, and methods. For businesses, the more restrictive an environment becomes, the more expensive it is to perform business in that context. In military contexts, the higher the intensity for violence and the collapse of society within that conflict region, the more expensive it becomes for nearly all business to occur. However, an inverse tends to occur for most security forces in that the lower the intensity of violence, the greater the restrictions become for even routine activities. International peacekeeping activities represent some of the most restrictive military contexts where a security force is usually under the oversight and control of multiple local, national and even international authorities. Yet the higher the intensity of conflict become, security forces subsequently see a dramatic expansion of options coupled with a reduction of control and restrictions. If a military finds itself fighting for the very survival of a nation's existence, nearly all restrictions are removed.

Militaries are unlike all commercial industry in that they are, by their established nature as state-sanctioned instruments of power, able to be directed into extremely dangerous contexts by political decision-makers. Further, only in the military context does a security organization tolerate the distinct possibility of severe injury, death, and destruction within not only the application of military power in real conflict environments, but also in the rehearsal, practice, and training environments that exist only to simulate and prepare the organization to operate in actual conflict contexts.[115] While many commercial enterprises chose to operate in hostile, austere, or otherwise dan-gerous contexts in pursuit of profit and advantage against competitors, the key operating term is "choice" in that a military instrument of power does not have final authority on where or when it chooses to operate. There are no required oaths or ceremonies of alliance in commercial enterprise outside of legal contracts and non-discloser matters for competition bounded once more by commerce and not national policy and matters of foreign policy.[116]

Militaries, particularly those western industrialized ones addressed in this research, enjoy industrialized state wealth, power, and abilities, and are the single expeditionary entity for their national government. This means that they can express organized and state-sanctioned violence as an extension of that national will to act. Military philosopher Carl von Clausewitz framed this as follows: "war is an extension of policy by other means".[117] Militaries in the modern context exist solely to project the threat or application of organized violence out into hazardous, hostile, lawless, or otherwise crisis-stricken geographic part of the world where military presence provides that nation's policymakers with distinct advantage. These militaries will operate in these contexts at the direction of their political superiors and attempt to accomplish strategic goals on behalf of political policymakers regardless of the level of risk, danger, or other notion of threat that otherwise would influence all commercial enterprises differently. Security forces can inform political leadership of the extraordinary costs associated with high-risk security activities, but they do not get to decline the directive to act. They exist to serve at the disposal of political leadership in service of the nation.

This raises one last distinction for why military design contexts differ from commercial endeavors. While all businesses can go out of business, nearly all militaries are immune from elimination, *even at the defeat of a nation*. Armies are defeated and although the individuals within that army may be removed, some security force will be restored and used for the nation or territory for the security and defense of that area. Many senior leaders in the Afghanistan security forces in the mid-2000s were earlier working with Soviet forces prior to being defeated by Taliban insurgents in the 1990s, only to return to power after the post-9–11 American invasion in 2001. Militaries are not businesses, and they are not-for-profit enterprises that happen to cost quite a bit of money to run. There are business aspects within militaries, as well as many business-like behaviors, qualities, and relationships. One might argue the modern military industrial complex is indeed a business, yet within this convoluted mass of companies, lobbyists, politicians, and military services, the security forces themselves are still bound by noncommercial rules and regulations. *Militaries are not businesses*, and consequence the direct application of business-inspired design done without addressing these uniquely military contexts will be insufficient for military-specific challenges that security organizations must address.

There is also the matter of control and whether one can choose to cease to continue unprofitable or undesired activities. Security forces, as non-profit entities operating as monopolistic instruments of state power, frequently become victims of maintaining and continuing projects that are not profit-driven, especially long after any commercial enterprise would have abandoned them or gone bankrupt. Israeli Defense Forces Commando and postmodern designer Gal Hirsch in his autobiography would offer the following on how militaries frame the concept of "profit" as distinct and unrelated to commercial contexts: "Unlike the business world, in the military realm, the 'profit' is reflected by the ability to create an effective and timely operational response".[118]

As profit is perhaps one of the most fundamental characteristics defining commercial enterprise, this dissimilar military interpretation of it makes for profound consequences when a commercial design methodology is applied (or more likely misapplied) to security challenges where the application of organized violence is paramount. Militaries can continue self-destructive behaviors where profit is not a factor (in the same regard) because they conduct activities through political, societal or institutional consensus instead of profit. Historian Carl Builder explored this in his study of the U.S. Army, Navy, and Air Force and how they often pursue institutionally self-serving activities at the expense of other armed services and even at the detriment of national interests.[119] This occurs through policy, public demand, culture, internal organizational self-interests, or often a combination of all of these. Rarely can any commercial enterprise do this and survive for long in a competitive marketplace.

Militaries, even if defeated, do not usually disappear, and frequently maintain the same form and function. Often, militaries extend outdated practices and behaviors well overdue into future wars, intending to win the next war by reimagining the last one. Often, militaries uphold their shared belief systems and values that feature strong elements of ritualization, where a military defines itself by what is has historically done for a nation, instead of what it might need to design away from. Many marine organizations remain wedded to amphibious operations and airborne infantry units to mass-tactical vertical envelopments despite the future feasibility and effectiveness of such activities. Air forces struggle with the possibility of unmanned aviation systems pushing out the need for human pilots. Militaries struggle with cyberspace, and institutionally often frown upon missions or tasks that prevent them from doing what they institutionally value or cherish. Commerce rarely functions this way, with monopolies usually dismantled, and slow adaptors often put out of business. Yet security forces can continue to lurch forward in self-serving, often wasteful or unproductive ways, provided they secure resources and direction from policymakers and the society they support to continue such activities.

Conclusions

There are clear distinctions in what design is depending on the *context*. Designers use different theories, models, and methods complete with unique language and underlying belief systems of that institution or community. This chapter established the primary distinctions of how design differs in two broad yet dissimilar areas of practice: that of war and commerce. These communities of design practice emerged over the last few centuries in often overlapping and symbiotic ways where the industrialization of modern society also drove warfare to new horrific heights of destruction, devastation, death, and transformation. Without designers, neither would have occurred. Yet today many design practitioners remain confused or unaware why such a distinction exists in designing toward security affairs or that of commerce, and whether the methodology harnessed in one might not simply translate over to the other one with minor adjustment. We have, as an interdisciplinary community of practice, forgotten how we arrived at this point, and we lack awareness of how commercial and military design would evolve interdependently. The chicken and the egg would manifest together, along with the designed hand grenade as both an economic construct and a technologically needed tool for war. Design appears to fold, unfold, and refold infinitely through multitudes of human wants and needs, paradoxically creating differences and distinctions as well.[120]

The next chapter addresses the development of security forces over the centuries where premodern (both of antiquities and the European Feudal Age) militaries transformed into modern security forces, and how the challenges of contemporary security contexts are ushering in some postmodern world that requires different thinking in warfare. Postmodernism is a controversial topic, yet one that is essential to any deep discussion of what military design is, and how it formed in the 1990s in the Israeli Defense Forces and spread around the globe. The force required to apply organized violence is a brutal, dangerous, usually horrifying action deeply rooted in all of human history. To design within these contexts takes particular cognitive requirements, attributes, skill and imagination. To understand how these create entire design methodologies of warfare will also require continuing an epistemological study and reflection of how the application of organized violence unfolds through human interactions.

Notes

1 War is the overarching human construction of organized violence to achieve some societal, political, or related desire, while warfare is how societies understand to go about waging war in specific contextual, geographical, technological, and rule-based manner. The philosophical inquiry executed in this book seeks not only the "what" and "how", but also the "why" of both how humans understand and create war and warfare.

2 Philippe Beaulieu-Brossard and Philippe Dufort, "The Archipelago of Design: Researching Reflexive Military Practices," The Archipelago of Design: Researching Reflexive Military Practices, 2017, www.militaryepist emology.com; Aaron Jackson, "Towards a Multi-Paradigmatic Methodology for Military Planning: An Initial Toolkit," *The Archipelago of Design Blogs* (blog), March 4, 2018, http://militaryepistemology.com/multiparadigm2018/

3 Victor Papanek, *Design for the Real World: Human Ecology and Social Change* (New York: Pantheon Books, 1971), 21.

4 Bruce Archer, "The Structure of Design Processes" (thesis/dissertation/report, Royal College of Art, London, 1968), 5.

5 Kees Dorst, *Frame Innovation: Creating New Thinking by Design*, Design Thinking, Design Theory (Cambridge, Massachusetts: MIT Press, 2015), vii.

6 Buckminster Fuller, "Introduction," in *Design for the Real World: Human Ecology and Social Change*, by Victor Papanek (New York: Pantheon Books, 1971), vii–viii.

7 Richard Boland Jr. and Fred Collopy, "Designing Matters for Management," in *Managing as Designing*, ed. Richard Boland Jr. and Fred Collopy (Stanford, California: Stanford Business Books, 2004), 4.

8 Richard Buchanan, "Wicked Problems in Design Thinking," *Design Issues* 8, no. 2 (Spring 1992): 18.

9 Donella Meadows, *Thinking in Systems: A Primer*, ed. Diana Wright (White River Junction, Vermont: Chelsea Green Publishing, 2008), 106.

10 Cara Wrigley, Genevieve Mosely, and Michael Mosely, "Defining Military Design Thinking: An Extensive, Critical Literature Review," *She Ji: The Journal of Design, Economics, and Innovation* 7, no. 1 (Spring 2021): 117–119; Ken Friedman, "Editorial," *She Ji: The Journal of Design, Economics, and Innovation* 7, no. 3 (Autumn 2021): 303–308.

11 Klaus Krippendorff, "Propositions of Human-Centeredness; A Philosophy for Design," in *Doctoral Education in Design: Foundations for the Future: Proceedings of the Conference Held 8–12 July 2000, La Clusaz, France* (Staffordshire (UK): Staffordshire University Press, 2000), 2.

12 Jean-Pierre Protzen and David Harris, *The Universe of Design: Horst Rittel's Theories of Design and Planning* (New York: Routledge, 2010), 23.

13 Siniša Malešević, *The Sociology of War and Violence* (Cambridge, United Kingdom: Cambridge University Press, 2010), 259.

14 Peter Rowe, *Design Thinking*, Seventh (Massachusetts: MIT Press, 1998), 41; Stephen Waring, "Taylorism and Beyond: Bureaucracy and Its Discontents," in *Taylorism Transformed: Scientific Management Theory since 1945* (Chapel Hill, North Carolina: The University of North Carolina Press, 1991), 9–19.

15 Anthony Giddens, *The Consequences of Modernity* (Stanford, California: Stanford University Press, 1990), 60.

16 Ricardo Herrera, "History, Mission Command, and the Auftragstaktik Infatuation," *Military Review*, August 2022, 53–66.

17 Justin Kelly and Michael Brennan, "Alien: How Operational Art Devoured Strategy" (Carlisle, Pennsylvania: Department of the Army's Strategic Studies Institute, September 2009), 61–62.

18 Malešević, *The Sociology of War and Violence*, 126.

19 There are many more design theorists and thought leaders in the last two decades, however for the research focus of this chapter, the first design theory from the late 1950s–1990s represents the primary sources for how and why contemporary commercial design exists today. Primary theorists are subsequently identified and cited within this chapter.

20 Papanek, *Design for the Real World: Human Ecology and Social Change*.

21 When designers do address military or paramilitary organizations, it is usually in how design research can explain various universal phenomena. See Weick

and Roberts on naval aircraft carrier culture, and Weick on U.S. Forestry Service smoke jumpers: Karl Weick and Karlene Roberts, "Collective Mind in Organizations: Heedful Interrelating on Flight Decks," *Administrative Science Quarterly* 38, no. 3 (September 1993): 357–381; Karl Weick, "Drop Your Tools: An Allegory for Organizational Studies," *Administrative Science Quarterly* 41 (1996): 301–313.

22 Malešević, *The Sociology of War and Violence*; Siniša Malešević, "The Organization of Military Violence in the 21st Century," *Organization* 24, no. 4 (2017): 456–474.

23 Chris Gray, *Postmodern War: The New Politics of Conflict* (New York: The Guilford Press, 1997), 230.

24 Peter Gossel, ed., *The Bauhaus: 1919–1933; Reform and Avant-Garde*, trans. Maureen Roycroft Sommer, Basic Art Series 2.0 (Slovakia: TASCHEN, 2019), 15.

25 Magdalena Droste, *Bauhaus: 1919–1933*, ed. Angelika Taschen and Nicole Opel, trans. Jane Michael and Karen Williams, Second Edition (Poland: TASCHEN, 2019), 30–40; Papanek, *Design for the Real World: Human Ecology and Social Change*, 23–24.

26 Ellen Lupton and J. Abbott Miller, eds., *The ABCs of [Triangle, Square, Circle Shapes]: The Bauhaus and Design Theory*, Second (New York: Princeton Architectural Press, 2019), 40.

27 Droste, *Bauhaus: 1919–1933*, 26–27.

28 Lupton and Miller, *The ABCs of [Triangle, Square, Circle Shapes]: The Bauhaus and Design Theory*, 43.

29 Gossel, *The Bauhaus: 1919–1933; Reform and Avant-Garde*, 92.

30 Anatol Rapoport, *The Origins of Violence: Approaches to the Study of Conflict* (New Brunswick, New Jersey: Transactions Publishers, 1995), 142–147.

31 Gossel, *The Bauhaus: 1919–1933; Reform and Avant-Garde*, 15.

32 Eric Shaw, "Marketing Strategy: From the Origin of the Concept to the Development of a Conceptual Framework," *Journal of Historical Research in Marketing* 4, no. 1 (2012): 32, https://doi.org/10.1108/17557501211195055

33 Archer, "The Structure of Design Processes," 3.

34 Klaus Krippendorff, "Principles of Design and a Trajectory of Artificiality," *Product Development & Management Association* 28 (2011): 411–412; Buchanan, "Wicked Problems in Design Thinking," 9; Archer, "The Structure of Design Processes," 5.

35 Rapoport, *The Origins of Violence: Approaches to the Study of Conflict*, 144.

36 Daisuke Wakabayashi and Scott Shane, "Google Will Not Renew Pentagon Contract That Upset Employees," *The New York Times*, June 1, 2018, www.business-humanrights.org/en/latest-news/google-will-not-renew-pentagon-contract-that-upset-employees/

37 Molly Wood, "Some Tech Employees Don't Want Their Work Used by the Military," *Marketplace Tech Blog* (blog), February 27, 2019, www.marketplace.org/2019/02/27/a-lot-of-tech-employees-dont-like-the-of-idea-their-work-being-used-by-the-military/

38 Paul Virilio and Sylvere Lotringer, *Pure War*, trans. Mark Polizzotti, "Twenty Five Years Later" edition (Cambridge, Massachusetts: MIT Press, 2008), 19.

39 Bernard Tschumi, *Architecture and Disjunction* (Cambridge, Massachusetts: MIT Press, 1997); Bruno Munari, *Design as Art* (New York: Penguin Books, 1966); Papanek, *Design for the Real World: Human Ecology and Social Change*.

40 Robert Chia and Robin Holt, *Strategy without Design: The Silent Efficacy of Indirect Action* (New York: Cambridge University Press, 2009).

41 This does not include postmodern criticism of militaries or conflicts, which were extensive in the 1970s–1990s but ignored or dismissed by military theorists, doctrine writers and leadership.

42 Archer, "The Structure of Design Processes," 42.

43 Christopher Paparone, *The Sociology of Military Science: Prospects for Postinstitutional Military Design* (New York: Bloomsbury Academic Publishing, 2013), 13–15.

44 Dorst, *Frame Innovation: Creating New Thinking by Design*, 42.

45 Archer, "The Structure of Design Processes," 66.

46 Meadows, *Thinking in Systems: A Primer*, 146.

47 Krippendorff, "Principles of Design and a Trajectory of Artificiality," 413.

48 Rapoport, *The Origins of Violence: Approaches to the Study of Conflict*, 97–108; Robert Chia, "From Modern to Postmodern Organizational Analysis," *Organization Studies* 16, no. 4 (1995): 579–604.

49 This is indeed a very brief and potentially insufficient summarization. A deeper analysis of postmodern theory and how it influenced design methodologies for security affairs is addressed in Chapter 2.

50 Martin Kilduff and Ajay Mehra, "Postmodernism and Organizational Research," *Academy of Management Review* 22, no. 2 (1997): 469–470.

51 Aaron Jackson, "Civilian and Military Design Thinking: A Comparative Historical and Paradigmatic Analysis, and Its Implications for Military Designers" (IMDC 2019 Conference, Lancaster, UK: unpublished, 2019), 1–17.

52 Dorst, *Frame Innovation: Creating New Thinking by Design*, 1–10; Papanek, *Design for the Real World: Human Ecology and Social Change*, 137–142.

53 Malešević, "The Organization of Military Violence in the 21st Century," 460.

54 Buchanan, "Wicked Problems in Design Thinking"; Jeff Conklin, "Wicked Problems and Social Complexity," in *Dialogue Mapping: Building Shared Understanding of Wicked Problems* (CogNexus Institute, 2008), www.cognexus.org; Conklin.

55 Krippendorff, "Propositions of Human-Centeredness; A Philosophy for Design," 2–4.

56 Papanek, *Design for the Real World: Human Ecology and Social Change*, xxii.

57 Protzen and Harris, *The Universe of Design: Horst Rittel's Theories of Design and Planning*; Chanpory Rith and Hugh Dubberly, "Why Horst W.J. Rittel Matters," *Design Issues* 23, no. 1 (Winter 2007): 72–74.

58 Herbert Simon, *The Sciences of the Artificial*, Third Edition (Cambridge, Massachusetts: The MIT Press, 1996).

59 Rittel in multiple original lectures addresses his disagreements with Simon's theory. See: Protzen and Harris, *The Universe of Design: Horst Rittel's Theories of Design and Planning*; Chia and Holt, *Strategy without Design: The Silent Efficacy of Indirect Action*, 45.

60 Krippendorff, "Principles of Design and a Trajectory of Artificiality," 413.

61 Protzen and Harris, *The Universe of Design: Horst Rittel's Theories of Design and Planning*, 64.

62 Protzen and Harris, 24–25; Archer, "The Structure of Design Processes," 70–78.

63 Papanek, *Design for the Real World: Human Ecology and Social Change*.

64 Krippendorff, "Propositions of Human-Centeredness; A Philosophy for Design," 1.

65 Haridimos Tsoukas, "A Dialogical Approach to the Creation of New Knowledge in Organizations," *Organization Science* 20, no. 6 (December 2009): 946.

66 John Gero and Udo Kannengiesser, "An Ontology of Donald Schön's Reflection in Designing" (Key Centre of Design Computing and Cognition, University of Sydney, November 26, 2015); Donald Schön, *Displacement of Concepts* (London: Tavistock Publications (1959) Limited, 1963).

67 Krippendorff, "Propositions of Human-Centeredness; A Philosophy for Design," 1–4.

68 Buchanan, "Wicked Problems in Design Thinking," 18.

69 Archer, "The Structure of Design Processes," 7.

70 Harold Nelson and Erik Stolterman, *The Design Way*, Second (Cambridge, Massachusetts: The MIT Press, 2014).

71 Rowe, *Design Thinking*, 41.

72 Jesper Simonsen et al., eds., "Situated Methods in Design," in *Situated Design Methods* (Cambridge, Massachusetts: MIT Press, 2014), 1–19; Henry Mintzberg, "Patterns in Strategy Formation," *Management Science* 24, no. 9 (May 1978): 944.

73 Naomi Stanford, *Guide to Organisation Design: Creating High-Performing and Adaptable Enterprises* (London: Profile Books Ltd, 2007), 47–50.

74 Haridimos Tsoukas, *Complex Knowledge: Studies in Organizational Epistemology* (New York: Oxford University Press, 2005), 290–293.

75 Papanek, *Design for the Real World: Human Ecology and Social Change*; Protzen and Harris, *The Universe of Design: Horst Rittel's Theories of Design and Planning*; Rith and Dubberly, "Why Horst W.J. Rittel Matters."

76 Frank Gehry, "Reflections on Designing and Architectural Practice," in *Managing as Designing*, ed. Richard Boland Jr. and Fred Collopy (Stanford, California: Stanford Business Books, 2004), 19–35; Papanek, *Design for the Real World: Human Ecology and Social Change*; Tschumi, *Architecture and Disjunction*; Krippendorff, "Principles of Design and a Trajectory of Artificiality," 415.

77 Krippendorff, "Principles of Design and a Trajectory of Artificiality," 415.

78 Krippendorff, "Propositions of Human-Centeredness; A Philosophy for Design," 2.

79 Rowe, *Design Thinking*, 39.

80 Aaron Jackson, "Introduction: What Is Design Thinking and How Is It of Use to the Australian Defence Force," *Australian Journal of Defence and Strategic Studies*, Joint Studies Paper Series, no. 3 (December 2019): 3.

81 Figure 1.1 and 1.2 are demonstrations of how these HCD methodologies operate, but they are not precisely an exact representation of how many companies present their design processes. Additionally, many schools and companies change the aesthetics of their design format for marketing reasons, yet the methodological function remains unchanged.

82 Seth Johnson, ed., *IBM Design Thinking Field Guide Version 3.1* (IBM Corporation, 2015), ibm.biz/idt_fieldguide.

83 Willemien Visser, "Schön: Design as a Reflective Practice," *Collection*, Art+Design & Psychology, no. 2 (2010): 21–25; Karl Weick, *Sensemaking in Organizations*, Foundations for Organizational Science (Book 3) (California: Sage Publications, 1995).

84 Krippendorff, "Propositions of Human-Centeredness; A Philosophy for Design," 2.

85 Bruce Nussbaum, "Design Thinking Is a Failed Experiment. So What's Next?," *Fast Company* (blog), April 5, 2011, www.fastcompany.com/1663558/design-thinking-is-a-failed-experiment-so-whats-next

86 Nussbaum.

87 Natasha Iskander, "Design Thinking Is Fundamentally Conservative and Preserves the Status Quo," *Harvard Business Review*, Entrepreneurship, September 5, 2018, https://hbr.org/2018/09/design-thinking-is-fundamentally-conservative-and-preserves-the-status-quo

88 Protzen and Harris, *The Universe of Design: Horst Rittel's Theories of Design and Planning*, 151–158; Archer, "The Structure of Design Processes," 8–9.

89 Protzen and Harris, *The Universe of Design: Horst Rittel's Theories of Design and Planning*, 7.

90 Protzen and Harris, 116.

91 Buchanan, "Wicked Problems in Design Thinking," 15.

92 Rowe, *Design Thinking*, 37.

93 Buchanan, "Wicked Problems in Design Thinking," 15–16.

94 Protzen and Harris, *The Universe of Design: Horst Rittel's Theories of Design and Planning*; Horst Rittel and Melvin Webber, "Dilemmas in a General Theory of Planning," *Policy Sciences* 4 (1973): 155–169.

95 Mary Jo Hatch, *Organization Theory: Modern, Symbolic, and Postmodern Perspectives*, Third Edition (Oxford, United Kingdom: Oxford University Press, 2013), 230.

96 Rowe, *Design Thinking*, 37.

97 Majken Schultz and Mary Jo Hatch, "Living with Multiple Paradigms: The Case of Paradigm Interplay in Organizational Culture Studies," *Academy of Management Review* 21, no. 2 (1996): 529–557; Dennis Gioia and Evelyn Pitre, "Multiparadigm Perspectives on Theory Building," *Academy of Management Review* 15, no. 4 (1990): 584–602.

98 Papanek, *Design for the Real World: Human Ecology and Social Change*.

99 The author deployed for 12 months as a strategic planner and designer for NATO Training Mission: Afghanistan (NTM-A) in 2012, and the entire Afghan Police strategy for training and equipping centered exactly on these parameters. They remained as such until the collapse of the Afghan government in 2021.

100 Saeed Shah, "Taliban Commander Who Launched Bombings in Kabul Is Now a Police Chief in Charge of Security," *The Wall Street Journal*, October 20, 2021, www.wsj.com/articles/taliban-commander-who-launched-bombings-in-kabul-is-now-a-police-chief-in-charge-of-security-11634740097

101 Natasha Jen, "Design Thinking Is B.S.," *Fast Company* (blog), April 9, 2018, www.fastcompany.com/90166804/design-thinking-is-b-s

102 Katharine Schwab, "IDEO Breaks Its Silence on Design Thinking's Critics," *Fast Company* (blog), October 29, 2018, www.fastcompany.com/90257718/ideo-breaks-its-silence-on-design-thinkings-critics

103 Schwab.

104 GK VanPatter, "Double Diamond Method: Understanding What Was Missed," digital online blog, *Humantific.Com* (blog), April 24, 2019, www.humantific.com/post/double-diamond-method. VanPatter alleges that despite Double Diamond advocates tracing its origin to Bela Banathy's 1996 book *Designing Social Systems in a Changing World*, Banathy allegedly took some inspiration from earlier creative problem solving methods such as the "Five Diamond Model" in practice in the 1970s.

105 Buchanan, "Wicked Problems in Design Thinking," 20.
106 Protzen and Harris, *The Universe of Design: Horst Rittel's Theories of Design and Planning*, 25–28.
107 Select elements of the First Earth Battalion concept paper demonstrate several design concepts as well as numerous "New Age" philosophical ideas that were popular in the 1960–1970s and coincided with significant military introspection and frustration as the Vietnam War came to a close. The group was highly controversial and eventually disbanded and rejected by the larger force.
108 Jim Channon, "The First Earth Battalion: Ideas and Ideals for Soldiers Everywhere" (declassified U.S. Army internal concept paper, 1979), www.scr ibd.com/doc/21926670/The-First-Earth-Battalion-Field-Manual
109 Gray, *Postmodern War: The New Politics of Conflict*, 205.
110 Malešević, *The Sociology of War and Violence*, 90–98.
111 This observation is value neutral. Whether one is a blood-thirsty dictator bent on world domination or the President of a small nation defending itself against an aggressor, both control the power and authority to apply organized violence.
112 The matter of military contractors is not addressed here. Contractors typically hold to these design distinctions, in that the service provided is a technical or specific activity/purpose that does not directly involve acts of warfare for legal purposes.
113 Brian Bloomfield, Gibson Burrell, and Theo Vurdubakis, "Licence to Kill? On the Organization of Destruction in the 21st Century," *Organization* 24, no. 4 (2017): 2; Jeremiah Monk, "End State: The Fallacy of Modern Military Planning" (Research Report, Air War College, Maxwell Air Force Base, Alabama, April 6, 2017).
114 Santi Suthinithet, "Laotian Refugees Reach out to Aid Their War-Torn Country," *Hyphen*, 2010, http://legaciesofwar.org/resources/books-docume nts/land-of-a-million-bombs/
115 Morris Janowitz, *The Professional Soldier: A Social and Political Portrait*, Paperback Edition (New York: Free Press, 2017), xv.
116 Many commercial enterprises hold to national security protocols, clearances, and handle extremely sensitive government materials or support classified missions and activities. These relationships do blur the boundaries here slightly, but the framework for orienting toward commerce and profit and that of security affairs and war are still visible.
117 Peter Paret, "Clausewitz," in *Makers of Modern Strategy: From Machiavelli to the Nuclear Age*, ed. Peter Paret (Princeton, New Jersey: Princeton University Press, 1986), 186–213.
118 Gal Hirsch, *Defensive Shield: An Israeli Special Forces Commander on the Front Line of Counterterrorism, the Inspirational Story of Brigadier General Gal Hirsch* (Jerusalem: Gefen Publishing House, Ltd, 2016), 132.
119 Carl Builder, *The Masks of War: American Military Styles in Strategy and Analysis* (Baltimore: John Hopkins University Press, 1989).
120 David Pick, "Rethinking Organization Theory: The Fold, the Rhizome and the Seam between Organization and the Literary," *Organization* 24, no. 6 (2017): 800–818.

2   Premodern, Modern, and Postmodern War Designs

Throughout human history, militaries have been organized and employed by leaders of civilizations to accomplish a range of different security goals. In most contexts, militaries accomplished war objectives using methods that were indeed reliant upon some flawed theories, incomplete or inefficient models, or using language and metaphoric devices that would be later criticized as limited or counter-productive from those using other perspectives or belief systems. Across the centuries, some militaries proved more effective than others, and humanity would gain and lose abilities through the chaotic and emergent pathways of societal development. Regularly, practical wisdom combined with ideological or cultural guidelines would sharpen the point of most war machines, while in the last few centuries, that machine would become scientifically far more destructive than ever before. Mental models, metaphors, as well as theories came from many different design paths, collecting in the heads of those tasked with waging organized violence against cunning adversaries. Often too, those adversaries battled using other combinations of war concepts inside of their skulls, and frequently the underdog could shock and surprise, whether disadvantaged in numbers, resources, technology, or knowledge. From premodern warfare through present day, many ways of thinking and acting in human struggles would manifest, with rarely a best or perfect solution to endure the test of time.

Indeed, many successful violent groups today defeat larger, more technologically resourced and educated opponents routinely, while hybridization of primitive warfighting concepts with advanced ones creates even stranger yet strikingly effective configurations. The Taliban stormed Kabul in 2021 as American forces struggled to evacuate citizens in a sloppy, strategically muddled withdrawal. Cunning drug cartels operate remote submersibles to move shipments past networked border security efforts. Decentralized, anonymous hacker groups can act with near impunity while striking powerful nations amid Russia and the Ukraine's more conventional armed conflict. The world's deadliest terrorist group in 2015 was not the sophisticated and well-resourced al-Qaeda or Hezbollah, rather a regional Nigerian group referred to as Boko Haram.[1] The group, declaring a desired return to a feudal age, low-technology, Islamic lifestyle, has since 2009 displaced over two

DOI: 10.4324/9781003387763-3

million people, killed thousands, uprooted modern education while successfully holding off against African, NATO and American military and security
efforts.[2] The fact that they could do this using simple technology, terror, and
information warfare against vastly more resourced and equipped opponents
demonstrates this pattern that war is never a sure thing, nor are any rules
hard and fast. Even the most astoundingly advanced militaries stumble and
fail today, regardless of how closely they follow the best designed methods
available.

In the last few decades, sufficient change and complexity have increased
in myriad aspects of human civilization so that a growing minority of
designers within security studies propose a redesign of not only methods,
models, and practices, *but even the normally unquestioned theories of warfare and
institutionalized belief systems themselves.* This is not necessarily a new idea, as
the transformation of war (including form, function) over time has been
argued in many ways, in different eras or periods by theorists from security
studies, military history and other disciplines.[3] In this chapter, we will consider how that the military design movement is a new manifestation of
disruptive, novel conceptualization that puts nothing off limits, including
the last few centuries of scientific ordering of how humans understand war.
This chapter outlines how humans have understood and applied organized
violence within a pre-modern, modern, and arguably now a *postmodern*
structure. On this controversial point, various groups offer positions that
society is entering a postmodernity, war is becoming postmodern, or the
organizations waging warfare are themselves transforming into postmodern
developments. Each of these require investigation, as war is designed differently whether one is using a premodern, modern, or postmodern frame
for warfare.[4]

The application of state-organized violence for political, societal, and religious interests has preoccupied generations of soldier-statesmen, theorists,
and leaders while the epistemological underpinnings of war's nature,
behavior, and meaning has usually left to the philosophers. The concepts
addressed in this chapter set up the subsequent examinations of different
design constructs and address the formalized need for design as an emergent
subset of the security studies field where in previous military periods it did
not exist except tacitly and informally. This is best done philosophically, as
otherwise we may not consider much beyond our current belief system in
operation. How war would be designed for action in premodern times would
draw from observations, logical rationalization, inspiration from ideologies, and patterns of successful past behaviors. War would be redesigned in
modernity to assume (or mimic) the precision and rigor of natural sciences,
reformed within shifting societal understanding of national and international
orderings for life. Over the last seven decades, postmodern thinkers have
challenged whether society is now in some postmodern reality, and in the last
few decades, military design communities have carried those controversial
ideas directly into military organizations.

Defining Premodern War: Ritualized, Localized, and Brutalized

The premodern movement concerning war begins in the antiquities, where the first human records of war become available. This extends into the Dark Ages where parts of the world such as Europe slid backward, losing, or forgetting much of what had been developed and curated.[5] From Greek military practices to that of the High Feudal Age, we will focus upon several centuries leading up to European modernization coinciding with rapid industrialization of warfare. European feudalism, stretching from the ninth through mid-sixteenth centuries, is selected for deeper inquiry due to how the changes in this premodern period directly link to the rise of a modern, professional, and unavoidably European-centric model for warfare. This does not discount amble material on premodern warfare thinking such Sun Tzu's seminal works around 500 BC in ancient China or others outside the European continent.[6] This summarization of premodern warfare orients toward how Europe would rise into a dominant, international, military colossus in modernity, and now teeters in the uncertainty and confusion of what can be called a postmodern shift.

To grasp the premodern framework of theories, models, methods, and language concerning *their design of war*, we need to consider how feudal societies organized and functioned differently than modern civilization. The feudal model would position a minority of powerful, wealthy elites that nonetheless were dependent upon the limited scope and skill of their vassals. Both groups were fixed in perpetual class roles, with the minority elites passing their status, wealth, and power to their offspring. The majority poor would be unable to escape passing the same onto their children too. This was set within a land-based economy that helped consolidate power, authority, and wealth to an elite minority, dependent upon a majority to work upon that land and provide blood to protect it.[7] Eventually, inventions such as the printing press, a money-based economic system, and political developments to challenge the autocratic structure of premodern societies would transform these things. The world changed slowly, with one century having much in common with the next, until rapid developments transformed European societies into Imperial Empires of fantastic power and reach. Such change would occur through war and conquest, as well as through new ideas, technological breakthroughs, and paradigm shifts in shared belief systems.

People thought of war differently in premodern times than today, which is difficult in that we often struggle with cognitive biases that insert modern sensibilities and values against what was entirely dissimilar, strange, and often lost in historical records. War in a premodern sense would not be rationalized with nation-state bureaucracies or any systematic application of science, technology, or international rules and norms. Nor would it be bounded in ways that declared who could participate in war with those that could not.[8] Ritualized, premodern war by design defined how organized violence could be applied, almost always involving some autocratic, deity-mandated leader to declare and wage war using nobles to organize armies of conscripts, slaves, and mercenaries.[9]

Modern ideas of warfare again fail to illuminate how different premodern wars were designed to occur. Jaynes highlights the shocking reality that largely illiterate, rural peasant girls (potentially hallucinating with drugs or even schizophrenic) were recruited to serve as Oracles and be the central method of decision-making for over a thousand years in ancient Greece. Most wars required oracle decisions as necessary precursors. While the masses of Greeks faithfully accepted Oracle divine guidance, historians show evidence that much of this was performance art with powerful elites bribing the girls in so that divine declarations aligned to their own designs.[10] This perform-ance, illusionary and ritualized as it was, nonetheless defines a central part of the most advanced civilization at that time for decisions involving policy and war. How information existed in premodern society differed, and for most of the poor or professionalized military involved, challenging such matters could be fatal. Premodern war was designed for elites to steer toward adver-saries and competitors, which also held numerous risks and rewards.

Whether a premodern society could go to war, defend against invasion, or otherwise secure itself reliably would be nested in cost, availability of amateur farmer-soldiers, seasonal and weather concerns, geography, cul-ture, and the significant technological and informational limitations defining the premodern world. Van Creveld describes the typical fighting season for Ancient Greek through feudal age armies as follows: "An army on campaign normally spent no more than a handful of days in actual combat. Much of the largest part of the season was always taken up by something best described as a mixture of tourism and large-scale robbery."[11] War could be justified or decried in many ways, from the drug-fueled babbles of oracles to the religious inspiration of a king, religious leader, or wealthy nobleman. War moved slowly and frequently would be sporadic affairs where actual fighting need not occur at all to accomplish limited political, tangible, or symbolic objectives.[12] Much of premodern war was indeed a sometimes-bloody per-formance art, in that superior positioning could force an army to flee, force terms of negotiation, or create lengthy siege conditions to starve a population into surrender.[13] Often the military force able to occupy a powerful position could defeat the premodern opponent without a strike upon a combatant or avoid pitched battle unless all advantages ceded to the aggressor.[14] This also does not ignore the exhaustive history of brutal, destructive acts of warfare well chronicled in that same history.

Premodern Warfare: Dangerous Games, Rules, and Rituals

The premodern military amateur could wage war and apply organized violence with horrific, gruesome effects.[15] This design of graphic violence required brutal methods, in that the impact of such activities could in turn influence those populations that learned of the actions and therefore might change their behaviors to support those capable and willing to repeat such actions. Foucault would liken premodern punishments with the atmosphere of a festival, where rituals and societal expectations of the act of punishment were public and despite the extraordinary brutality of many punishments, they were carried out with a festive, socially engaged manner.[16] Indeed, many societies demonstrate awareness of death with holidays, rituals, and festivals where the living interact with the dead, with these dating back to premodern origins.

Malešević explains the premodern pattern of conducting gruesome violence at a local level due to the feudal limitations of time, space, and resources.[17] A savage application of organized violence could scale upward and influence behaviors of surrounding populations. A horrific yet localized act could help that army achieve larger effects to circumvent the limitations of the feudal age itself. Word can travel fast, and a gruesome story might travel even faster. Amateurs were not necessarily less violent or less capable than modern professional soldiers if one discounts scaling war technology. Rather, they represent an earlier period where the slow speeds and high costs of warfare meant that a concentrated act in one location, done properly, could influence surrounding areas into suing for peace without as much resistance. Contextually, premodern acts of brutality would be local due to technological reach and the limitations of muscle and natural power. The longer arm of modern militaries would not equate to less brutal actions either.

Premodern militaries consisted of massed forces of ignorant, illiterate, poor or enslaved forces, enhanced (or at times, substituted) by trained mercenaries, and led by elite amateurs of title and status. The term *amateur* is not pejorative, in that military leaders would assume command of militaries briefly, with rarely any special military education, training, or experience. Successful premodern leaders appear to be "self-made", ambitious and cunning as well as creative in war. This pattern extends into today, where many participants in contemporary warfare are raw conscripts, untrained, irregular, or otherwise unorthodox combatants lacking formal military education or development. This of course does not prevent them, and in many examples does not appear to hinder them in succeeding against professional adversaries.

Premodern amateurism permeated Medieval armies, in that a campaign often consisted of titled elites gathering up peasants, workers, and supply trains while relying on family and friends to orchestrate the adventure. Modern concepts such as a general staff, military academies, doctrine, and even a "military profession" did not exist yet, and cannot be confused with how premodern, especially feudal armies functioned. This does not suggest premodern militaries were hapless, disorganized fools either. Greek and Roman soldier-statesmen during the classical period mastered the principal skills and knowledge of their day, many demonstrating exceptional intellect and wisdom. However, none of these military leaders received any military-technical training "beyond what was normally provided by the Greek [later

Roman] city-state".[18] Premodern experts in war were self-made, with no real continuity across the premodern world. Keegan summarizes this premodern dynamic as follows:

> How did Alexander form his military judgements? It is dangerous in any age much before our own, to speak of a "general staff", because to do so is to imply a bureaucratization of society quite at odds with reality. The general staff, officered by men selected and trained to perform intelligence, supply and crisis-management tasks, was a nineteenth-century Prussian invention. The Romans, via the *cursus honorum,* anticipated something akin to it. But mediaeval armies knew it not at all, while even the Renaissance and dynastic armies of early modern Europe were staffed at best by gifted amateurs, usually the friends or favorites of the commander.[19]

Early warfare functioned upon a simpler design of contextuality, intuition, reason, as well as the unavoidable happenstance to what class or status a person might be born into. Nepotism along with bribery and flattery would remain a pattern of military selection and promotion from the antiquities up through military reform efforts of the eighteenth century, particularly in Europe.[20] The premodern war era combined fraternal or political networks, often along class or blood lines. Some elites, if so inclined, might have drawn military knowledge from existing manuscripts, elite tutoring, or the hard school of experience on the battlefield. Others pursued sport, rituals of the court, and other affairs that in no way prevented them from assuming powerful leadership roles when the occasion called them to battle.

Premodern war was designed so that organized violence, directed by the privileged elites, could exploit opportunities, and cede risks to expand an empire through land conquest, population seizure, as well as the detainment and negotiation of wealthy opponents for plunder and enhanced terms for peace. Greek and Roman designs for war became sophisticated and highly destructive, given the limitations of technology, knowledge, and resources of the time. Even at the height of Roman military power, premodern war remained a fragmented, self-maintained, and rather fragile institution. This is highlighted with the collapse of the Roman Empire and the subsequent loss of ancient and effective military knowledge. Foucault describes feudal age society as "made up of an unstable mixture of rational knowledge, notions derived from magical practices, and a whole cultural heritage whose power and authority had been vastly increased by the rediscovery of Greek and Roman authors".[21] Premodern war designs rose and fell with the societies and empires that used them, often at the mercy of victors who lacked any particular interest in preservation of military knowledge, rituals, or valuable techniques of a fallen enemy.

With the collapse of the Roman Empire, premodern warfare entered a significant decline, where much of the earlier skill, sophistication, and existing military knowledge became lost or forgotten. The Dark Ages were aptly named as much of the millennium between the fall of Rome and the commencement of the Medieval Renaissance was cast in the darkness of ignorance and retrograde of societal progress.[22] Premodern war thus spans the antiquities of Greek and Roman military might and that of a more limited, ritualized, and somewhat simplistically violent feudal age setting. Some of the peculiar contexts of the European feudal age would be the fractioning of society into small fiefdoms and domains, steep autocratic growth of kings and churches, a rigid class system, and a steep decline in military education, professionalization, and warfare innovation.

Feudal battles occurred sparingly, with most war activities being a combination of the mobilizing of forces and plundering expeditions that might descend into small-scale battles of individual fights between the knights.[23] Knights were sociologically speaking, harboring predisposition against military discipline on the battlefield as seen earlier in Greek and Roman formations of infantry. Feudal knights would attempt to act in battle purely on personal honor, reputation, and in keeping with tightly ritualized behaviors of the gentlemen class.[24] Employing chivalry and bravery on the field of battle, the actions of a valiant knight could resolve the outcome of the battle and even bring about the strategic end of the entire campaign in some instances.[25] Premodern war design centered not on creative, cunning, or progressive military leaders but on rote memorization of drills, automatic and conditioned responses, and a rather brutal process of institutionalization upheld by cultural, religious, and strict hierarchical class structures. The breakdown of Roman society and centuries of significant retrograde of knowledge, security, and quality of life during the Dark Age would produce a feudal age military institution that would be violent and destructive, but only to immediate, simplistic, and limited aims.

Designing toward Modernity: Feudal Age Developments

Premodern war was designed to be limited in size, scope, duration, particularly after the collapse of the Roman Empire and a long period of feudal stagnation across Europe. Rules of war were fleeting, with some upheld across noble houses and politically aligned groups, yet in other contexts of foreign invasions, piracy, and raiding parties, the only rules were that of force. Throughout the ancient and later feudal ages, it was expensive and risky to wage war. Leaders dependent upon their largely agrarian based societies considering the immense costs and time required to raise, train, move, and employ an army, not to mention the significant risks if one was not exceedingly sure of a successful outcome in advance.[26] Rival nobility competed fiercely through violence, trade, and marriages through constant preservation or expansion of their family wealth would agree to hold to declared rules for war that were often intertwined with religious edicts and regulation.[27] The feudal system also prevented any one king, prince, count, or bishop to gather enough power and military resources to establish absolute domination over their peer competitors. Additionally, the independent nature of vassals under the king meant that rulers needed to perpetually manage their networks through material such as land or treasure and ideational means such as titles and promotions.[28]

Finding quality military officers was difficult as well, due to premodern emphasis on class and title over merit or qualification. Premodern military officers did not attend any sort of school for war, nor were they required to study books or take examinations for advancement in rank or position.[29] As the money-based economy replaced the outdated land-based one, wealthier families could purchase officer commissions to strengthen status and prestige, while a rising merchant class could seek purchase of titles, positions, and status for a price. This threatened the earlier stable model where roughly no more than two percent of the population had the proper blood lines to assume command. Societies could never exploit the full intellectual potential of the population this way, and most of the eligible officers had no military education, training, or experience in warfare.[30] Elites believed a gentleman that could hunt, ride, and lead drills of common men had everything necessary to win in any battle, provided they had proper blood lines as well.[31] Aristocratic elites of feudal Europe saw war more as a sport of the nobility, albeit with dangers that demanded cultured refinement befitting a gentleman.[32]

The design of military leaders would shift from entitled nobility to upstarting middle class families desiring access to the previously denied elite attributes, primarily through defeat and humiliation when Napoleon would upset the legacy systems of European warfare. Elite families themselves would also need to shift if only to maintain sufficient status and power within changing military forms. Yet depending on culture and geography, noble status, and an affinity for sport and hunting as primary officer skills would persist well through World War I such as in the British army.[33] Amateur warmakers of the premodern world would gradually fade away into obscurity or transform with the times. The metaphoric devices of "war hero" and "war manager" are relevant here and offer insight into a shift in war theories and conceptual modules as the premodern war paradigm transformed. Societies saw a gradual replacement of traditional battlefield heroes in war with an increasingly scientific manager of technologically more complex organized violence. War managers would expand the original limited leadership of the war hero across increasingly more vast time and space. As Keegan described, war heroes would provide their leadership prior and after battles, but once the fray commenced, the hero would need to fight and survive or not be around to continue leading afterward.[34]

Heroes, as the elite leaders were at the tip of the spear, and tragically sometimes under it for centuries. Nobles may not have had formal military education or development, but the pressures of demonstrating gentlemen status, chivalry, as well as maintaining a family's reputation proved motivational in encouraging heroic behaviors on the limited battlefields of the feudal age. In the gradual shift to an industrialized, scientifically oriented modernity in war, military managers would learn to lead and decide differently, as would the warfare they administered. The military leader would gradually move further and further behind the front lines, as technology and industry permitted faster communication over the horizon and the dangers of war weapons required greater safety for senior commanders. The demands of the growing complexity in war through technology, science, time, and space would transfer war heroes toward largely tactical and symbolic needs while creating a new form of military decision-making and command that operated upon a far more complex and technologically advanced design.

The Industrial Revolution would force the displacement of heroic military architypes for that of modern managers, if only because the increased scale, scope, and information density of modern warfare could no longer afford to have the leader at the front edge of battle. Heroes would remain popular, particularly in desired leadership styles and egos, but the rise of military managers created a new requirement for understanding the growing bureaucracies.[35] Amateurism in military leadership roles declined, with the reforms of seventeenth century leaders such as Frederick the Great, theorists like French nobleman, the count de Guibert, German educator Gerhard von Scharnhorst, and minor aristocrat Freiherr Heinrich Dietrich von Bulow who all would advocate for formal education and merit-based selection through examinations.[36]

Not only did the leadership roles and functions transform over time, but so did the methods in which militaries curated knowledge, experimented, and adapted to change. Whether in the ancient world or within most all European conflicts through the late seventeenth century, soldiers learned their drills through rote memorization, mimicry of farming movements replaced with war weapons, or for the literate minority, by reading non-doctrinal concepts in literature or through oral history. Those elites and select individuals able to become literate or be tutored by literate military scholars would gain access to a fragmented range of ancient, medieval, Renaissance, and contemporary war texts.[37] Prior to the sixteenth and seventeenth centuries, "antiquity was still the great teacher in all that concerned the broader aspects of military theory and the secrets of military genius".[38] Militaries were designed to be simplistic, reliable and brutally disciplined using clear, rote memorization of drills for largely illiterate participants.[39] Standardization only existed in localized contexts where an army would mechanically repeat certain tasks in uniform ways, dependent on many things that might produce different military standards and drills just a mountain range away. One followed clear, simple orders and moved as part of a larger unit, whether holding shields and pikes or firing muskets in rigidly choreographed volleys.

Premodern militaries also had different forms and functions for policy documents and institutional regulation. Modern militaries operate predominantly through self-published, institutionalized doctrines, checklists, rules, and processes that are explicitly trained and enforced across the organization both to reinforce past experience and to attempt to predict future war patterns.[40] Premodern militaries had no such standardization, nor did military officers have any requirement or incentive to study anything outside of the expected general knowledge requirements of their given society and status. Doctrine, way to produce uniformity, convergence, and predictability in group behavior, first originated in the Roman Catholic Church.[41] The tenets of obedience, conformity, repetition, reliability, and discipline became the cognitive models that drove premodern war methodologies. One designed toward warfare to steer forces of uneducated, illiterate, often untrained participants into organized violence.

The Transformation of Doctrine from Church to Army

Modern militaries assume that their doctrinal publications are highly scientific, based upon proven, objective, and analytical constructs representative of a modern professional curation of knowledge. Few realize that modern military doctrine developed out of religious texts and practices, and today retains far more premodern, feudal, and *ascientific* constructs than one might realize.[42] Doctrine was designed for designating the body of correct beliefs and behaviors; early religious scholars would produce religious texts to clarify, eliminate dissent, or differentiate between compliant members and heretics.[43]

Religious doctrine provided a scalable and uniform way to establish similar baselines for explicit knowledge across an organization or institution, particularly in the overlap of autocratic leadership and the vast power of the Church. Naveh et al. remarked on this church and military parallel as follows: "Just as literacy facilitates bureaucratic, administrative centralization, it also makes possible the codification and logical centralization of doctrine."[44] How premodern, particularly feudal societies regulated behavior, codified culture, justified law, and order would extend from ideological, political, and social forces directly into how those same societies waged war. Military assimilation of doctrinal practices would come directly from observed success of religious institutions and their doctrine for designating the body of "correct" beliefs.[45]

According to Malešević, military discipline blended rituals and ideological discourse from religion and mythology in premodern periods to not only justify conquests and coercive governmental forms, but to forge modern bureaucratic structures.[46] Feudal militaries would parallel the steep hierarchical organizational structure of the Catholic Church as would many feudal establishments. The authority of the ruler coincided with divine decry, and the rituals and customs of feudal society partnered war with religious justification. Power structures of one mirrored the other, as large armies and navies would often feature ranks, customs, titles, authorities, as well as mannerisms that were found both in the royal court and within the pews. Doctrine encompassed everything from the minute details of when to sit, stand, or kneel in service, to the vast bureaucratic functions of the Vatican. Often, both religion and military would combine, in religious wars, divine justifications for acts of war, forced conversion of captured populations, destruction or conversion of rival religious structures, and control of specific territory deemed "holy" or otherwise symbolic for a people to control. Military doctrine would draw heavily from religious content and form, assimilating technical, managerial, and tactical concepts toward the design of warfare.

In the transformation from premodern to modern form, late Medieval militaries would retain the religious inspired ordering of their doctrine while expanding their depth and specialization using mimicry of natural science metaphors and concepts. This commenced with the Renaissance and accelerated in the sixteenth through eighteenth centuries as premodern militaries had to adjust to significant technological, informational, and industrial developments that were changing how warfare needed to be fought. Militaries would need to incorporate new ways of understanding reality if only to properly govern the application of technologically advanced tools of war, such as artillery, siege warfare formulas, engineering challenges, and how to navigate fleets across oceans.[47]

Gerhard von Scharnhorst, first Chief of the Prussian General Staff as well as teacher of Carl von Clausewitz, focused on the importance of military students studying mathematics

> because it provides the mental discipline and firm foundation upon which all other subjects build. Mathematics deals with the use of reason in an orderly, cogent manner ... Scharnhorst stated that there is a direct link between the study of mathematics and the art of war. Both had to be learned following a scientific or systematic approach.[48]

Scharnhorst would also champion military history as a critical topic for students, providing a new emphasis upon the specialization of warfare as an area of historical inquiry and enlightenment that later would profoundly impact Prussian military reformers such as Helmuth von Moltke (the Elder), who studied under Scharnhorst's famous pupil, Carl von Clausewitz.[49]

This feudal age transformation would start with a rigid, mathematical reinterpretation of warfare where chance, chaos, and uncertainty could be cast away, just as the new scientific disciplines and Age of Enlightenment would sweep away the ritualized backwardness and superstitions of the Dark Ages. Early theorists argued that war could be organized into a machine-like precision of steps, rules, principles, and formulas. Greek and Roman classics would be rediscovered or "reborn", while an entirely new mode of inductive reasoning would revolutionize European society from a relatively obscure collection of cultures and ethnic groups into the dominant imperial powers.

Within a few centuries, European powers equipped with sophisticated, scientifically aligned and technologically advanced militaries would topple previously competitive rival empires, surge past the curated knowledges of ancient regimes, and capture and enslave indigenous populations through colonization and assimilation. With mathematical precision in war, soon geometrical, geological, and chemical reasonings would be incorporated, yet these advancements would not entirely displace the premodern constructs, belief systems, or institutionalized rituals. Modern warfare designed with premodern origins, and subsequently any systemic inquiry into why modern militaries think about war to engage in technologically exquisite (and destructive) warfare requires acknowledgment of both.

Defining Modern War: A Rise of a Scientific Way of Organized Violence

The transition from premodern to modern would span centuries of human progress and development, yet once European militaries began to reform and professionalize through scientific discoveries, political reforms, and cultural shifts, they soon moved away from slow, limited, ritualized acts of warfare to something much bigger. One might pluck an infantryman, cavalryman, or an archer from the Roman period of warfare and throw them onto the field of battle in fourteenth century Europe, and aside from clothing and language differences, the warfare tactics and techniques would hardly differ. Once early natural sciences engineered new technological weapons such as cannon and early firearms, the days of knights deciding battle results in one-on-one duals was over.

Premodern war spanned millennium where few innovations promoted radical, catastrophic change. In roughly four centuries, modern militaries would advance from large, cumbersome cannon[50] and musket ball, smooth-bore firearms fired in tight, regimented formations to become able to employ radar, orchestrate mechanized armor attacks involving hundreds of thousands of soldiers across different time zones, as well as develop atomic weapons and the first computers. This transformation represents the most profound redesign of war in human history and lays the foundation for examining whether another redesign of war is underway today, as well as why such a question is troubling to the military institution.

From a contemporary political and societal perspective, the treaties of Westphalia in 1648 marks the gradual transformation of societies from premodern to modern ones with the rise of new states spanning multiple populations but clearly defined with geographical boundaries, often a common language as well as some fundamental cultural and ethnic qualities. Every state, once established and accepted by the international community (of similar nation states) had an equal right to sovereignty. This birth of the modern state coincided within decades of the establishment of the first military professional academy, as well as the first modern published military doctrine.[51] Although France, Prussia, Great Britain, and a few other nations would create academies and construct formal military doctrines, many other nations lagged far behind. Often, dramatic military defeat at the hands of an opponent led by professional, educated officers appointed by merit would stimulate this important shift. Rapoport offers important summarization of how nationalism and modern warfare would transform Europe together:

> By "country" we understand here a nation–state, that is, a politically unified region, whose inhabitants share certain characteristics, in the first instance a common language and usually common knowledge about the history of the region, especially of events that reflect a common fate of the inhabitants, for example, struggles against invaders or foreign rulers. The concept of nationalism so defined is a comparatively recent notion. It originated in Europe following the decline of the feudal system and of the power of the Church ... [The] power of the secular princes was greatly increased and their domains were more firmly welded into integrated political units ... However, it is not until the near absolute power of the princes was challenged by the French Revolution that nationalism emerged as a major force in European history.[52]

The Westphalian state would centralize the multiplicity and general inefficiency of feudal era war machines into state-sanctioned, analytically optimized, and legitimate monopolization of public force.[53] Giddens reinforces this with: "The successful monopoly of the means of violence within territorially precise borders is distinctive to the modern state".[54] Premodern societies typically fell short of this or could only do so briefly and at significant cost, while many sprawling empires would suddenly collapse with the death of a signature unifying authority. While premodern military ritualization and powerful social foundation shares many attributes of religious institutions concerning knowledge management and curation, the incorporation of scientific analysis and inductive thinking presents the modern military transition where the promise of greater prediction, control and precision in war is quite attractive. Yet premodern war functioned in physical states and planes of existence that were, if compared to contemporary conflicts, far simpler.

Premodern warfare was designed within the clear limitations of available technology and knowledge, human and animal labor, and the boundaries set by time and space. Through muscle power or wind, premodern armies moved at slow speeds, at enormous costs, and communication remained largely localized to immediate areas at best. Modern warfare could design novel ways to expand or even jump to entirely different planes of existence to do battle, sometimes in ways that differ from historical patterns profoundly.[55] Premodern warfare occurred largely on land or sea, with only minimal activities in the air (proximate, low altitude). The rise of airpower at the turn of the twentieth century, the development of deep undersea naval capabilities, mechanized vehicles using internal combustion engines, chemical warfare in World War I would transform warfare. The first Quantum Revolution would introduce in World War II radar, atomic weapons, the first computer concepts, and atomic weapons. Later in the same century, development of a virtual, digitized, and human accessible cyberspace created an entire new plane of existence where future warfare could occur in ways incompatible with premodern concepts. Humans would become able to operate in space itself around Earth, marking the first species able to escape the planetary confines. In the last decades, society witnessed a rise of social information and networked societies. This second Quantum Revolution of blockchains, quantum encryption, quantum attack presents even more examples where war can now be exercised in ways unimaginable even a few centuries ago.

The shift to modernity spanned centuries, yet theorists often point to key events, conflicts, or technological developments for milestones in this transformation. Tschumi would posit 1750 as the turning point from a premodern world in architecture to the modern form, representative of the hardened structures and living spaces that humanity would experience this shift.[56] Foucault, in defining Napoleon's arrival as another proposed beginning of modernity and war, saw his desire to organize and arrange around him

> a mechanism of power that would enable him to see the smallest event that occurred in the state ... a meticulous observation of detail ... for the control and use of men ... and from such trifles, no doubt, the man of modern humanism was born.[57]

Most of these developments and events span the mid-sixteenth through the early eighteenth centuries, marking a powerful transformation of societies and war. War would become scientific, precise, machine-like where men became disposable, mass-produced parts of the war machine. Antoine Bousquet also frames this "scientific way of warfare" as follows:

> The successes of modern science in uncovering seemingly external laws of nature and developing or perfecting technological contraptions to take advantage of them has unsurprisingly proved highly attractive to military thinkers and practitioners seeking to dominate the battlefield and render their activity as predictable as possible.[58]

The incorporation of scientific rationalized logic into military thought on the application of organized violence would shift the creation process of military doctrine into a manufactured, pseudo-scientific form that would have one foot in the premodern world of canonized lore and codified hierarchical laws, and the other in a modern mindset of experimentation, testing, validation, and subsequent community of shared practice and theory. Militaries would think about warfare in scientific terms but believe in war itself through a combination of philosophy, culture, and an intended mimicry of other professions undergoing similar modernization processes. War would be framed within a new Westphalian political model, while warfare would be waged within a blending of older church-inspired forms with new inspiration from natural science origins. War would also become universalized in the tested laws of classical physics and Newton's laws of motion, with technological developments viewed as powerful game-changing opportunities to inflict greater destruction. With scientific methods, one need only unlock the core principles of war so that formulas and patterns could aid the military in realizing what truths found in one battle could be applied to any other one, provided the science supported such rationale.

Clausewitz would inject the supernatural and the fantastic through German Romanticism that chipped away at Jominian efforts to render all of warfare into formulaic checks and balances,[59] yet even Clausewitz would seek to produce tenets, patterns, maxims, and laws that offered some hope of the clever strategic leader. He offered new ways to conceptualize the destruction of opponents, how to secure goals, and outwit cunning adversaries in meaningful battlefield accomplishments framed in a scientific theme that highlighted chaos, uncertainty, fog, and friction in western, natural science metaphoric devices and associations.[60] Paparone, also citing David Lai's military study of Chinese strategic concepts, notes that for Clausewitz, "the metaphor of 'friction' [was like] walking through water, the Chinese perspective contextualizes very differently: '[Water] has no constant shape. There is nothing softer and weaker than water, yet nothing is more penetrating and capable of attacking the hard and strong'".[61] War was chaotic in a *westernized* sense, rendered understandable in scientific framings complete with theories, conceptual models, and terminology taken from the natural sciences.[62] War became increasingly larger, total mobilizations of national will and resources set to a Clausewitzian logic of offensive concentration of force toward critical points.[63]

Paparone charges that modern militaries attempt to interpret war not necessarily through scientific rationalism, but by using "pseudo-scientific logic" reliant upon an assimilation of selective parts from natural science discourse and engineering language.[64] Newton would define gravity and the laws of motion for physics, while Clausewitz would explain a nature of war where a center of gravity followed his philosophical framing of warfare between passionate people using concentrations of destructive force, not any natural science formulations themselves. Vauban and Jomini would, among other military theorists in the sixteenth through eighteenth centuries, attempt to remove all uncertainty and chance through positivist, formulaic principles of warfare that mimicked natural science stability.[65] Jaynes, while not directly addressing militaries, provided another example of how militaries and other professions might take from natural scientific findings. "The first half of the nineteenth century was the age of great geological discoveries" where geology applied the metaphoric device of "layers of the earth's crust" toward new models of understanding reality.[66] Other disciplines quickly absorbed these models, including psychology (levels of consciousness, subconsciousness, and unconsciousness) and, as this author suggests, militaries may have adapted this into levels of warfare (tactical, strategic, and grand-strategic levels of war) as framed in contemporary military doctrine. Hundreds of other modern military examples include "spectrum of war", "lines of effort", "principles of war", "kinetic/non-kinetic", "fog and friction", and "gray zone" which is yet another optical/visual metaphor.

Why might early modern militaries design themselves in such a way that they retain some feudal qualities, shed others, and assimilate portions of natural science concepts while mutating them beyond their original purpose? Militaries were under a great deal of pressure, desperate to modernize and attracted to the uniform, objective, and analytically optimized rigor of natural science. Indeed, militaries developed formal professions well behind earlier medical, legal, engineering, and mathematical professionalization.[67] Military educators such as Scharnhorst "hoped to develop a cult of objectivity within the army", to question "all received 'truths' in a disciplined fashion, responding correctly to a critical, independent analysis of the facts".[68] If natural sciences seemed intoxicating to the military on how one could quickly professionalize and rationalize in clear, quantitative, and presumably predictive ways, they were hardly the only ones to do this. The race to industrialize, professionalize, and modernize spanned all fields, disciplines, and guilds, where the sooner one could correlate their particular focus area to a scientifically sound, logical, and technocratic framework, the better.

Militaries, being late to the professionalization race, quickly drew inspiration from established natural sciences and civilian professions. Jaynes would order geology as the first "fashionable science" to transform much of society through geological metaphors, language, and subsequently models and methods nested in new geological theory. Next would come chemistry, and physics, biology, and psychology. With each of these new scientific fields impacting how society made sense of reality and adapted new constructs, militaries too went through a professionalization to modernize and readily drew from these natural science cognitive models, metaphoric devices, and language to produce their own military theories on warfare that *appeared* scientific too. Through the new theories provided by Machiavelli, Jomini, Scharnhorst, Clausewitz (and later still, Mahan, Liddell Hart, Douhet, Schlieffen, and Svechin) to mention just a few, militaries would begin to render activities within war and the nature of war itself understandable primarily through natural science constructs, terminology, and principles.[69]

However, war is of human design. It did not exist before humans, as did the primary focus of physics, chemistry, biology, and geology in that their inductive reasoning produced theories that extend before and after human existence. In the desire to associate a natural scientific frame to what is socially constructed and maintained within human consciousness, modern militaries fixated on the tangible and the objective aspects of warfare. They would marginalize the subjective and social, or merely assume theoretical superiority of hard science over all else. This produced a scientific rationalism where war could be chaotic, but within the areas of human precision, focus, and organization, one could accomplish goals and defeat adversaries. This technocratic, scientific rendering of war would define the modern military war paradigm, which today is under assault by a postmodern antagonistic perspective.

Gray would critique this shift toward a mechanistic, scientifically rationalized, and technocratic way of warfare as follows: "That faith in technology assured military technophiles that bombing cities would lead to victory, despite all the evidence to the contrary".[70] This rationalized scientific mode of inquiry would transform military language, building on top of the still enduring premodern (feudal) model of rank structures, rigid order and discipline and narratives of heroic lore with precise terms and Newtonian physics-based metaphors.[71] Gibson, in critiquing the American military's systemic failure in the Vietnam War attributes much of it in his concluding chapter to how modern warfare prioritizes certain knowledge over others, with scientific rationalism firmly atop the pile:

> A basic conceptualization of the relationships between knowledge and social stratification has been present throughout this analysis. War-managers are at the top of the stratification system. They think in instrumental categories taken from technology and production systems, and the business accounting rationales of debit and credit ledger. Those at the top of the stratification system had a virtual monopoly on socially accepted "scientific" knowledge. Conflict among different war-managers was quite common, yet those conflicts all occurred within the paradigm of Technowar and its technical knowledge about war. Never was the "otherness" of the foreign Other really questioned, nor was the social world of the Vietnamese peasantry examined, nor were the terrible contradictions and double-reality facing U.S. soldiers in the field ever confronted. Debates at the top were only debates and struggles concerning the direction of the Technowar, not a questioning of its basic assumptions.[72]

The modernization of thought and action in war prioritized scientific theory, analytic logic, and a uniform categorization of all actions associated with the conduct of war into a manufactured, sequenced, and manageable affair of organized violence, resources, and risk.[73] At least, this was the ideal that militaries strove to achieve in battle. Yet could complex warfare really become a bureaucratic, managerial affair of nation-states focused on efficiency, destruction of the enemy, and total war mobilization? War could, in this Newtonian styling, primarily be interpreted through the methodologies underpinned by scientific models and theory and be measured by natural scientific rules that alter any war debate to a foregone conclusion.[74] Virilio would charge this techno-scientific obsession of modernity to be if anything "eclipsing ... the real, in the aesthetics of scientific disappearance ... science is now less concerned with truth than with the effect created by the announcement of a new discovery".[75] Critics, particularly after 1945 and throughout the military challenges of the late twentieth century, would lash out at the pseudo-scientific logic, modeling, language, and methods that modern militaries had invested into over the last few centuries of modernization.[76]

In terms of epistemology, the modernist military movement largely presents one of "rules and maxims".[77] It corresponds with a movement that established a Weberian World Order "with the *rational-legal nation-state* as the center of gravity".[78] Nation-states wage war between states, using formal military instruments of power while seeking formal declarations of war and strategic closure through some ceasefire and metric of victory against a framed enemy force.[79] Asymmetric or irregular war would be possible and in reality quite frequent in the modern military era, however security forces would educate, equip, and act with an orientation towards high-intensity or total war considerations against near-peer and existential threats. Threats that fell below these high thresholds would also be dealt with; however, any military organization that is configured and prepared to wage high-intensity war was expected to also be qualified and competent to act in low-intensity war efforts including humanitarian and peacekeeping efforts that increased dramatically after the end of the twentieth century's decade-long Cold War between nuclear superpowers.[80]

Politically, the rise of the Westphalian Order would create a

> rule-based system of states, based on the concept of sovereign equality
> ... this resulted in a framework of widely respected norms for interstate
> behavior – to be sure, never for long unchallenged – that permitted states
> to turn greater attention toward their internal development.[81]

Here, *war was redesigned* so that it could only be actioned by states against other states, or by a state against a non-state entity if necessary. Giddens explains:

> the nation-state system was forged by myriad contingent events from the
> loosely scattered order of post-feudal kingdoms and principalities whose
> existence distinguished Europe from centralized agrarian empires ...
> nation-states concentrated administrative power far more effectively ...
> a massive leap forward in economic wealth and also in military power.[82]

In premodern war, the idea of war was flexible, and these actions might be taken by any number of actors or groups, and also feature profound supernatural and existential qualities where divine decrees and scripture provided significant influence. Yet in this modern war redesign, states created an exclusive club where war became a tightly controlled state activity and expressed in international politics and policy. One state could wage war against another, while other nations could determine in elaborate rules and laws whether they wished to enter this war, remain neutral, or collaborate in other ways exclusive to nation-state abilities. By doing so, proponents of this Westphalian redesign anticipated that in state hands, war should become less frequent, more predictable, with strict boundaries and rules to govern the application of organized violence. As the world erupted into two World Wars in quick succession ending in the arrival of the atomic age of warfare, this clearly did not go as designed.

Hierarchical models for decision-making and linking strategy to operations and tactics in war continue to dominate the modern military form.[83] Earlier, the nobility, king, bishop, or shogun might direct war activities down to the lowest level where serfs, indentured servants, slaves, and mercenaries obeyed orders. Yet the industrialization of warfare would apply machines to magnify man's physical capacity; the cognitive parallels would suggest analytic science could aid in controlling and predicting the natural world as well.[84] Armies would be described as a "large and very complicated machine" where the leader "must only steer" and where "[general] rules and standard operating procedures must be the foundation" for warfare.[85] Militaries grew larger, expanding in greater sophistication and intricacies, yet a centralized hierarchy of command and control remained firmly in place. To compensate for this massive expansion in scale, scope and intricacy, militaries would redesign how they developed and curated knowledge within the force. Modern military education would promote "an environment of learning that embraces uniformity and enhances scenario based pre-planned drills as ways of conducting military operations".[86] If natural sciences could unlock universal principles of reality to make the world more predictable, understood, and therefore controllable, proponents reasoned so should a "military science" do the same for the chaos and uncertainty of war.

The seventeenth through nineteenth centuries witnessed an emphasis on engineering and the mathematics of artillery for war study.[87] Indeed, the founding of major military academies such as West Point was based primarily on the need for new technologically skilled military specialists in engineering and artillery of American origin and allegiance. Modern war became a pursuit toward a methodology of maintaining order even within the chaos of war, predominantly through mechanistic and what became a scientific worldview where war had to be managed in specific, often quantifiable ways. Scharnhorst, as an early educator and military reformer of the Hanoverian army, in 1796 ran their artillery school.

> Classroom instruction in geometry and arithmetic took place Monday through Thursday mornings, from six to noon. Scharnhorst taught "artillery" from 10 to 12. Afternoons were spent in "laboratories" in which students had the opportunity to put into practice the theory they had learned in the morning.[88]

Bousquet provides the metaphors of the "clock" for the Newtonian period for what he frames as the first metaphoric development within what is considered modern war.[89] The "steam engine" metaphor followed the clock, with an assimilation of thermodynamics, flows of energy clashing in battle. This occurred whether "ballistic, motorized, industrial or moral nature, as nation-states clashed in ever wider conflicts that drew on all the resources at their disposal".[90] Jomini's principles of warfare demonstrate more of a machine-like emphasis of the "clock" warfare concept, whereas the Prussian war philosopher Clausewitz and his concepts of fog, friction, ambiguity, and the psychological factors within minds experiencing war align with the "steam engine" association. In this metaphoric alignment, massive nation-state war machines attempted to destroy one another in exchanges of swirling, chaotic energies, and forces using natural sciences, quantitative analysis, as well as social sciences such as psychology to consider fundamental laws and maxims on the human psyche.[91] Bousquet's delineation between these two provides added context to the gradual shift from premodern to modern war paradigms in organizational form, doctrine, and practice.[92] Liddell Hart writing on Clausewitzian applications in both World Wars offers: "[Clausewitz] was a *codifying* thinker, rather than a *creative* or *dynamic* one … In defining the military aim, Clausewitz was carried away by his passion for pure logic".[93]

Another critical additional contribution to the modernist military form developed in the early twentieth century addressed standing armies. Nation-states moved away from temporary, small-scale, and peacetime armies out of fears of an industrialized, total-war oriented aggressor overwhelming them. Instead of ramping up conscription or drafts to rapidly expand a small military force in times of war and then downsize accordingly, a new movement for permanent "total war" preparation and improvement began where nations sought to maintain massive, costly military instruments of power at the ready. The nation-state would be able to increasingly mobilize vast segments of its population quickly, train and arm them sufficiently, and deploy them rapidly *en masse* through modern transportation and logistics in ways unprecedented in human history. To do this required a sophisticated, technologically enabled, industrialized defense framework. The rise of military bureaucrats, commercial industry oriented toward supplying expensive and ever-developing war tools of increased sophistication and power, and the international system of competition would bring forth the rise of an industrial–administrative complex as the zenith of this modern war paradigm.

Redesigning war required new ways of thinking about how to improve, expand, and sharpen more powerful state instruments of power. Militaries are costly, but their ability to project power and enforce policy could mitigate or prevent war or potentially enable a military victory over an adversary with all the added benefits coming to the victor. This fostered a cottage industry for militarization within every nation-state large enough to deal with existential threats permanently. Once this occurred, the modern national security apparatus would exist not just for any immediate perceived threats to the nation, but as a permanent destructive capacity for all possible future security considerations through the rise of the military industrial complex.[94] Unavoidably, the design of industry and commerce became wedded to this development, where massive companies and entire fields of research and experimentation became dedicated to war technology and an expansion of new products and user experiences nested in a security context.

As the new methods expressed these new cognitive models and exercised new war theories, the very language of articulating the "what", "how" and "why" of warfare would transform as well. Metaphoric devices such as "front line" would be developed in war's modernity because prior to industrial tools such as rail systems and mass conscription, *such contexts did not exist in war* nor did military language require such ideas to be articulated. Modern military goals became larger than the limited gaze of the tactical commander, and strategic desires of senior politicians could not peer down into the broad conflict zones in "total war" scenarios without significant confusion.

Designing Modern Managers of War

While earlier wars might be waged by champions and despots in somewhat episodic or sporadic affairs of conquest, empire building, and gore, modern wars were meant to be managed and executed in precise scientific means with clean, efficient, and decisive results.[95] Heroes could still inspire tactically or locally, but it became increasingly difficult with a dilution of impact the larger the scale or size of the conflict. Strategic and national heroes would be rare, and their symbolic qualities often transcended specific military methodologies.

This did not eliminate heroic war leaders or the occurrence of exceptionally gory conflicts outright; rather the twentieth century harbored the arrival of military scientists and managers to offer novel combinations and hybrid outcomes done in a scale and scope previously unreachable by even the largest pre-modern militaries. The premodern war hero could still exist, but this character was joined and often replaced (or made subordinate) to the military manager that wielded scientific analysis, calculations of risk, and the factory-style approach to efficiencies and optimization of the modern war machine. Even a heroic national leader needed an army of managers, bureaucrats, scientists, and engineers to support them, otherwise they risked sudden elimination or defeat by better organized adversaries.

Despite the promise of such cold, scientific precision, war would in the twentieth century remain chaotic and often untamed by pseudo-scientific design efforts. From the horrific bureaucratic genocide managed by Germany through creating emotional distance between Jewish victims and Nazi executioners to the misguided self-inflicted damage the British did in their strategic bombing efforts against Germany where they lost more British aviators than destroyed enemy resources, modernist rational war logics created perhaps *an illusion* of scientific certainty despite often producing unscientific results.[96]

The modern military manager of war would dominate and displace the premodern heroic or divinely chosen leader, but this did not necessarily make warfare more effective in accomplishing policy goals through organized violence. It would expand the scope, scale, and sophisticated horror to a level unprecedented in human history. Military managers use modern military doctrine in a manner unlike premodern texts and teachings. Modern doctrine seeks to demonstrate scientific processes, theories drawing from natural sciences, and mental models that equate to a "technical rationalism" of war itself and how one conducts warfare successfully. This modern way of understanding reality (including war) ordered the world in a scientific way to conceptualize, articulate, and describe everything:

> In the Western world, a particular regime of order emerged with modernity ... Order came to be increasingly justified and organised on the basis of a scientific and technical rationality. Knowledge produced through the inductive methods of scientific enquiry gained ascendancy over deductive theological and scholastic claims about the world.[97]

Subsequent decision-making methodologies become process-oriented, sequential, and quantitatively oriented so that they can attempt to gain in efficiencies, risk reduction, uniformity, predictability, and collective understanding of the many participants. The very language of modern militaries and the doctrinal narratives used by them are filled with precise terms, conceptual models, and metaphoric devices that desire objectivity, rational analytical thinking, and the uniformity of declared "laws of warfare".[98] This reinforces a scientific way of modern warfare through training, education, practice, and institutionalization.

Schön, in framing how modern high-technology societies approach contexts such as warfare, states: "practitioners solve well-formed instrumental problems by applying theory and technique derived from systematic, preferably scientific knowledge".[99] Asrid Nordin and Dan Oberg support this as follows:

> Doctrinal text relies heavily on abbreviations and contains an impersonal and administrative language. It is as if the potential for symbolism − for the text to mean more than itself − has been subtracted from the paragraphs in favour of Orwellian "newspeak".[100]

Military leaders would thus become modern managers of organized violence, using technological rationalization of war through a lens of natural science metaphors while also insulating the organization from major changes in strategy.[101] James Gibson, in explaining the failures of the Vietnam War, cites Henry Kissinger who along with Robert McNamara would be most associated with this modern managerial approach to warfare:

> Kissinger writes that since 1945, American foreign policy has been based "on the assumption that *technology plus managerial skills* gave us the ability to reshape the international system and bring domestic transformations on 'emerging countries ...'" The West, in Kissinger's view,

had been committed to this hard epistemological work since Sir Isaac Newton first formulated his laws of physics ... The West is deeply committed to the notion that the real world is external to the observer, that knowledge consists of recording and classifying data — the more accurately the better. Cultures which have escaped the early impact of Newtonian thinking have retained the essentially pre-Newtonian view that the real world is almost entirely internal to the observer ... [and] are therefore totally unlike the West and its leading country. Those who are totally unlike us and live in their own delusions are conceptualized as foreign Others. The foreign Other can be known only within the conceptual framework of technological development and production systems.[102]

Thus, war managers of modernity can direct vast amounts of manpower, technology, and logical methods of analytic optimization toward all aspects as understood in this modern war design that establishes cognitive models upon Clausewitzian theories oriented toward Westphalian nation-state warfare, supported in somewhat paradoxical ways with Jominian and additional modern war theorists. The war manager would become a critical metaphoric device in this redesign of premodern to modern war paradigm. Gray would cast McNamara as a key example of a war manager where in World War II, McNamara as a young Harvard professor specializing in bargaining theory would calculate strategic bombing missions and operations against Imperial Japan. Later, as Secretary of Defense under President Kennedy and tasked to manage the Vietnam Conflict, "he would run a war of his own",[103] one that Gray declares the first "postmodern war" waged. McNamara and Kissinger would become the top war managers associated with this shift, as they would direct generals and military forces with executive level authority and designs.

Kissinger, according to Gray's analysis, placed overemphasis on what was considered a rational opponent (rational actor theory, game theory) and attempted to render war and policy into computer models of analysis and rational decision-making a limited structure of modernity, to subsequently be challenged with alternatives in postmodernity.[104] The new philosophy of modern science and warfare would "be a masculine project in two senses ... adapting logic and emotionless calculation from the scholastics and marrying them to objective clinical experimentation ... [and] in its rhetorical equation that woman is nature and science is torture and domination".[105] War managers would move beyond the original heroic leaders of premodern wars, able to channel the energies and resources of the Westphalian state model toward mechanized, scientific, sterilized, and analytically optimized acts of organized violence, all done from the safety of a far-away office complex, underground bunker, or the other side of the globe surrounded by computers, simulations, and analysts.

Military Modernization: Technological Machines of Total War

Modernization of military professional education occurred over centuries where military academies would spring from mimicry of earlier scientific schools and universities across Europe.[106] Natural science sought academic publication of new theories and standardization of how the discipline would function, and militaries would also follow suit. By combining earlier premodern and largely ideological inspired doctrines for institutional norms, practices, and behaviors, military theorists and educators would blend scientific constructs for engineering, siege warfare, artillery applications, and other new warfare activities that depended upon education, scientific training, and rigid, uniform practice. Doctrine and education would become the primary tools for professionalizing militaries into their modern forms.[107] Vauban's early doctrine on siege warfare and artillery bombardments demonstrate a formulaic, mathematical, and geometric logic of reducing many war activities and decisions into precise, regulated checklists and recipes to follow.[108] Militaries used science to turn war into a bureaucratic form of managed violence, which Sookermany critiqued with: "[running] the risk of fostering a reductionist view where the human body is reduced to mere biology … consequently, the objectivized and alienated body is easily associated with the image of a machine."[109]

Chris Gray observed that this scientific style of military doctrine is "itself the style of nonstyle. It is not supposed to become, or even influence the content of what is asserted".[110] It dictated compliance, never inquiry or curiosity on testing doctrinal standards against something else. Doctrine existed to standardize, to control and coordinate the contributions of large numbers of units, mirroring the conditioned beliefs on war through historical interpretation.[111] Those that obey doctrine are "trained, rewarded, and promoted according to a particular way of doing business".[112] Doctrine retained its earlier religious epistemological origins in that militaries would obey, study, and repeat the processes contained within them without question. If the institution published it, there would be no other justification other than to follow it as closely as possible and seek ways to further reinforce it. While centralized hierarchical organizations like the Catholic Church and their doctrine would influence feudal and modern militaries, in the classical period of Greek and Roman military soldier-statesmen, military-technical treatises were in circulation but never became doctrinal, nor even required for promotion or qualification to lead military forces.[113] Only in the late feudal and early modern period of warfare would scientific understanding create the need for formal education, certification, and evaluation so that military officers could specialize in artillery, engineering, siege warfare, and other technical and emerging fields for warfare.

This in turn caused the establishment of formal military doctrine that could accomplish the distribution and reinforcement of war knowledge so that such goals of uniformity, reliability, repetition, and craft mastery could be accomplished. Unlike the ritualistic, symbolic battles of the feudal age where two knights might fight to decide an outcome, an artillery officer needed to thoroughly understand how to fire their weapons correctly and avoid catastrophe of individual error or whimsy.[114] Military doctrine would be composed in the language of scientific *inspired* methodology as well as the assimilation of classical mechanics concepts and metaphors into war-oriented methods, while also retaining the hierarchical structure and institutional conformity of religious doctrine. The military would draw from natural science theory to erect parallel military theories on warfare that echoed similar scientific models, methods, and shared language (with metaphoric devices intact). The birth of a military science thus began in the Age of Enlightenment and was enhanced through the arrival of the Industrial Revolution.[115] Increased technological industry begets an expansion of military prowess and lethality, subsequently joining both design of commerce and design of war together and demanding a scientific educational construct to harmonize this union.

Modern military doctrine would morph from religious origins to assume a bureaucratic, managerial focus toward convergence, uniformity, and predictability of prescribed military behaviors. Highly tactical, technical, and later managerial doctrines centered on numerous mathematical, geometrical, and mechanistic constructs assimilated from natural sciences (hence a "Newtonian style"),[116] which in the twentieth century would expand into directing specific theoretical, strategic, and operational constructs.[117] Paparone and others critique this as "pseudo-scientific" in that the production of military doctrine mimics scientific communities and academia, but largely remains set in the centralized hierarchical orderings of religious doctrine.[118] Content is exclusively controlled, managed, edited, and published by authorized training organizations where committees of anonymous authors present final versions to be signed by a senior authority. Peer review is non-academic, controlled in a bureaucratic, opaque manner of staffing actions and timelines by the doctrine writers.[119]

Military doctrine rarely has citations or sources identified, nor are alternative theories, models, and methods ever presented. There are no authors declared, only the signature of an authority declaring the new truth to the force. All new doctrine must first comply with all other existing doctrine, and once published the institution must no longer consider the ideas in any outdated or replaced doctrines. This emulates not scientific discovery and testing, but the orthodoxy of a steeply hierarchical religious entity. Just as religious organizations declare doctrinal findings as authoritative and truthful, modern militaries converge upon these publications that are considered *proven* not in any actual scientifically rational manner, nor evaluated by outside peers, but largely through institutionalized belief systems seeking *process compliance and obedience*. Feudal practices, rituals, and institutional biases would therefore extend into modernity in how militaries seek to curate knowledge on warfare, as well as declaring what war is.[120] Modern militaries would often feature a strong bias toward traditions and continuity over new designs and innovation, illustrating deep patterns of premodern behavior extending into today.[121]

War also became more sophisticated, with the infusion of science creating new tools of war that demanded scientific study of mathematics, physics, and chemistry to effectively employ, maintain, and improve these new powerful tools. From naval improvements in propulsion and ranged artillery to coastal defense and batteries attempting to counter ship developments, a science of war took hold. The increased technological requirements of modern military weapon systems as well as significantly larger and increasingly complex force structures required a shift away from the premodern military demand for heroic military warriors toward an educated military manager and engineer of more destructive and also more precise war machines.[122] While premodern militaries might conducted excessively gory acts of localized violence in order to control populations, the modern military scientist and manager could be far less "brutal" while also destroying and killing at a vastly larger scope and scale.[123] This redesign from feudal kingdoms and a more flexible (and ritualized) form of war would be a transformation not just of militaries and national power structures, but of entire societies.

Vulnerabilities with the Modern Military Form and Function

If premodern wars consisted largely of expensive mercenaries, slaves, and the temporary conscription of amateur soldiers for limited political, ideological, and security goals,[124] the modern era introduced professionalization of war at a scope and scale unlike anything previously. With the American Civil War which framed as one of the first likely examples of modern military management at the operational level of warfare,[125] earlier relationships between strategic goals and tactical victories became problematic. Gone were the days of a tactical victory that could immediately accomplish strategic aims. Modern war grew too large, with too many activities, actors, effects, and emergent possibilities.

Military desires for the organized application of violence were vastly exceeding previous cognitive limits in scope and scale. In proposing the modern to postmodern war transition, Gray would also refer to the U.S. Civil War as "the first industrial war, that ushered in the last phase of modern war".[126] The American Civil War would have one foot planted firmly in earlier pre-industrialized, even feudal aged constructs of earlier warfare forms while the other foot would step into new Westphalian "total-war" strategies, fueled by technology, science, and modern management of vastly larger conflicts. Historian French highlights this as follows:

> The French made war an affair of the whole nation and drastically altered its objectives. Instead of using the limited resources of the *ancient régime* to

achieve limited objectives, the French republic and its imperial successor sought to destroy their enemies' means and will to resist.[127]

Modern wars arrived on larger stages, with many more set pieces, actors, effects, and the resources and energies of entire nations invested. Tactical thinking alone was too localized and limited, and strategic orchestration became too removed from immediate and pressing decisions to properly exploit opportunities and adapt to change. Modern war manifested a new area of conceptualization between the two.

The formalization of "operational art" and an "operational level of war" would come from post-World War I theoretical publications of Soviet military professionals, such as Alexander Andreyevich Svechin. Svechin and a few others attempted to fill this conceptual gap in modern war thought, although they faced opposition from Clausewitzian advocates arguing for an offensive-oriented form of warfare.[128] Due to Stalin's military purges in the 1930s and his, as well as other Soviet military leaders', fixation on a Clausewitzian emphasis on offensive warfare, these ideas on operational art would slowly spread. The Soviets would rediscover Svechin's ideas on operational art later in World War II, long after Stalin had purged him, and it would take until the 1980s for the American Army to realize the importance of thinking operationally, between tactics and strategy.

While modern warfare did expand the scale, scope, sophistication and clearly the lethality of state-sponsored organized violence for political or societal designs, did it redesign for everything that premodern war was lacking? Bigger does not necessarily mean better, nor does faster, more efficient, or more lethal. The modernization of warfare brought with it numerous paradoxes such as an escalation of violence by multiple participants to a point where none can deescalate or slow the destruction. Nor did the technological sophistication and high education and training of modern military forces deter lesser developed adversaries, or in combat guarantee the clear defeat of the weaker opponent. The high costs of warfare as well as defense would balloon budgets, create economic and societal stresses, and in some instances trigger the collapse of a nation even without a shot fired. In other instances, some nations raced to gain forbidden or exclusive weapon technology such as nuclear arms to accomplish some existential protection against perceived adversaries. The rise of the permanent and increasingly expensive military industrial complex in most every developed or developing nation promises that with politics and industry come lobbyists, special interest groups, bloated military budgets, never-ending new weapon investments, and what Builder surmised as a military service's own self-interests and identity sometimes taking over even at the expense of national security interests.[129] Doctrine would be used this way too, for service self-relevance and interests.[130] Modern war machines were larger, more lethal, yet they did not become solutions to longstanding issues regarding war and peace. In some ways, they exacerbated the potentiality of *more* warfare, whether in frequency or intensity.

In another paradox of industrialized modern war design, premodern or low-tech approaches still worked, even positioned within a larger, modern, technologically advanced world able to respond. The late-modern genocides in Africa were accomplished with primitive weapons such as machetes and axes coupled with radio broadcasted propaganda and instructions.[131] Even in the modern world, groups armed with steel-edged weapons and low-cost communication systems could slaughter entire groups of people in the ways of premodern war while the rest of the world watched at a distance. In the collapse of Afghanistan to the Taliban in 2021, we find yet another inconsistency where the most technologically advanced nation in the world is unable to decisively defeat one of the poorest and low-technology opponents, nor halt the defeat of the Afghan military in record time. The American Afghan War from 2001 to 2021 presents one of the most resource and technologically lopsided conflicts in the last century. It illustrates that advanced technology, wealth, and massive military force are not always sure bets in war. The fact that the Taliban's precursors did the same to the Soviet occupiers for over a decade with mostly the same long-term results further emphasizes this point.

The modern war era also has ushered in a more chaotic, potentially unmanageable form of "total war" that has no real premodern equivalent. Premodern and modern militaries are problematic to compare outright. Premodern war despite low-technological capabilities could be highly intensive and extraordinarily violent and gruesome. The limited, rule-based, and highly politicized "cabinet wars" of feudal and early modern periods illuminate this difference. In the "limited war" of Europe's feudal age,

> the object of the campaign frequently was to reach a situation (by proper maneuvering) in which it could become clear that one's own side had a strategic or tactical advantage ... At that point, the side that was likely to lose could capitulate with honor, after which everyone would go home.[132]

There was no operational layer for war until Soviet theorists would conceptualize its design after World War I. Instead, strategic thinking usually nested into immediate tactical applications, and in some conflicts the tactical battle would determine the strategic outcomes immediately.[133] Total war would not be realized until the combination of technology and political objectives using the application of organized violence allowed for the creation of a "state within a state" or a permanent economy of war.[134]

The 1870s are another period where theorists such as Virilio and Lotringer saw a shift from limited wars to one of total war preparation and posturing.[135] Societies would develop and maintain entire war economies oriented toward total war preparation in peace, or complete application of the national potential through logistical machination. He would posit the arrival of the "post-industrial era" for western society beginning in the 1920s with the rise of Hollywood and "the de-realization of the world" therein.[136] This climaxed in the arrival of the nuclear weapon, the technological solution to the modernist military staff problem of how to achieve decisive victory. Modern war designed not toward moderation in organized violence through greater technological ability, but often to achieve greater destructive ends. This culminated in 1945 with nuclear weapons, where "ironically enough, that very fireball marked the end, as well as the apotheosis, of total war. Limited war has returned to central stage since 1945, although moderation certainly has not".[137]

Malešević referred to this shift as the "cumulative bureaucratization of coercion" in war, where military violence could change its form and utilize new technological modes of expression, all the while increasingly gaining greater control of the process to monopolize violence more effectively. While nuclear weapons were terrifying, they could potentially achieve select political goals through a rather deliberate and objective application of violence without the traditional risks to one's own military in the execution of the enemy's destruction, at least initially and against non-nuclear adversaries. Further, the use of advanced war technology such as a nuclear device offers the bearer of that war system a level of precision (complete city-sized destruction) and sterility (total annihilation) unreachable with earlier or more primitive means. Societal outrage notwithstanding, nuclear weapons essentially provide what hundreds of years of previous military organizations had been seeking; the ultimate city destroying capability that a military could employ to force the adversary to surrender or definitively be destroyed. With technological developments open to one's adversary, the hasty arrival of "mutually assured destruction" and nuclear stalemate in the mid-twentieth century was also an expected and unavoidably violent development for technologically advanced humanity.

Malešević also suggested support of this transition from premodern to modern military concerning the application of violence, where

> there was a shift from the Medieval type of violence defined by excessive cruelty and a very low military efficiency to the wars of modernity characterized by the gradual displacement of gruesomeness and greater military proficiency resulting in enormous human casualties (i.e. WWII).[138]

What took thousands of soldiers over the course of months of campaigning using hand-wielded edged weapons hundreds of years ago could now be done quickly and cleanly by mere hundreds of aviator crews in bombers flying over a targeted city. Today, the ratio gets even smaller where a single soldier might wield a city-destroying device of technology, particularly from behind a computer half a world away. Furthermore, smaller groups of adversaries capable of what previously could be state-controlled instruments of power can now be independent of states themselves. This deconstruction, or perhaps the fragmenting and challenging of the Westphalian monopolization of organized violence, leads to the last section in this chapter and opens

Pandora's box of new war designs, as well as ways to design *a conceptualization of war itself* differently.

Entering a Postmodernity, a Postmodern Shift in War, or Postmodern Militaries?

This last section treads carefully in that there are three areas of concern. First, the area of postmodern theory represents probably the most antagonistic, misunderstood field of possible study for most military professionals, security affairs experts, and defense academia. Second, most postmodern theorists disagree among one another on postmodern concepts, and potentially would find any postmodern adaptation of their ideas by the military itself as an abomination. Third, postmodern concepts require dense language that relies on entirely dissimilar metaphoric devices and operates under radically different theories and models than that underpinning the modern war paradigm. In terms of commercial design, there are few if any examples of any postmodern overlap with mainstream design praxis except for a select group of design theorists such as Alexander, Tschumi and perhaps Papanek.[139] Most commercial designers do not indulge in postmodern language, ideas, or models. Few military designers understand and articulate what postmodern concepts they design with, how they function and why it matters. Many more prefer to reject postmodern ideas and substituting modern ones in military design, as the next few chapters will detail.

While most modern military scholars and practitioners readily agree that the character of war has changed so much that one can differentiate between a premodern and modern war era, few will venture into *whether the nature of war* is also up for discussion. Premodern military philosophers mused upon universal laws concerning war and satisfied most inquiries using logic, wisdom, and collective experiences. With the rise of the modern scientific era of war, a natural science-inspired framing for warfare took hold. Most all security affairs, international strategic efforts, and associated war paradigms of western industrialized (and those still developing) nations subscribe to a natural order of war based upon the theories and models of Carl von Clausewitz, Antoine-Henri Jomini, Gerhard Scharnhorst, Basil H. Liddell Hart, and others that reinforce this scientific rendering of warfare and war itself. Today, the modern war paradigm maintains this epistemological framework in how it formulates decision-making, organization, administration as well as rationalizing how warfare must (and must not) be expressed.[140]

Just as postmodern theorists in the 1960s–1990s challenged the foundations of modern industrialized, technologically advanced (largely western) societies, the first designers operating within security contexts would follow suit and apply some postmodern arguments, ideas, models as well as language to deconstruct, disrupt, challenge, and reform the modern war paradigm. Military design would shift from a modern, managerial, and organizational extension of established military plans and strategies to one of postmodern deconstruction, reframing, and novel conceptualization beyond previous barriers of imagination. Such design becomes heresy to those defenders of the modern military orthodoxy, and these divisions would play out across military forces, schools, and even within doctrine.

The first way in which postmodern concepts have been applied to war and the modern military profession is certainly the most antagonistic to modern military identity. Has war itself transformed from a modern to postmodern form and function? This would mean that many of the beliefs, models, theories, and methodologies that were previously employed during modern war are vulnerable to evaluation and possible elimination for something new, paradoxical, and highly disruptive to how modern militaries think, act, and organize. This postmodern stance threatens the very bedrock of military doctrine, theory, and practice in that everything Clausewitz, Jomini, Liddell Hart, or others declared about war in the modern sense.[141] Virilio and Lotringer, for example, invert Clausewitz's maxim of "war is politics by other means" into "Total Peace of deterrence is Total War pursued by other means".[142] Deleuze and Guattari would invert Clausewitz as well where the war machine takes charge of the original aims of the state (politics), "appropriates the States, and assumes increasingly wider political functions".[143] The war machine, as something beyond the control or even awareness of the nation-state, consumes all of the Westphalian, Clausewitzian constructs entirely in this postmodern deconstruction.

Postmodernists that argue the transformation of modernity into a postmodernity must include war itself, as war is a human construct and postmodern societies will transform war along with everything else. Postmodernists suggest war went from a modern to postmodern transformation at the zenith of World War II, or immediately following the use of atomic weapons in 1945. Others choose the Vietnam War as the first postmodern war, some place the shift at the collapse of the Berlin Wall, while still others offer the First Gulf War as that changing point into postmodernity. This is a wonderful example of how fragmented the postmodern field is, why it is so controversial in security affairs and defense academia. Postmodernists themselves rarely seem able to agree on their own ideas on deconstructing and disrupting modernity. This does not discount the utility of the postmodern perspective, as the essence of postmodernism will not be readily understood or diagnosed using modern concepts.

Gray positions the start a postmodern war era at 1945 with the first atomic bombs used offensively.[144] Others posit that postmodernity occurring later with the Vietnam War,[145] the end of the Cold War,[146] or later still with the terror attacks of September 11, 2001.[147] Gray views the Vietnam War as the first "postmodern war", implying that conflicts between the end of World War II and through the Korean War are transitional incidents of organized violence.[148] In a subtle precursor to this notion of postmodern war, Virilio would frame World War I as the first "total war" of humanity *against* man, where nation-states employed the entirety of "the scientific complex ranging from physics to biology [to] psychology",[149] injecting a stylized sequencing of postmodernity that would lead to what he saw as a post-industrial reality. This was ushered in through Hollywood in the 1920s, nesting with Virilio's emphasis on signs, information, and narratives that are broadcasted into everyone's living room over radio, newspapers, later movies and then television.[150] There is a blend between postmodern war dealing with extremely, even potentially unstoppable weapons of technological terror and the rise of socialized information that would also move at terrifying speeds to audiences unprecedented in any earlier society.

Often, the idea that Westphalian nation states rose provides postmodernists with the desire to announce their downfall and with it, the transformation of war as an application of state-sponsored organized violence for political desires. Lockhart and Miklaucic refer to the last twenty-five years of academic acknowledgement that failing and failed states present a clear and significant risk to international security, yet a majority position on the plight of the Westphalian Order remains one that it is the "best of all worst possible options" and that "the best hope to retain and build on this progress is to preserve the system of effective sovereign states, ensure that states can perform required functions, and adapt this system to the needs and realities of the twenty-first century".[151] Thus, they want to preserve the legacy state of modernity to include modern war designs, but make progressive, incremental improvements that do not destroy the scientific foundations.

This stance represents a preservationist argument where the modern Westphalian Order for war and the application of organized violence ought to be retooled and modified to remain relevant within this emerging postmodernity. Gorka in cataloging the last two centuries of declared/categorized wars asserts that due to the vast preponderance of wars being irregular, our militaries are rather misoriented to what war even is now, or how to adapt to these changes.[152] Gorka joins Lockhart and Miklaucic in promoting improvements to the current Westphalian system if only to improve its relevance to emergent alternative competitors that violate the original, natural science inspired war rules. McFate offers "new rules for war" to include the title of his book for militaries now operating in this post-Westphalian world, but he also argues that a nature of war remains unchanged and enduring to these same natural science constructs.[153] War in this postmodernity is considered unlike the traditional state-on-state conflicts of recent centuries, yet western industrialized nations such as American, British, Canadian, and Australian militaries use doctrine and military analytical methodologies to prepare and understand warfare in the legacy, outdated frame. Gorka declares this as fundamental to why the modern war frame is failing:

> As a nation, we must move beyond outdated and Clausewitzian understandings of war as solely a functional operation of the nation-state. This is not to denigrate the Prussian's genius. However, his description

of war as a continuation of politics by other means was an idealized description of state-on-state war and as such is fit fine for describing and understanding World War II or the Gulf War, but definitely lacking when we face groups that are not motivated by politics as we understand them, like al-Qaeda or [the Islamic State]. Whether fighting Shaka Zulu in Africa in the 19th century, the Taliban in Afghanistan, or [the Islamic State] today, our adversaries do not play by the Clausewitzian rule book. His concepts of friction and fog still apply, but the idea that our enemies will make rational cost-benefit analyses about the reasons for going to war in ways that serve the *raison d'etat* really does not apply in the irregular domain, especially one in which our main enemy is transcendentally and apocalyptically motivated.[154]

Thus, some postmodern war advocates are not at all in camp with postmodern philosophers such as Baudrillard, Deleuze, Derrida, or Virilio. Some critique the theoretical foundations of the modern war paradigm as inadequate or outdated. Gorka takes issue with how Clausewitzian logic dominates most all modern military logic, doctrine, and even terminology in the twentieth century, along with Paparone, Rapoport, Jackson, Naveh, and others. Gray reinforced this by casting the Vietnam War as the first postmodern one where some of Liddell Hart's earlier critiques of Clausewitzian logic in both World Wars were misapplied. This works in parallel with Michel Foucault's view that military over-infatuation with technological knowledge production, the influence of institutional power structures upon a "rules of the discourse system", and the disruption of what knowledge regime produces the real or valued "truth" manifest.[155] Rappaport would, in his 1968 introduction to Clausewitz's *On War,* make numerous alternative war frames for the Soviet Union and other adversaries that also seem to reject the modern war frame so readily assumed as universal and unquestionable by western industrialized forces.[156] There are many strange bedfellows in the postmodern war camp, yet they all tend to rely on different perspectives, theories, models, and even paradigms to argue their points. This in turn once more makes postmodern arguments concerning war a difficult and confusing affair.

In terms of postmodern philosophers willing to directly confront the modern war paradigm, Baudrillard did so quite aggressively in his writings as well as a series of newspaper articles during preparation for the First Gulf War and the war itself.[157] On the shift to postmodernity, he stated:

The true revolution of the nineteenth century, of modernity, is the radical destruction of appearances, the disenchantment of the world and its abandonment to the violence of interpretation and of history ... I analyze the second revolution, that of the twentieth century, that of postmodernity, which is the immense process of the destruction of meaning, equal to the earlier destruction of appearances.[158]

His broad brush for this shift between modernity and postmodernity brackets most other nuanced postmodern positions as well.[159] Baudrillard argued that a complete destruction of reality is now replaced by a simulacra or "copy without any original form", in that societies no longer realize they have substituted a false world that they now exist in, to include war. This false reality is composed of symbols and meanings maintained by a social construction that becomes an external force to any individual thought. We become trapped in a false way of understanding war itself and are forced to conform our actions in warfare within this simulacrum of reality.

Other postmodernists frame the primary security threat in modernity as that of a physical invasion of a nation's sovereignty or that of an ally ... that this became the complete purpose of modern war. However, in what some term a "late modern period" during the nuclear arms race of the Cold War, the core purpose of war pivoted toward a nuclearized brinksmanship over annihilation.[160] Moskos et al. discuss in *Towards a Postmodern Military* that: "The power to destroy civilizations was the defining quality of the Cold War".[161] Once the Cold War ended, some postmodernists saw the arrival of a new period where multipolarism, international systems, networks, and dynamic transformations created new contexts for the application of organized violence.[162] Gray summarized this profound shift in the very logic of war itself as follows:

> I call it postmodern war. Why choose "postmodern" over the other possible labels? There seem to be two good reasons. First, modern war as a category is used by most military historians, who usually see it as starting in the 1500s and continuing into the middle or late twentieth century. It is clear that the logic and culture of modern war changed significantly during World War II. The new kind of war, while related to modern war, is different enough to deserve the appellation "postmodern." Second, while postmodern is a very complex and contradictory term, and even though it is applied to various fields in wildly uneven ways temporally and intellectually, there is enough similarity between the different descriptions of postmodern phenomena specifically and postmodernity in general to persuade me that there is something systematic happening in areas as diverse as art, literature, economics, philosophy and war.[163]

If indeed a new war paradigm (whether postmodern war, war in postmodernity, etc.) is exercising across the world now, a case should be made by these postmodernists that previous modern methods of warfare (that were once successful) *are no longer so*. Contemporary military practitioners, political leaders, strategists, and theorists seem to be aware that the existing dominant and hierarchical modes of waging war in state-on-state "Great World Wars" style may be losing relevance or changing so that previous power relationships no longer work as before. No adversary, whether rational in a modern context or operating in some postmodern frame, seeks to fight opponents where the strengths and advantages are stacked against them. Nuclear superpowers wield the most sophisticated, advanced military forces in existence, yet cunning terrorist groups, cartels, and guerrilla movements find the best ways to sidestep, mitigate, or avoid these strengths. Atran, in addressing revolutionary movements and their resistance to previously well-engineered, modern military solutions to security challenges, observed that contemporary terror networks such as the Islamic State[164] were able to exist and even thrive under conditions that would have previously obliterated earlier opposition:

> During the surge of American troops in Iraq, up to three-fourths of the fighters were neutralized in al-Qaeda's Iraqi affiliate, which would become ISIL, and an average of about a dozen high-value targets were eliminated monthly for 15 consecutive months, including its top leader, Abu Musab al-Zarqawi. Yet, the organization survived and the group went on to thrive beyond all expectations amidst the chaos of Syria's civil war and Iraq's factional decomposition.[165]

With reference to postmodernity and whether the earlier state-centric stability of conflicts, the dominance of the centralized hierarchical form for military organizations (as well as criminal entities) may also be faltering or under some postmodern redesign. Dishman in studying criminal and terrorist organizations notes that analysts used to be better at determining the goals, motives, and impacts of terror groups and criminal entities like drug cartels because they used to follow a modern centralized hierarchical form – one that was familiar to those using the modern war paradigm and seeking similar structured adversaries.[166] Farah argues that the new relationships between adversarial nation-states and criminal as well as terror groups have changed, and along with it, the dynamics for understanding and acting for security challenges in this new world (of postmodernity). He observes:

> In the construct of the new rules they are writing for their game, none of the state-sanctioned or state-sponsored activities with transnational organized crime (TOC) groups or terrorist groups are illegal or questionable – they are revolutionary tools to obtain a strategic objective.[167]

Along with nation-states, non-state actors are increasingly able to assume state-like capabilities and characteristics for applying organized violence despite not being "states". Dishman goes on to say: "Ungoverned by hierarchical rules, today's networked actors are increasingly polymotivated and pursue a spectrum of criminal and terrorist activities".[168] Antagonist groups such as Hezbollah, previously understood as a "proxy" under the Iranian nation-state and a regional threat tied to particular geopolitical and cultural frameworks, is increasingly reinterpreted as a new sort of threat requiring new thinking, new terminology, and a change in how militaries understand and act in order to deal with it. Levitt posits:

> The challenge Hezbollah poses has become global in nature … [it] can no longer be seen as an Iranian proxy and terrorist organization alone; it is now a powerful military force, a globally lethal terrorist organization, and a complex criminal and money laundering network.[169]

The modern interpretations of proxy terror elements are falling short within the postmodernity context for war, although many of the modern war effects such as targeting Hezbollah leadership still have significant results notwithstanding.

The analytic processes championed in the modern war frame are increasingly fragile or potentially irrelevant within postmodernity, advocates suggest.[170] Scientific knowledge, along with the technological rationalism that modern militaries employed in conjunction, does not represent the totality of knowledge for postmodernists.[171] The scientific way of warfare in modernity would, from the 1950s onward, increasingly be challenged as would the nation-states and the technocratic militaries under their control. This indicates another significant aspect of a shift away from the nation-state-centered modern war construct toward something different and requiring new ways of thinking about war. While the atomic bomb, a technological marvel and output of total-war orientation, had ended World War II, the promises that future security concerns would be rapidly dealt with through rapid, precise, overwhelming military superiority did not seem to work.

The decade-long Vietnam War showcased this tension with advanced American forces unable to dislodge or defeat Viet Cong adversaries. A generation later, after the spectacular invasion of Iraq by the most technologically advanced, resourced, and trained military force arguably in human history, a low-technology, poorly resourced adversary did not crumble. U.S. Special Operations Forces should be most able to conduct analytic optimization of war beyond most any other rival on the planet, yet

> by the end of the year, there had been more terrorist attacks in Iraq alone than there had been in the entire world in 2003. And it only got worse. The forces that comprised the Special Operations Task Force had clear and undeniable points of superiority … despite all of these advantages, it was clear by 2004 that AQI was somehow outpacing some of the world's most highly trained and well-funded units.[172]

The very best in the world at war were only able to keep pace with the growth and adaptation of one of the poorest, least trained, and under-resourced rivals employing a different operating logic and defying the rules of the established game. Terror groups like AQI, ISIS, Boko Haram, and others would be decimated or otherwise diminished by 2022, but did this occur in the context of modern war paradigms demonstrating competitive value, or is it possibly a last gasp of an outdated and extremely expensive approach to warfare in postmodernity?

Despite nearly unlimited resources and almost comical technological overreach against these low-tech, poorly equipped, and informally trained adversaries, the U.S. military elite could only strike at an increasingly illusive and rapidly expanding enemy movement that seemed immune to virtually every indoctrinated method in the book. There is little doubt that in the two decades since the 9–11 attacks, legions of military staffs and units have churned out strategic plans, campaign designs, operational and tactical plans in complete process compliance, often well beyond previous efforts of earlier generations. Individual failings aside, how can so many military enterprises design warfare with such doctrinal adherence, procedural effectiveness, and technological integration yet still systemically fail at reaching even modest, original strategic goals of their policymakers and societies that sent them to war? Few generals or senior leaders were ever fired or relieved, nor were subordinate officers held accountable. If operators were not to blame, and the planned deliverables of modern military methodologies not faulty, was the modern military frame potentially at fault?

Postmodernism and Military Design

Whether war is now postmodern, or we are entering some "late-modernity" that demands difference in thinking, there are plenty of examples of military and strategic failings to go round. Perhaps the question is not whether we are entering some postmodernity or that war itself is undergoing some radical postmodern transformation, *but whether this design movement can provide security organizations with something new and advantageous* versus repeating our legacy decision-making frameworks for addressing warfare. Are military designers proposing something different and disruptive to the existing, legacy frame? If so, military design as an emerging community of practice needs to be understood in how the theories, models, methods, language, and overarching belief system concerning war philosophy (or philosophies) differ, and why it matters.

This is hardly a promise that design might be some silver bullet or magical cure for the madidities of today's security dilemmas. Designing for defense functions differently than traditional military decision-making logic, and that implementing design requires careful and deliberate education and leadership. It is unclear whether any new methodology for warfare is emerging to replace the traditional, linear, and mechanistic modes for thought and action in modern warfare, with many military design groups moving in quite different directions upon diverging paths. Yet this also is in keeping with the fragmentary nature of postmodernism writ large. In this military design movement, some deserve the term of "postmodernist", others might better be described as "designing in postmodernity", and yet others could be seeking redesign of modernity to compete with postmodern adversaries.

Debates between modernists and postmodernists concerning war are often fixated on ideas; meaning and belief systems seem nested into the war paradigms themselves so that a challenge to one method, doctrine, or war theorist causes the wagons to circle and an institutional defense of everything sacred. This is where perhaps postmodernists and modernists cannot see eye-to-eye as they see the world and that of warfare through clearly different epistemological and ontological frameworks. This makes it exceptionally hard to have one influence the other, or even engage in useful discourse. They talk past one another as they both interpret many of the same words and ideas in entirely dissimilar ways.

Lyotard, in explaining this divide between modernists and postmodernists offers:

> The rules of the game of science, are immanent in that game, that they can only be established within the bonds of a debate that is already scientific in nature, and that there is no other proof that the rules are good than the consensus extended to them by the experts.[173]

Modernists reject postmodernism by employing the hallmarks of scientific rationalism to discount anything not scientific outright. Sookermany notes: "the postmodern describes a world in continuous change where existing knowledge is under constant pressure and meaning 'floats'".[174] In these paradoxes, one can apply postmodern theory to deconstruct why in Vietnam "to save the village, we had to destroy it" and other appropriate contradictions. Virilio illustrates this in *The Information Bomb* where he describes an inversion of modernity due to globalization, technology, speed of information, and social networks where reality still attempts to apply organized violence through military forces:

> For the US general staff, then, the pips are no longer inside the apples, nor the segments in the middle of the orange; *the skin has been turned inside out.* The exterior is not simply the skin, the surface of the Earth, but all that *is in situ,* all that is precisely localized, wherever it may be. There lies the great Globalitarian transformation, the transformation which extraverts localness – all localness – and which does not now deport persons ... but deports their living space.[175]

Postmodern proponents thus declare that all modernity is over, including the legacy frame for modern military doctrine, methodologies, theories, and organizational form. Moskos, in taking a postmodernity stance for emerging war contexts, posits that the nation-state security concerns shifted from earlier nuclear and existential ones in the 1990s to that of ethnic violence and terrorism.[176] Here, postmodern acts of war are not clearly aligned with specific nation-states or prone to isolation and prediction, and earlier Westphalian enabled security concerns have (at least when written) decreased. This reinforces the ideas of the last section where individual actors and groups now are intruding upon what had previously been monopolized *exclusively* by nation-state capabilities for organized violence. Non-state entities now operate in space alongside advanced, space-faring nations, while individual actors can hack with exceptional precision in cyberspace in parity with national security abilities. Terrorism now blurs into narco-cartels, human trafficking, counterfeiting, and hybrid combinations that elude national security efforts. Yet the modern Westphalian war paradigm already encompassed non-state actors buzzing and stinging larger, formal nation-state entities. Postmodernism folds this upon itself.

Virilio targets the Israeli attack on Beirut in 1969 as the shift where nation-states adapted terrorist behaviors to defeat perceived postmodern threats, ushering in a postmodern era of perpetual state-crimes, economic war machines, and the disappearance of politics in any traditional format able to arbitrate concerning organized violence.[177] For postmodernists, the nation-state that, in realizing the flexibility of non-state terror activities, can seek to invert the dynamic so that a nation-state gains the abilities of the nimble non-state actor, or in an effort to slow down a nimble adversary, encourage the non-state actor to become more state-like.[178] Deleuze created the postmodern concept of a fold, where "the world and everything in it is in a constant state of folding, unfolding, and refolding … Deleuze illustrates this … by describing how a caterpillar envelops a butterfly (i.e. it is folded inside of it)."[179] When the caterpillar grows into a butterfly, it is "unfolding", and when it dies and decomposes into content for the larger ecosystem, it "refolds" back into its constituent parts. This postmodern perspective of folding provides a different way to conceptualize the ever-changing dynamic of nation-state warfare and non-state actor innovation. The non-state actor is folded inside a nation-state, it unfolds in organized violence through terrorism, the nation-state refolds by assimilating novel terror advantages, and the cycle continues, ever changing and adapting. Virilio and Lotringer too would play upon this theme, morphing the nation-state away from "classical war" in the nuclear age to one where states become terrorists in how they wage limited wars.[180] The deterrence of nuclear annihilation would, in a postmodern shift, force nation-states to assume previously undesired or unnecessary warfare methods and ideas.

McFate later theorizes a similar argument evoking some postmodern perspectives where: "Traditionalists cannot contemplate wars without states, even though such wars surround us".[181] The postmodern adversary could apply instruments of state power without needing to be a nation-state in the modernist Westphalian model.[182] Williams in his conclusion in "The Global Crisis of Governance" proposed a post-Westphalian model for the twenty-first century where "variations on state-building based on a Westphalian ideal … is increasingly passé … in the twenty-first century, the only forms of governance that are likely to be sustained in large swaths of the world are those that are, in effect, post-Westphalian".[183] Gilman, sharing a similar postmodern perspective, suggested that contemporary insurgencies reflect a post-Westphalian postmodernity where "the replacement of the Westphalian ideal of uniform authority and rights within national spaces [occurs] by a kaleidoscopic array of de facto and de jure microsovereignties".[184] Postmodern thinking designs toward shifting our focus to concerning ourselves with understanding processes and the ways in which these understandings contribute to changing reality. Pick summarized this as follows:

> Our attention [as postmodernists] is drawn to modes of change, expansion, propagation, occupation, contagion, and peopling … as a result, possibilities arise to open up new registrars of thought, action and speed – for example, in the form of new metaphors, ideologies and paradigm assumptions.[185]

While the modern war paradigm orients upon a Clausewitzian emphasis on destruction of another adversary's ability to fight and resist, postmodern thinking deconstructs this and suggests alternatives. The postmodern argument includes concepts where destruction is not necessarily of pure physical ends and means, but also where the social construction of reality and rich networks of information and relationships are now exchangeable with tangible bombs and territory. Ideas become as lethal and dangerous as bullets.

> The world is crisscrossed by networks, some based on ties like ethnicity or nationality, others on shared ideas, concerns or ideology … advancements in communication technology, especially the ability to meld reality and fantasy … confuse the young and the unstable who then feed each other's delusions via virtual communications.[186]

Reality exchanges with fantasy, and the virtual with the tangible to blend society into hybrids that may lose track of where one ends and the other begins. The fantastic aspects of postmodern war perspectives create a most difficult ground for debate or discussion. However, in 2022 where notions of "fake" news, disinformation, social media wars, ambiguity over whether one is human or AI online, cultural wars of identity, meaning, and shifting values, as well as the myriad ethical, moral, and legal ambiguity in emerging warfare contexts[187] present unique examples where a postmodern perspective may have significant value.

Enter the Military Design Movement

The military design movement is recent, materializing in a formal community of practice only within the last three decades, despite informal design occurring throughout the history of warfare. This growing community of practice remains even at the time of this writing a minority group within a much larger discipline of security affairs, foreign policy theory, and general military education and military practitioners. Military designers remain mostly unknown to the general commercial design community, with formal acknowledgement slowing developing in recent years. One instance of this is a rigorous military design literature review published in 2021 and represents the first formal study of military design curated knowledge.[188] Across the globe, only a handful of mostly western, largely Anglo-Saxon security forces have experimented with some military design methodology either through inclusion into formal military doctrine, military education, or some grassroots applications at unit level. Many of these design communities operate independently, often within a specific military service or national identity, with only loose frameworks of design networks beginning to form and communicate.

Yet in the span of less than three decades this movement has grown from a singular origin in the Israeli Defense Forces to an international and increasingly influential yet diverse community of design practitioners. Many of the design methods, theories, and cognitive models are unlike commercial design counterparts as well. Some military designers seek to enhance existing, institutionalized practices in strategy and planning, while others propose to deconstruct the entire military system of decision-making and build an alternative unlike the original. Others are in-between, with significant plasticity and ambiguity defining the movement overall.

Most military designers seek change and innovation, with some preferring incidental, evolutionary design risk and others radical paradigm shifts in war itself. These designers are often declared heretics, mavericks, and sometimes worse by the mainstream military institution. This is in part due to how institutions are largely unable to assimilate innovation into doctrine,[189] and acts of innovative thinking typically force radical revisions or change of institutionally sanctioned, historical knowledge that may no longer be relevant. Paparone articulates how the design movement is often associated with deviancy, abstract thinking, and radical conceptualization outside the norms of mainstream military thought as follows:

> The design school is arguably a Lyotardian *petit recit,* introduced by a somewhat deviant segment of the military community led by military intellectuals such as Brigadier General (Ret.) Shimon Naveh, the founder and former head of the Israel Defense Forces' Operational Theory Research Institute. This school of thought is particularly countercultural and orients practice on paralogical communicative processes.[190]

The following chapters expand on this tension between modernity and postmodernity, the nation-state mechanistic, industrialized war frame and that of a postmodern alternative, and how military designers have formed a movement that uses different theories paired with unique conceptual models, requiring new terminology and dissimilar metaphors than what the modern military institution expects. Across the globe, for various reasons, American, Canadian, Australian, British, NATO, and other European forces have witnessed this rise of a military design movement. It does not correlate directly with commercial design theory and practice, nor does it spring directly from how established military beliefs, theories, and practices are constructed. The design movement is eclectic, counter-cultural, disruptive, imaginative, innovative, and from the perspectives of institutional defenders, potentially destructive and dangerous. This may originate ultimately in the attitude and spirit of the first military design theorist, Dr. Shimon Naveh, who is the subject of the third and sixth chapters.

Naveh, building upon a wide and eclectic range of disciplines and fields previously unexplored or utilized in any military organization, sought to change his own Israeli Defense Force so that his nation could win the next conflict. Thus, he presents a paradoxical fusion of Westphalian ideals and nation-state perspectives with those that are decidedly postmodern in design. Earlier in this chapter, we discussed how historians and theorists proposed that certain conflicts had one foot in the premodern, the other in the modern, and again for the modern–postmodern shift. Many theorists and military scholars before Shimon Naveh proposed ways to change, innovate, design, and transform militaries to prepare for upcoming conflicts. Yet few presented entirely novel war paradigms, complete with new theoretical combinations, alternative models for conceptualization, and an independent decision-making methodology that did not collaborate or support the legacy frame for war. Naveh, in producing his "systemic operational design" construct, was able to stand with one foot as a Brigadier General and combat veteran of the Israeli paratroop infantry in modernity and extend his other foot into a postmodern design unlike anything seen previously in military thought or action.

Notes

1 This translates to "Western education is forbidden" and is a common reference to the terror organization. The group calls itself "Jamā'at Ahl as-Sunnah lid-Da'wah wa'l-Jihād", meaning "Group of the People of Sunnah for Dawah and Jihad" in Arabic.
2 Madiha Afzal, "From 'Western Education Is Forbidden' to the World's Deadliest Terrorist Group: Education and Boko Haram in Nigeria," *Security, Strategy, and Order* (Washington, D.C.: The Brookings Institution, April 2020), www.brookings.edu/wp-content/uploads/2020/04/FP_20200507_nigeria_boko_haram_afzal.pdf
3 The postmodern war section of this chapter expands on this with primary sources and examples.
4 Markus Mäder, *In Pursuit of Conceptual Excellence: The Evolution of British Military-Strategic Doctrine in the Post-Cold War Era. 1989–2002* (Bern, Germany: Peter Lang AG, 2004), 70–77.
5 John Keegan, *The Mask of Command* (New York: Penguin Books, 1988), 117.

6 Sun Tzu, *The Art of War: Complete Text of Sun Tzu's Classics, Military Strategy History, Ancient Chinese Military Strategist Deluxe Collection Edition, 1*, trans. Lionel Giles (Las Vegas, NV: Amazon, 2022).

7 Morris Janowitz, *The Professional Soldier: A Social and Political Portrait*, Paperback Edition (New York: Free Press, 2017), 60–61; Alfred Vagts, *A History of Militarism: Civilian and Military (Revised Edition)*, Revised (New York: The Free Press, 1959), 42–43.

8 Chris Gray, *Postmodern War: The New Politics of Conflict* (New York: The Guilford Press, 1997), 111.

9 David French, *The British War in Warfare 1688–2000* (Cambridge, Massachusetts: Unwin Hyman Ltd, 1990), 8. This does not undercount standing, professional armies of the ancient world such as Sparta.

10 Julian Jaynes, *The Origin of Consciousness in the Breakdown of the Bicameral Mind*, Third (Boston, Massachusetts: First Mariner Books, 2000), 321–323.

11 Martin van Creveld, *The Training of Officers: From Professionalism to Irrelevance* (New York: The Free Press, 1990), 20.

12 Basil H. Liddell Hart, *Strategy*, Second revised edition (New York: Meridian Book, 1991), 61, 71.

13 Henry Guerlac, "Vauban: The Impact of Science on War," in *Makers of Modern Strategy: From Machiavelli to the Nuclear Age*, ed. Peter Paret (Princeton, New Jersey: Princeton University Press, 1986), 73–74, 85; Amos Fox, "The Reemergence of the Siege: An Assessment of Trends in Modern Land Warfare," *Institute of Land Warfare Publication* 18, no. 2 (June 2018): 1–2.

14 Liddell Hart, *Strategy*, 59–61.

15 Siniša Malešević, "The Organization of Military Violence in the 21st Century," *Organization* 24, no. 4 (2017): 456–474.

16 Michel Foucault, *Discipline & Punish: The Birth of the Prison*, trans. Alan Sheridan (New York: Vintage Books, 1995), 111.

17 Malešević, "The Organization of Military Violence in the 21st Century."

18 van Creveld, *The Training of Officers: From Professionalism to Irrelevance*, 8.

19 Keegan, *The Mask of Command*, 40.

20 Charles White, *Scharnhorst: The Formative Years, 1755–1801* (Warwick, England: Helion & Company, 2020), 209–210, 266.

21 Michel Foucault, *The Order of Things: An Archaeology of the Human Sciences*, Vintage Books Edition, April 1994 (New York: Vintage Books, 1994), 32.

22 Keegan, *The Mask of Command*, 117.

23 van Creveld, *The Training of Officers: From Professionalism to Irrelevance*, 20; Siniša Malešević, *The Sociology of War and Violence* (Cambridge, United Kingdom: Cambridge University Press, 2010), 108–109.

24 Malešević, *The Sociology of War and Violence*, 107.

25 Vagts, *A History of Militarism: Civilian and Military (Revised Edition)*, 41–46; Justin Kelly and Michael Brennan, "Alien: How Operational Art Devoured Strategy" (Carlisle, Pennsylvania: Department of the Army's Strategic Studies Institute, September 2009), 12–13.

26 French, *The British War in Warfare 1688–2000*, 2.

27 This does not include invaders from other regions, cultures, or otherwise unwilling to agree to such rules.

28 Malešević, *The Sociology of War and Violence*, 105.

29 van Creveld, *The Training of Officers: From Professionalism to Irrelevance*, 13.

30 R. R. Palmer, "Frederick the Great, Guibert, Bulow: From Dynastic to National War," in *Makers of Modern Strategy: From Machiavelli to the Nuclear Age*, ed. Peter Paret (Princeton, New Jersey: Princeton University Press, 1986), 92.

31 van Creveld, *The Training of Officers: From Professionalism to Irrelevance*, 19.

32 White, *Scharnhorst: The Formative Years, 1755–1801*, 147, 229; van Creveld, *The Training of Officers: From Professionalism to Irrelevance*, 49; Vagts, *A History of Militarism: Civilian and Military (Revised Edition)*, 48–52.

33 van Creveld, *The Training of Officers: From Professionalism to Irrelevance*, 49–51.

34 Keegan, *The Mask of Command*; Brian Linn, *The Echo of Battle: The Army's Way of War* (Cambridge, Massachusetts: Harvard University Press, 2007).

35 Janowitz, *The Professional Soldier: A Social and Political Portrait*, 262–263.

36 Palmer, "Frederick the Great, Guibert, Bulow: From Dynastic to National War," 91–119.

37 Guerlac, "Vauban: The Impact of Science on War," 71–74; Keegan, *The Mask of Command*, 18–19; Vagts, *A History of Militarism: Civilian and Military (Revised Edition)*, 52.

38 Guerlac, "Vauban: The Impact of Science on War," 71.

39 French, *The British War in Warfare 1688–2000*, 11.

40 Herbert London, *Military Doctrine and the American Character: Reflections on AirLand Battle*, National Strategy Information Center Agenda Paper Series 14 (New York: National Strategy Information Center, Inc, 1984), 15–16.

41 Elizabeth Kier, *Imagining War: French and British Military Doctrine Between the Wars* (Princeton, New Jersey: Princeton University Press, 1997), 29; Shimon Naveh, Jim Schneider, and Timothy Challans, *The Structure of Operational Revolution: A Prolegomena*, A Product of the Center for the Application of Design (Fort Leavenworth, Kansas: Booz Allen Hamilton, 2009).

42 Ascientific is not necessarily unscientific. It is not based on or in accordance with the scientific methodology.

43 C. C. Pecknold, "How Augustine Used the Trinity: Functionalism and the Development of Doctrine," *Anglican Theological Review* 85, no. 1 (2003): 127–141.

44 Naveh, Schneider, and Challans, *The Structure of Operational Revolution: A Prolegomena*, 23.

45 Mäder, *In Pursuit of Conceptual Excellence: The Evolution of British Military-Strategic Doctrine in the Post-Cold War Era. 1989–2002*, 28–29.

46 Malešević, *The Sociology of War and Violence*, 8–14.

47 Christopher Paparone, *The Sociology of Military Science: Prospects for Postinstitutional Military Design* (New York: Bloomsbury Academic Publishing, 2013), 90–97; Guerlac, "Vauban: The Impact of Science on War."

48 White, *Scharnhorst: The Formative Years, 1755–1801*, 123.

49 Gunther Rothenberg, "Moltke, Schlieffen, and the Doctrine of Strategic Envelopment," in *Makers of Modern Strategy: From Machiavelli to the Nuclear Age*, ed. Peter Paret (Princeton, New Jersey: Princeton University Press, 1986), 296–299.

50 Cannon first were developed in China in the 12th century. The author considers the standardization of cannon and the publication of siege warfare theory and doctrine in 16th century Europe for this framing.

51 Aaron Jackson, *The Roots of Military Doctrine: Change and Continuity in Understanding the Practice of Warfare* (Fort Leavenworth, Kansas: Combat Studies Institute Press, 2013); Mäder, *In Pursuit of Conceptual Excellence: The Evolution of British Military-Strategic Doctrine in the Post-Cold War Era. 1989–2002*, 28–29.

52 Anatol Rapoport, *The Origins of Violence: Approaches to the Study of Conflict* (New Brunswick, New Jersey: Transactions Publishers, 1995), 68.

53 Éric Alliez and Maurizio Lazzarato, *Wars and Capital*, trans. Ames Hodges (South Pasadena, California: Semiotext(e), 2016), 19.

54 Anthony Giddens, *The Consequences of Modernity* (Stanford, California: Stanford University Press, 1990), 58.

55 Submarines, mass-tactical paratrooper operations, encryption and codebreaking, chemical warfare, wireless radios, naval aviation, and atomic weaponry are but a few of the radical developments in the first half of the twentieth century. These new ways to wage war would make earlier forms of strategy and tactics obsolete, or at best inconsequential.

56 Bernard Tschumi, *Architecture and Disjunction* (Cambridge, Massachusetts: MIT Press, 1997), 48.

57 Foucault, *Discipline & Punish: The Birth of the Prison*, 141.

58 Antoine Bousquet, "Chaoplexic Warfare or the Future of Military Organization," *International Affairs (Royal Institute of International Affairs 1944–)* 84, no. 5 (September 2008): 919.

59 John Shy, "Jomini," in *Makers of Modern Strategy: From Machiavelli to the Nuclear Age*, ed. Peter Paret (Princeton, New Jersey: Princeton University Press, 1986); Bousquet, "Chaoplexic Warfare or the Future of Military Organization," 920–921; Azar Gat, *A History of Military Thought: From the Enlightenment to the Cold War*, 1st edition (Oxford: Oxford University Press, 2002), 158–256.

60 Paparone, *The Sociology of Military Science: Prospects for Postinstitutional Military Design*, 30–31; Astrid Nordin and Dan Oberg, "Targeting the Ontology of War: From Clausewitz to Baudrillard," *Millennium: Journal of International Studies* 43, no. 2 (November 3, 2014): 392–410.

61 Paparone, *The Sociology of Military Science: Prospects for Postinstitutional Military Design*, 79; David Lai, "Learning from the Stones: A GO Approach to Mastering China's Strategic Concept, SHI" (Strategic Studies Institute and U.S. Army War College Press, May 2004), 4, www.carlisle.army.mil/ssi/

62 Antoine Bousquet, *The Scientific Way of Warfare: Order and Chaos on the Battlefields of Modernity* (London: HURST Publishers Ltd., 2009).

63 Liddell Hart, *Strategy*, 183, 208–212, 338–352.

64 Paparone, *The Sociology of Military Science: Prospects for Postinstitutional Military Design*, 13–14; Christopher Paparone, "On Metaphors We Are Led By," *Military Review* 88, no. 6 (December 2008): 55–64; Christopher Paparone, "How We Fight: A Critical Exploration of US Military Doctrine," *Organization* 24, no. 4 (2017): 516–533, https://doi.org/o0r.g1/107.171/1773/51035058048147167699933853

65 Mäder, *In Pursuit of Conceptual Excellence: The Evolution of British Military-Strategic Doctrine in the Post-Cold War Era. 1989–2002*, 29.

66 Jaynes, *The Origin of Consciousness in the Breakdown of the Bicameral Mind*, 2–4.

67 van Creveld, *The Training of Officers: From Professionalism to Irrelevance*, 18.

68 White, *Scharnhorst: The Formative Years, 1755–1801*, 389.

69 Bousquet, *The Scientific Way of Warfare: Order and Chaos on the Battlefields of Modernity*; Bousquet.

70 Gray, *Postmodern War: The New Politics of Conflict*, 149.

71 Paparone, "On Metaphors We Are Led By."

72 James Gibson, *The Perfect War: Technowar in Vietnam*, First Edition (Boston: The Atlantic Monthly Press, 1986), 462.

73 Lorraine Daston, *Rules: A Short History of What We Live By* (Princeton, New Jersey: Princeton University Press, 2022), 233–234; Bousquet, "Chaoplexic Warfare or the Future of Military Organization," 920–923.

74 Jean-Francois Lyotard, *The Postmodern Condition: A Report on Knowledge*, trans. Geoff Bennington and Brian Massumi, Theory and History of Literature, Volume 10 (Minneapolis, MN: University of Minnesota Press, 1979), 29.

75 Paul Virilio, *The Information Bomb*, trans. Chris Turner, paperback, Radical Thinkers (New York: Verso, 2005), 3–4.

76 Lyotard, *The Postmodern Condition: A Report on Knowledge*, 37.

77 Anders Sookermany, "On Developing (Post)Modern Soldiers: An Inquiry into the Ontological and Epistemological Foundation of Skill-Acquisition in an Age of Military Transformation" (dissertation for the degree of Dr. Philos, University of Oslo, 2013), vii.

78 Sookermany, 10.

79 Sean McFate, *The New Rules of War*, First Edition (New York: William Morrow, 2019), 28–33.

80 Brian Bloomfield, Gibson Burrell, and Theo Vurdubakis, "Licence to Kill? On the Organization of Destruction in the 21st Century," *Organization* 24, no. 4 (2017): 9–10.

81 Clare Lockhart and Michael Miklaucic, "Leviathan Redux: Toward a Community of Effective States," in *Beyond Convergence: World Without Order*, ed. Hilary Matfess and Michael Miklaucic (Washington, D.C.: Center for Complex Operations; Institute for National Security Studies, 2016), 298.

82 Giddens, *The Consequences of Modernity*, 62–63.

83 Kelly and Brennan, "Alien: How Operational Art Devoured Strategy," 63–65.

84 Christopher Alexander, *Notes on the Synthesis of Form* (Cambridge, Massachusetts: Harvard University Press, 1964), 1–11.

85 White, *Scharnhorst: The Formative Years, 1755–1801*, 377–378.

86 Anders Sookermany, "Military Education Reconsidered: A Postmodern Update," *Journal of Philosophy of Education* 51, no. 1 (2017): 310.

87 Linn, *The Echo of Battle: The Army's Way of War*, 23; van Creveld, *The Training of Officers: From Professionalism to Irrelevance*, 15.

88 White, *Scharnhorst: The Formative Years, 1755–1801*, 295. White references original letters and manuscripts by Scharnhorst translated from German by the author for this quote.

89 Antoine Bousquet and Simon Curtis, "Beyond Models and Metaphors: Complexity Theory, Systems Thinking and International Relations," *Cambridge Review of International Affairs* 24, no. 1 (2011): 45.

90 Bousquet, "Chaoplexic Warfare or the Future of Military Organization," 921.

91 Liddell Hart, *Strategy*, 340.

92 See also: Kelly and Brennan, "Alien: How Operational Art Devoured Strategy," 14–16.

93 Liddell Hart, *Strategy*, 340–341.

94 Bloomfield, Burrell, and Vurdubakis, "Licence to Kill? On the Organization of Destruction in the 21st Century," 5.

95 Linn, *The Echo of Battle: The Army's Way of War*, 6–7.

96 Bloomfield, Burrell, and Vurdubakis, "Licence to Kill? On the Organization of Destruction in the 21st Century," 3–6.

97 Bousquet, *The Scientific Way of Warfare: Order and Chaos on the Battlefields of Modernity*, 13.

98 Paparone, "On Metaphors We Are Led By"; Alex Ryan et al., "Full Spectrum Fallacies and Hybrid Hallucinations: How Basic Errors in Thinking Muddle Military Concepts" (Land Warfare Conference 2010, Brisbane, Australia: Australian Land Warfare Centre, 2010), 237–253; Jackson, *The Roots of Military Doctrine: Change and Continuity in Understanding the Practice of Warfare*; Paparone, "How We Fight: A Critical Exploration of US Military Doctrine."

99 Donald Schön, *The Reflective Practitioner: How Professionals Think in Action*, 1st Edition (New York: Basic Books, 1984), 3–4.

100 Nordin and Oberg, "Targeting the Ontology of War: From Clausewitz to Baudrillard," 401–402.

101 Henry Mintzberg, "Patterns in Strategy Formation," *Management Science* 24, no. 9 (May 1978): 944.

102 Gibson, *The Perfect War: Technowar in Vietnam*, 15–16.

103 Gray, *Postmodern War: The New Politics of Conflict*, 135.

104 Gray, 78.

105 Gray, 80.

106 Jackson, *The Roots of Military Doctrine: Change and Continuity in Understanding the Practice of Warfare*; David Lindberg, *The Beginnings of Western Science: The European Scientific Tradition in Philosophical, Religious, and Institutional Context, 600 B.C. to A.D. 1450* (Chicago: University of Chicago Press, 1992), 190–191.

107 van Creveld, *The Training of Officers: From Professionalism to Irrelevance*, 14–18; Jackson, *The Roots of Military Doctrine: Change and Continuity in Understanding the Practice of Warfare*.

108 Sebastien le Prestre de Vauban, *The New Method of Fortification, as Practised by Monsieur de Vauban, Engineer-General of France. Together With a New Treatise of Geometry. The Fifth Edition, Carefully Revised and Corrected by the Original*, Fifth (London: S. and E. Ballard (reprint by Creative Media Partners), 1722); Guerlac, "Vauban: The Impact of Science on War."

109 Sookermany, "On Developing (Post)Modern Soldiers," 69; Gray, *Postmodern War: The New Politics of Conflict*, 111.

110 Gray, *Postmodern War: The New Politics of Conflict*, 96.

111 London, *Military Doctrine and the American Character: Reflections on AirLand Battle*, 37, 60–61; Mäder, *In Pursuit of Conceptual Excellence: The Evolution of British Military-Strategic Doctrine in the Post-Cold War Era. 1989–2002*, 22.

112 Barry Posen, *The Sources of Military Doctrine: France, Britain, and Germany Between the World Wars* (Ithaca, New York: Cornell University Press, 1984), 44.

113 van Creveld, *The Training of Officers: From Professionalism to Irrelevance*, 9.

114 Vagts, *A History of Militarism: Civilian and Military (Revised Edition)*, 44–45.

115 Max Boisot and Bill McKelvey, "Integrating Modernist and Postmodernist Perspectives on Organizations: A Complexity Science Bridge," *Academy of Management Review* 35, no. 3 (2010): 418; Jackson, *The Roots of Military Doctrine: Change and Continuity in Understanding the Practice of Warfare*.

116 Naveh, Schneider, and Challans, *The Structure of Operational Revolution: A Prolegomena*, 35–36.

117 Haridimos Tsoukas, *Complex Knowledge: Studies in Organizational Epistemology* (New York: Oxford University Press, 2005), 213–216. James Der Derian,

"Virtuous War/Virtual Theory," *International Affairs (Royal Institute of International Affairs 1944-)* 76, no. 4 (October 2000): 786.

118 Paparone, *The Sociology of Military Science: Prospects for Postinstitutional Military Design*, 20.

119 In academia and scientific review, theories are presented to external reviewers, and if published, later tested, critiqued, and debated by peers along with consideration of competing theories and ideas.

120 Modern military doctrine assumes a Clausewitzian, but also a Jominian inspired hybrid theorization of what war is and is not. There are no alternative theories presented, nor are any statements underpinned with primary sources, citations, or additional explanation.

121 Mäder, *In Pursuit of Conceptual Excellence: The Evolution of British Military-Strategic Doctrine in the Post-Cold War Era. 1989–2002*, 297–302.

122 Paparone, "How We Fight: A Critical Exploration of US Military Doctrine"; Paparone, *The Sociology of Military Science: Prospects for Postinstitutional Military Design*, 90–97.

123 Malešević, "The Organization of Military Violence in the 21st Century," 464; Mäder, *In Pursuit of Conceptual Excellence: The Evolution of British Military-Strategic Doctrine in the Post-Cold War Era. 1989–2002*, 45.

124 French, *The British War in Warfare 1688–2000*, 2–3.

125 Liddell Hart, *Strategy*, 125–126; Naveh, Schneider, and Challans, *The Structure of Operational Revolution: A Prolegomena*, 47.

126 Gray, *Postmodern War: The New Politics of Conflict*, 116.

127 French, *The British War in Warfare 1688–2000*, 90.

128 Jacob Kipp, "General-Major A. A. Svechin and Modern Warfare: Military History and Military Theory," in *Strategy*, by Aleksandr Svechin, ed. Kent Lee (Minneapolis, Minnesota: East View Publications, 1991), 23–56; David Stone, "Misreading Svechin: Attrition, Annihilation, and Historicism," *The Journal of Military History* 76 (July 2012): 688–689.

129 Carl Builder, *The Masks of War: American Military Styles in Strategy and Analysis* (Baltimore: John Hopkins University Press, 1989).

130 Mäder, *In Pursuit of Conceptual Excellence: The Evolution of British Military-Strategic Doctrine in the Post-Cold War Era. 1989–2002*, 39.

131 Malešević, "The Organization of Military Violence in the 21st Century."

132 Anatol Rapoport, "Editor's Introduction to On War," in *On War*, by Carl Von Clausewitz, ed. Anatol Rapoport (New York: Penguin Books, 1968); Mary Ann Tetreault and Harry Haines, "Postmodern War and Historic Memory" (Annual Meeting of the International Studies Association, Montreal Quebec, Canada: International Studies Association, 2004), 2.

133 Kelly and Brennan, "Alien: How Operational Art Devoured Strategy," 12–17.

134 Paul Virilio and Sylvere Lotringer, *Pure War*, trans. Mark Polizzotti, "Twenty Five Years Later" edition (Cambridge, Massachusetts: MIT Press, 2008), 29.

135 Virilio and Lotringer, 30–31.

136 Virilio, *The Information Bomb*, 25.

137 Gray, *Postmodern War: The New Politics of Conflict*, 110.

138 Malešević, "The Organization of Military Violence in the 21st Century," 467.

139 Alexander, *Notes on the Synthesis of Form*; Victor Papanek, *Design for the Real World: Human Ecology and Social Change* (New York: Pantheon Books, 1971); Tschumi, *Architecture and Disjunction*.

140 Paparone, *The Sociology of Military Science: Prospects for Postinstitutional Military Design*, 90–97.

141 Mäder, *In Pursuit of Conceptual Excellence: The Evolution of British Military-Strategic Doctrine in the Post-Cold War Era. 1989–2002*, 21–23.

142 Virilio and Lotringer, *Pure War*, 39.

143 Gilles Deleuze and Felix Guattari, *A Thousand Plateaus*, trans. Brian Massumi (Minneapolis: University of Minnesota Press, 1987), 421.

144 Gray, *Postmodern War: The New Politics of Conflict*, 125.

145 Gibson, *The Perfect War: Technowar in Vietnam*.

146 Charles Moskos, John Williams, and David Segal, "Armed Forces after the Cold War," in *The Postmodern Military: Armed Forces after the Cold War*, ed. Charles Moskos, John Williams, and David Segal (New York: Oxford University Press, 2000), 4–7.

147 Mäder, *In Pursuit of Conceptual Excellence: The Evolution of British Military-Strategic Doctrine in the Post-Cold War Era. 1989–2002*.

148 Gray, *Postmodern War: The New Politics of Conflict*, 48.

149 Virilio, *The Information Bomb*, 55.

150 Virilio, 25.

151 Lockhart and Miklaucic, "Leviathan Redux: Toward a Community of Effective States," 299.

152 Sebastian Gorka, "Adapting to Today's Battlefield: The Islamic State and Irregular War as the 'New Normal,'" in *Beyond Convergence: World Without Order*, ed. Hilary Matfess and Michael Miklaucic (Washington, D.C.: Center for Complex Operations; Institute for National Security Studies, 2016), 353–368.

153 McFate, *The New Rules of War*.

154 Gorka, "Adapting to Today's Battlefield: The Islamic State and Irregular War as the 'New Normal,'" 354.

155 Michel Foucault, "Discourse and Truth: The Problematization of Parrhesia" (lecture, University of California at Berkeley, November 1983), http://fouca ult.info//system/files/pdf/DiscourseAndTruth_MichelFoucault_1983_0.pdf; Michel Foucault, "The Discourse on Language," in *The Archeology of Knowledge*, by Michel Foucault, trans. Rupert Swyer (New York: Pantheon Books, 1972); Foucault.

156 Rapoport, "Editor's Introduction to On War."

157 Jean Baudrillard, *The Gulf War Did Not Take Place*, trans. Paul Patton (Sydney, Australia: Power Publications, 2009).

158 Jean Baudrillard, *Simulacra and Simulation*, trans. Sheila Glaser (Ann Arbor: The University of Michigan Press, 2001), 160–161.

159 Baudrillard's assessment that modernity was born of the nineteenth century is, however, the exception to the vast majority of theorists for this project that align it with earlier sixteenth through nineteenth century developments such as the Industrial Revolution and the Enlightenment period.

160 Charles Moskos, John Williams, and David Segal, eds., *The Postmodern Military: Armed Forces after the Cold War* (New York: Oxford University Press, 2000), 14; Sookermany, "Military Education Reconsidered: A Postmodern Update," 314.

161 Moskos, Williams, and Segal, "Armed Forces after the Cold War," 2.

162 Sookermany, "On Developing (Post)Modern Soldiers," 14–23.

163 Gray, *Postmodern War: The New Politics of Conflict*, 22.

164 By 2021, the Islamic State is a far less dangerous terror threat in comparison to the height of their power in 2015. However, they are a useful example here in that surrogates and other emerging rival organizations learned from their successes and failures and continue to improve upon their model.

165 Scott Atran, "The Islamic State Revolution," in *Beyond Convergence: World Without Order*, ed. Hilary Matfess and Michael Miklaucic (Washington, D.C.: Center for Complex Operations; Institute for National Security Studies, 2016), 67.

166 Christopher Dishman, "Terrorist and Criminal Dynamics: A Look Beyond the Horizon," in *Beyond Convergence: World without Order*, ed. Hilary Matfess and Michael Miklaucic (Washington, D.C.: Center for Complex Operations; Institute for National Security Studies, 2016), 137–153.

167 Douglas Farah, "Convergence in Criminalized States: The New Paradigm," in *Beyond Convergence: World Without Order*, ed. Hilary Matfess and Michael Miklaucic (Washington, D.C.: Center for Complex Operations; Institute for National Security Studies, 2016), 181.

168 Dishman, "Terrorist and Criminal Dynamics: A Look Beyond the Horizon," 137.

169 Matthew Levitt, "Hezbollah's Criminal Networks: Useful Idiots, Henchmen, and Organized Criminal Facilitators," in *Beyond Convergence: World Without Order*, ed. Hilary Matfess and Michael Miklaucic (Washington, D.C.: Center for Complex Operations; Institute for National Security Studies, 2016), 173.

170 Paparone, *The Sociology of Military Science: Prospects for Postinstitutional Military Design*, 18–22.

171 Lyotard, *The Postmodern Condition: A Report on Knowledge*, 7.

172 Christopher Fussell and D.W. Lee, "Networks at War: Organizational Innovation and Adaptation in the 21st Century," in *Beyond Convergence: World Without Order*, ed. Hilary Matfess and Michael Miklaucic (Washington, D.C.: Center for Complex Operations; Institute for National Security Studies, 2016), 372.

173 Lyotard, *The Postmodern Condition: A Report on Knowledge*, 29.

174 Sookermany, "Military Education Reconsidered: A Postmodern Update," 312.

175 Virilio, *The Information Bomb*, 10.

176 Jeff Geraghty, "The Nature and Nurture of Military Genius: Developing Senior Strategic Leaders for the Postmodern Military" (For completion of Masters Level graduate requirements, Maxwell Air Force Base, Alabama, School of Advanced Air and Space Studies, 2010), 25; Moskos, Williams, and Segal, *The Postmodern Military: Armed Forces after the Cold War*, 14.

177 Virilio and Lotringer, *Pure War*, 41–42.

178 Ori Brafman and Rod Beckstrom, *The Starfish and the Spider: The Unstoppable Power of Leaderless Organizations* (New York: Penguin Books, 2006), 151–154.

179 David Pick, "Rethinking Organization Theory: The Fold, the Rhizome and the Seam between Organization and the Literary," *Organization* 24, no. 6 (2017): 803.

180 Virilio and Lotringer, *Pure War*, 61.

181 McFate, *The New Rules of War*, 182.

182 Hilary Matfess and Michael Miklaucic, "Introduction: World Order or Disorder?," in *Beyond Convergence: World Without Order* (Washington, D.C.: Center for Complex Operations; Institute for National Security Studies, 2016), ix.

183 Phil Williams, "The Global Crisis of Governance," in *Beyond Convergence: World Without Order*, ed. Hilary Matfess and Michael Miklaucic (Washington, D.C.: Center for Complex Operations; Institute for National Security Studies, 2016), 41.

184 Nils Gilman, "The Twin Insurgencies: Plutocrats and Criminals Challenge the Westphalian State," in *Beyond Convergence: World without Order*, ed. Hilary Matfess and Michael Miklaucic (Washington, D.C.: Center for Complex Operations; Institute for National Security Studies, 2016), 54.

185 Pick, "Rethinking Organization Theory: The Fold, the Rhizome and the Seam between Organization and the Literary," 814.

186 Steven Metz, *Armed Conflict in the 21st Century: The Information Revolution and Post-Modern Warfare* (Carlisle, PA: Strategic Studies Institute, 2000), 7.

187 For example, if a future autonomous weaponized system takes action that is declared a war crime, is the AI system held responsible, the remote operator or supervising human, or the computer programmers of the code that failed to produce the right decision?

188 Cara Wrigley, Genevieve Mosely, and Michael Mosely, "Defining Military Design Thinking: An Extensive, Critical Literature Review," *She Ji: The Journal of Design, Economics, and Innovation* 7, no. 1 (Spring 2021): 104–143.

189 Posen, *The Sources of Military Doctrine: France, Britain, and Germany Between the World Wars*, 54.

190 Paparone, "How We Fight: A Critical Exploration of US Military Doctrine," 11.

3 The Birth of Military Design

Heresy, Innovation, and Betrayal in Israel

Military design began out of fear and frustration. While one might assume that in the 1990s, militaries simply learned of well-established commercial design methods and adapted them into addressing complex security challenges and innovation on the battlefield, yet this did not occur except for direct technical or tactical "war tool" requirements that had no strategic or operational decision-making considerations. As covered in the first chapter, the military industrial complex already wove together the industry of designing for commerce with that of designing toward warfare. Each military service would quickly identify institutional relevance within specific warfighting abilities, skills, and mission specialization that required often sophisticated technical and tactical designs.[1] The Marines would associate their identity with amphibious assaults requiring unique equipment designs, while the 82nd Airborne Division perpetually existed to seize airfields and key terrain with paratroopers anywhere in the world again using different gear and skills. Army Special Forces could execute unconventional warfare with elite, multi-lingual teams of commandos, and the U.S. Air Force defined itself through advanced stealth bombers and fighter jets that continuously needed new, highly technological capabilities. Such war designs were never-ending, technologically rationalized within the modern military paradigm concerning how warfare occurs, and absolutely wedded to the linear, mechanistic, and systematic mode of decision-making found in military doctrine. Along the way, the war tools would be prioritized for design, yet somehow the ways that militaries conceptualized warfare and went about decision-making in war would seldom face similar redesign and innovation.

Modern militaries in the 1990s were undergoing lots of changes with downsizing after the end of the Cold War, a shift toward international peacekeeping and humanitarian aid around the globe, and many social and cultural reforms such as gender roles, inclusion of underrepresented groups, changing political winds, and societal expectations that militaries use advanced technology to accomplish faster, safer operations with fewer civilian and friendly casualties. American and Canadian militaries would struggle with significant political fallout in Somalia for different yet momentous reasons,

DOI: 10.4324/9781003387763-4

while Australia would enter a decade of military (and economic) stagnation. Aside from British counterinsurgency efforts in Northern Ireland, the IDF would be one of the few militaries that continuously operated in high stress, dynamic security contexts throughout this otherwise slow period of global curtailment. This is also why the military design story begins not at Fort Leavenworth or after the victory parades of the First Gulf War, but in Israel among a small population of Israeli military leaders and academics.

In Israel in the 1990s, small pockets of experienced military professionals grew increasingly disillusioned with how their security forces were failing to adapt to changing existential threats. Design and warfare would continue to demonstrate this pattern of military frustration and disappointment after various conflicts or defeats causing a ground swelling of reform and innovation efforts. The post-World War I development of the Bauhaus design movement from disgruntled German war veterans being one example, while individual military reformers such as Giulio Douhet, Billy Mitchell, B. H. Liddell Hart, and John Boyd are just a few of the innovators to protest institutionalized or outdated practices, often at risk to their own careers. Innovators and reformers are terms equally exchanged with mavericks and heretics, depending on which population one asks.

The first and most difficult to understand of military design methodologies came from the mind of Israeli Brigadier General Shimon Naveh. Great minds do not work in a vacuum, and Naveh drew from a small group of fellow frustrated military intellectuals to develop his design ideas, yet Naveh's name would become firmly attached to this first military design way of thinking and acting. Naveh would not simply add a new competing methodology or a new conceptual model to help the modern war paradigm perform better; he ushered forth an *entirely dissimilar and alternative war paradigm* complete with different models, methods, language, and theoretical constructs. This new war paradigm clashed with everything modern militaries believe war is, and how warfare is supposed to be understood and arranged to act within. Naveh and his team would call it "Systemic Operational Design" or SOD for shorthand. In the decades following this initial design, Naveh remains an enigma as a controversial, brilliant, sometimes despised and often combative character, and deservedly the founding father of the military design movement. Naveh will be the first to tell you so as well.

Naveh's ideas, despite meeting with controversy and often outright hostility over the years, have not been cast upon the military trash heap of fads, clichés, and experimental failures. Rather, they have multiplied and evolved, forming numerous branches and clones as well as some poor imitations and superficial facsimiles. This military design movement has in the span of roughly three decades grown from a single, eclectic application within Israel into a full-fledged international community of practice with tremendous variation therein. Military design currently exists in multiple formats, in doctrinal as well as practitioner-generated constructs, and across a diverse spectrum of epistemological, ontological, and methodological variations.

Some are good, others are bad, and others still are ugly. Yet such a diversity of designs is itself a powerful example of a thriving community of practice.

This chapter explains how military design first developed in Israel, why this occurred, and what would occur upon experimentation in real-world security affairs. For Naveh and his designers, this new military design would be used to challenge the deeply held, often ritualized belief systems that modern militaries maintain, such as the military hero construct, masculine roles in warfare, the prominence of a natural science order to war, and so on. One's biases, beliefs, culture, and philosophical designs of how to order complex reality would be tested by SOD. Nothing was off limits, and the postmodern ideas Naveh's SOD would use in their design process caused entirely new doors to open for military conceptualization and innovation. That said, opening those doors proved difficult, particularly for an institution conditioned to seek a universal key that was devoid of any challenging intellectual requirements, or the possibility that many different keys were needed depending on the door and the user.

This chapter starts the story of the military design movement. Military designers would challenge the deeply held beliefs on war and the conduct of warfare as explained in the second chapter. For Naveh and SOD advocates, a designer becomes some sort of strange jester, a disruptor … "the joker that delivers organizational reflection and reframe through seemingly eccentric intellectual activities"[2] and imagination concerning complex warfare. Design "changes the possibilities and makes things possible that were not possible before … it changes the needs systems and objectives of people".[3] At its very essence, designers seek to disrupt and deconstruct the organization using design constructs, systemic thinking, and alternative paradigms.[4] Naveh would over-deliver in this regard, as SOD would function like a conceptual wrecking ball upon the modern military paradigm, causing great introspection and reflection, but also producing confusion, frustration, and hostile rejection by the most devoted of the military legacy system for war. Israeli Brigadier General Gal Hirsch, one of Naveh's students, echoes this sentiment as follows:

> A conceptual-organizational change on such a scale is a matter for the highest leadership and necessitates a farsighted command … Primarily, we need the courage to enter new battlegrounds, step into dark new realms, and venture into a conceptual no-mans-land. A true leader will also retain the different, the nonconformist (even eccentric) individuals in his organization – there are many of those within the ranks of the special forces, and they could constitute an important asset in the staging of a breakthrough … Provided that one is aware of the conceptual fixation and restrictive paradigms, these nonconformist individuals can be the ones to embrace hardships as opportunities for leading to breakthroughs.[5]

Hirsch would be part of the first wave of new military designers willing to apply these concepts to real-world security contexts in the late 1990s through 2006, primarily in Israel in dangerous combat situations. Naveh and his design cadre would inspire students such as Hirsch to attempt radical transformation in the IDF, and later outward into American, Canadian, British, and Australian militaries. These military design ideas would move quickly and with intensity, but this would also usher in backlash and institutional banishment of designers to include Naveh. The fact that today the military design movement is larger than ever, expanding across the globe, and developing new theories, models, practices, and even military doctrinal revisions suggests that despite accusations of heresy, these designers are transforming the modern military institution.

Becoming a War Heretic: Naveh's Transformation through Intellectual Rebirth

During the 1990s in the small state of Israel, one visionary thinker would formally craft a new war paradigm complete with a newly designed methodology operating on new models while drawing from entirely different theories and language for security applications.

Many military innovators seek to modify the existing legacy system by producing some model, new terminology, or even a compatible methodology *that still supports* the overall modern military paradigm for decision-making in war. Boyd's "OODA Loop" model aids in experiential, reflective learning so that militaries can perform operational planning more effectively. Strange's "Center of Gravity Analysis Method" helps planners conceptualize the "center of gravity" model to enhance existing planning efforts. Constructs like "Effects Based Operations" and "Revolutions in Military Affairs" attempt to shift military thinking through new theories on how warfare ought to be waged, so that strategists and planners go about executing the institutionally promoted and regulated methods. SOD was designed not to enhance existing strategy, operational planning, or further the existing institutionalized war paradigm. SOD existed to deconstruct, challenge, destroy, and replace it as necessary in an ever-changing design of innovation, reflective practice, and experimentation in complex warfare against cunning adversaries.

Naveh, by creating SOD and making it completely unlike anything else, would in turn challenge the traditional decision-making methodology of an entire military organization,[6] which first occurred in the IDFs but would expand to many other organizations. Naveh would feature the insider perspective and credentials of the modern war leader yet operate in an outsider that had no qualms with destroying the sacred cows of modern military belief systems. Naveh wanted a security organization able to understand and apply organized violence *differently than previously, to design toward warfare unrestricted and in entirely novel ways.* Naveh cultivated a small band of heretics, intellectuals, dreamers, and disgruntled military professionals to foster the SOD invention.[7] This Israeli senior military officer and postmodern academic would usher in a new way of thinking about war, strategy making, and military organizational form in complex contexts that differed greatly from those of modernity and warfare.

Back in the 1990s many things had changed since the fall of the Berlin Wall, the collapse of the Soviet Union, and the end of the "Cold War" with only the United States remaining as a sole superpower. A majority of the industrialized west wandered in this post-Cold War landscape of downsized militaries and a rapid expansion of low-intensity peace-keeping around the world,[8] with many governments shifting to make their militaries a new testbed for social transformation and cultural redesigns for reasons outside of clear combat requirements.[9] While this happened across the west, the Israeli Forces experimented with a variety of novel concepts for organizing, creating defense strategies, and acting within conflicts for military application. Some new and competing constructs, such as "Effects Based Operations" (EBO) sought a highly centralized, nodal network where high-technology military systems could precisely target and collapse entire enemy organizational structures, enabled by the splashy arrival in that decade of "smart bombs", stealth fighters, and other amazing new weapons promising surgical precision.[10] Even within the Israeli military an intellectual battle would occur from the late 1990s through the 2006 Hezbollah War over whether military design, EBO, or niche applications such as "swarm theory" should be the new adapted frame for thinking and acting in war.[11]

Despite most military practitioners continuing towards technologically advanced solutions for applying organized violence using ever-greater degrees of precision, lethality, and efficiency, advanced industrialized militaries continued to encounter strategic failure and foreign policy quagmires against illusive and asymmetric opponents. This fixation on technical rationalism would create the opportunities Naveh needed to inject paradoxical concepts from postmodern philosophy, complexity theory, general systems theory, sociology, and more. In 1995, Naveh had just completed his doctorate at Kings College of London where he combined general systems thinking, the Soviet operational theories including theorists such as Svechin, and a proposed new way of conceptualizing complex warfare based on his ana-lysis of "operational shock" in World War II campaigns.[12] One critique of Naveh is that his writing is dense, filled with difficult concepts and language, requiring readers to leap from one discipline to another rapidly. His disser-tation would not disappoint, as would most of his later attempts at codifying SOD onto paper.

As a general officer and now an established academic, Naveh would gain access into a new Israeli endeavor to create a think tank to generate new ideas on warfare. A small number of rising combat leaders in the Israeli mili-tary would go to the newly formed Operational Theory Research Institute (OTRI) where Naveh and others had set up shop. They would study these new, radical ideas and later experiment with them in real-world Israeli security challenges, often growing their praxis through a grassroots sort of intellectual insurgency of sorts. Hirsch, a prominent student from this period, recalls how he discovered Naveh and OTRI:

> I had heard about OTRI and the team that taught the course, with their newfangled, weird concepts. I had also attended one of their classes at the brigade commanders' course and resented the aggressive, blunt and condescending attitude demonstrated by their lecturers ... [yet later when attending OTRI] something was beginning to shift my perception of the subject and this team, especially after hearing a lecture by Dr. Shimon Naveh. He was crude, rude, and impatient, but the effects of his words had on me was electric, challenging all my preconceived notions and generating a storm within me.[13]

Naveh was bold, often abrasive in his approach, yet within his radical ideas and new unfamiliar methodologies a design did exist that drew attention. SOD grew only within Israel at first, entirely within military affairs, and thus off the radar of any commercial designers except for a handful involved in development of the original concepts. Naveh would present his ideas to any military organizations interested in them, and as the SOD construct was theoretical and academic in origin, the Israelis did not restrict access as they do with most of their tactical, technical, and operational designs for war. Naveh would travel and present at any opportunity to expose new minds to these radical concepts. Along the way, he would continuously reform, experiment, and shape SOD further so that each time he presented, he would come with new and different material. While his ideas were radical, unorthodox, and even shocking to traditional military thinkers, some senior leaders quickly recognized what Naveh was trying to do. Naveh would visit the United States to present SOD in small sessions to senior leaders in the U.S. Department of Defense in the 2003–2004 period while he still was teaching SOD in Israel. U.S. Marine Lieutenant General (retired) Paul Van Riper recalls in an interview about his first encounter with Naveh:

> I attended a presentation on Systemic Operational Design that Naveh gave to a small group ... my impression, and I believe it was similar to that of others who listened to the brief, was that Naveh had an original insight, but his manner of presentation, curious use of English, and intense intellectual bent made it extremely difficult to understand SOD in any depth. I flew to Israel immediately following the wargame to attend a week-long course on SOD at the Israeli Defense Force's Operational Theory Research Institute (OTRI) founded by Brigadier General Naveh. I returned from this concentrated and demanding course with a greater appreciation of the design approach to planning, but still puzzled by Naveh's convoluted method of presentation. I was convinced that few American officers would have the patience to follow the intricate reasoning in his presentations, slides, and written materials. Brigadier

General David Fastabend (then Deputy Director, Futures, U.S. Army) and Colonel Robert (Bob) Johnson, his assistant also felt this way, which led to attempts to describe SOD in plain English, few of which were successful.[14]

From the beginning, Naveh and his OTRI team of intellectual disruptors would create intrigue, stimulate deep introspection of why militaries hold certain things as sacred or unquestionable, and usher in a new, provocative way to think and act in complex warfare unlike previous war paradigms offered. SOD would also create controversy, drama, as well as institutionalized political battles where egos, identity, and organizational norms would all be upset. Why was SOD so different than anything previously, and why would Naveh cause such a stir when introducing these ideas to military audiences? With myriad military concepts and models perpetually entering and exiting the modern defense zeitgeist, why did SOD cause so much controversy, and inspire so many innovators to adapt its ways despite clear risks to one's good standing and even career progression? To answer these questions, we need to return to Israel during the unsettling period of security unrest in the 1990s through early 2000s. The First and Second Intifadas span this period and represent major Palestinian uprises of complex violence, protest, and terrorism where the IDFs struggled to maintain dominance.

Designing New Strategies in War: The Israeli Defense Force 2000–2005

While the First and Second Intifadas played prominent roles in Israeli military introspection and frustration, they were not necessarily the proximate cause for SOD to emerge. By the 1990s, senior military professionals and academics were increasingly concerned with a lack of intellectual growth and organizational self-reflection within the Israeli military that they felt had decayed since the 1970s. This coupled with a growing fear that proxy enemies such as Hamas, Hezbollah, and hostile nation-states were quickly adapting to traditional Israeli military methodologies and decision-making models.[15] Yet skeptics within the Israeli forces found any proposed shift away traditional Israeli military concepts and methods as a risky, intellectually overindulgent distraction for a country under perpetual existential threat.[16] Regardless, OTRI would be established to foster debate, critical introspection, and create the conditions for innovative thinking on future warfare.

Naveh and his OTRI collaborators drew from an eclectic range of fields such as architecture, sociology, philosophy, poetry, non-western military theory, and systems thinking to produce an entirely different war paradigm. SOD would reconceptualize decision-making, organizational form, and military function *in an independent, comprehensive, and radically disruptive form.*[17] Thus, like the Bauhaus and Ulm School movements in commercial design covered in the first chapter, Naveh sought to break away from the institutionalized and the known, so that an entirely different and disruptive path could be plotted in any direction away from the norms and rituals of the status quo. The name "systemic operational design" was crafted by Naveh's designers to represent a new way to think in complex security contexts. Each of these words mattered and despite Naveh's verbose efforts to explain their significance, their meaning would frequently be distorted or misunderstood.

The term *operation/operational* in Naveh's position is synonymous with *experimentation*, in that every operation is unique and completely tailored to the synthesis of designing in complexity; they are not tactical plans.[18] In the earliest available draft SOD doctrine created by Naveh in English, he defines "operational" with "the act of working out the form of something (as by making a sketch or outline or plan)".[19] Naveh uses the term "operational" in how experiential learning theorists such as Donald Schön and Chris Argyris created reflective practice, double-loop learning, and related theories of management using "knowing-in-action" models.[20] Karl Weick's theory of organizing, enacted sensemaking, and organizational information theory all fall under social constructivism that Naveh would draw extensively from.[21] Social constructivists assume the world is not a fixed, static given and can only be interpreted as humans interact in an ever-expanding process of "becoming" that cannot be analytically optimized or reverse-engineered via reductionist principles and formulas.

When western military professionals hear "operation" or "operational", they immediately consider an abstract level of military order above the tactical and below strategic affairs. *Operational art* is defined in recent military doctrine as "the cognitive approach by commanders and staffs ... to develop strategies, campaigns, and operations to organize and employ military forces by integrating ends, ways, and means".[22] An earlier version of that same publication offers a variation that illuminates the confusion over the term *design* when SOD was first influencing American military thinking. *Joint Publication 3–0: Joint Operations* would later shed any mention of design, but the 2006 edition defined operational art as: "the application of creative imagination by commanders and staffs ... to design strategies, campaigns, and major operations and organize and employ military forces".[23] Designing was considered the development of established military activities at a higher level of abstraction, but still paired with "operational art integrates ends, ways, and means across the levels of war",[24] positioning military understanding of design as part of a specific war paradigm with a technically rationalized ordering of the world.[25]

Modern militaries posit an artificial hierarchical ordering of strategy-operations-tactics with "design" paired to the construction of military campaigns and operations. Naveh does not subscribe to this, and often acts paradoxically to such an assertion in how military design ought to function. SOD's use of "operation" is contextually dissimilar and likely antagonistic to the modern military frame, as SOD seeks to deconstruct, challenge, disrupt, and destroy irrelevant, obsolete, or incompatible constructs. Militaries attempting to shoehorn SOD into an abstract activity between strategic formation and orchestrated, tactical action misunderstand SOD's naming intent. Right from SOD's intentional title and meaning, Naveh and militaries using their modern warfighting frames set off in opposite directions.

SOD's use of "operational" does not translate to any traditional "level of war" modeling and links the activity to the term "planning" in a more abstract form instead of the linear, sequential mode of military plans. This lashing of complexity theory, military concepts, and postmodern ideas would lead to significant interdisciplinary tensions and fierce institutional resistance to how SOD rejects modern military givens. This is also where SOD's dense, exotic language and ideas cause confusion and frustration with novel apprentices. The postmodern ideas would prove problematic, yet their radically different stances were precisely what Naveh wanted to challenge the modern military mode of thinking. The postmodern perspective is provoking, antagonistic, and requires one to move into otherwise unexplored directions.[26] This new security design logic would demand "reflective practice" or "thinking about one's thinking" in war that would be difficult for many students to grasp.[27] Naveh used numerous previously well-established military terms in different ways as a postmodern military designer, triggering deep confusion for those expecting the terms to retain their legacy system content. That a military person might adapt postmodern ideas was equally frightening to many postmodernists that would discover what Naveh was doing.

Naveh's SOD model is oriented toward an Israeli framing (set in their Hebrew language) for an "operational *form* of warfare" that should not be taken literally in any English translation. Indeed, a better translation might be "operationalizing, in a drifting, emerging, systemic state *of becoming*" within complex warfare. Many concepts that Naveh and his team developed would later be reappropriated in English-speaking nations incorrectly or misinterpreted. "Operational" in Hebrew military thought became associated with the Anglo-Saxon, doctrinal "level of war linking strategic vision to tactical tasks" misinterpretation. This would spiral into decades of misguided attempts to fit design as a precursor to planning sequences in western military doctrine.[28] The security contexts facing Israel are different from that of the United States, and how the original SOD was designed for Israeli military needs might not necessarily translate into American requirements in the Department of Defense.

For most security applications, Naveh and his collaborators agree that SOD is intended for larger operational activities as well as certain strategic or grand-strategic (policy) level challenges. This does not preclude it from tactical or local military designs, nor is SOD limited to traditional military boundaries either. However, Naveh exclusively educated only IDF senior military leaders on SOD.[29] This may have led to expectations that SOD should be focused upon operational activities above the tactical fray.[30] Whether he intended it at that time or not, Israeli design under Naveh's hand went directly and only to the top ranked officers. In Israel, generals alone would need to learn to design in war.

In some ways, the relationship of general officers and education is difficult to appreciate, in that for many militaries, reaching the rank of general is the end of any formal or required educational development.[31] For design at least, the Israelis do not promote these ideas *until* the security professional achieves the rank of general officer and attends a general officer school where SOD is provided. American, Canadian, and other militaries would move to introduce design at subordinate levels, potentially in part not only due to the misunderstanding of "operational", but also due to different security contexts than those challenging Israel. There are many reasons as to why one military would invest design thinking into senior leadership only, while another might prioritize subordinate or even entry-level personnel instead. This correlates to commercial enterprise, where companies such as IBM shifted from educating design to a small population to a company-wide approach over the last decade.[32] Naveh's focus on teaching SOD only to Israeli Generals would have consequences too.

In a 2019 interview with the author, Naveh discussed the cultural differences between Israeli military design contexts and that of larger forces such as the American military: "There are some cultural differences between you guys and us, because in your culture … your system, the general never designs. The design is being done by his staff. In our case, the general does the design himself."[33] Naveh frequently states that *only a general can perform design*, and since Naveh himself was a general, he suggests that only generals such as he can properly mentor other generals in design. There likely is bias and ego at play here, which again are entirely in line with Naveh's brash persona and aggressive manner of articulating his core concepts.

This at times would be contradicted by the success of his co-educator, Dr. Ofra Graicer who is not a general (nor male in a male-dominant profession). Ofra has taught design with Naveh for decades, and while not a general, is an experienced member of the IDF. Naveh, over the years, has changed his position on who should learn design and when within their career and level of power and authority. Sometimes this appears to align with who is willing to invite Naveh to their schoolhouse to teach design, but Naveh has also suggested that complex warfare rarely presents any universal tenets or rules, including where design investment should be made.[34] Regardless, the only path to design education today in the IDF is through attendance at the General Officer School in Herzliya, part of the northern district of Tel Aviv as of 2021. Only generals learn design in Israel, although they certainly lead their units and disseminate design across their subordinates as they see fit. This remains a unique quality of Israeli security design and does not replicate elsewhere.

Abstractions Folding Upon Themselves: How SOD Challenges Everything

The rise of SOD seems unusual in that this original military design methodology (and philosophy) is considered quite intellectually demanding, difficult to implement into military units, and its own creator has declared it must not be put into military doctrine. As doctrine remains the preferred means of military knowledge curation and organizational convergence toward desired, uniformed behaviors, SOD appears from the onset to be an unlikely construct to gain any followership in what are rigidly hierarchical, rule-based, bureaucratic entities. Modern militaries, as the last chapter described, are risk averse, slow to change, and difficult to convince investment into something novel yet entirely unfamiliar.[35] Modern militaries are also often criticized as anti-intellectual, at least toward any new concepts that threaten institutionally sanctioned or ritualized beliefs such as the supremacy of technology and bureaucratic logic for management of war.[36] SOD would produce alarm for all the expected reasons within the mainstream military establishment bent on keeping everything in order.

Gaining access to SOD, even decades after its arrival, is exceedingly difficult. Part of this has to do with how Naveh has created the content (often in Hebrew), the strict classification levels and secrecy of Israeli military protocol, and the dense theoretical content Naveh used for establishing security design praxis. Naveh's work largely remains unpublished, existing predominantly in private correspondence, undated manuscripts, difficult to obtain PowerPoint presentations, as well as a few key student monographs and proprietary items done under military contract.[37]

A trove of SOD documents do exist in private hands, often given by Naveh or an associate to those able to gain access. Most are undated and often the only person able to properly date and explain their origin is Naveh himself. Only a few designers have published on Naveh and SOD, often in student research monographs during Naveh's time in Fort Leavenworth, through advocates promoting SOD in select military journals, and the occasional interviews Naveh has granted over the years. There are few case studies or clear examples of SOD used in warfare, with additional barriers of military classification and hesitancy to publish publicly also preventing more evidence from becoming available. Hirsch's autobiography on his Israeli military career does include numerous real-world examples of his struggles to apply SOD concepts inside the IDF with mixed results. Yet aside from Hirsch and a few other interviews and reports, there remains little evidence of SOD experimentation in war. This contributes to healthy skepticism on whether militaries ought to invest in it, but this hand is also overplayed by defenders of the military establishment terrified at the disruptive qualities of SOD itself. Hirsch himself was an early skeptic while attending OTRI in 1998–1999, yet would soon convert to a fierce advocate:

> I felt that I was dealing with advanced material, but I had my doubts as to its practical applications. The methodology was indeed suited for dealing with ever-changing and adapting enemies, especially those acting in a decentralized manner, organized in cells, and learning quickly, such as terror organizations and guerilla fighters. But how would I convey these ideas to regular military units, which are based on standard operating

procedures, checklists, orders and themes such as simplicity and conformity? How would they adapt to high-tempo changes such as these?
Did we even have a choice, if we wanted to remain relevant?[38]

Select documents have become primary sources on SOD. Weizman's
extensive study of Naveh and the Israeli design experiment from 1995
through 2006 is potentially the most well-known primary source material
outside of direct communication with Naveh or his small network of trusted
design agents and practitioners. However, there are concerns with Weizman's
assessment of Naveh due to ideological motives of his book within which
the IDF and design reflect a supporting role in topic and chapter location.[39]
Weizman's interviews are aging, with most dated to the earliest version of
Naveh's SOD in the early 2000s. Since then, SOD has undergone multiple developments. Naveh also wrote many unpublished documents and
drafts, yet these are often difficult to secure. He has created PowerPoint slide
presentations on SOD, yet most of these are incomprehensible without Naveh
present to explain them to you personally.[40]

SOD creates significant problems for modern militaries that seek to test
out new concepts, validate them, and insert them into standardized doctrine while also not contradicting or challenging any other existing doctrinal
practices. Many concepts end up in military doctrine, and publications over
the years seem like a revolving door of new terminology, fads, and brief
periods of particular interest until the military moves onto something else.
SOD could easily have been part of this immediately, yet doctrine writers
were held at bay for multiple reasons. Naveh was fiercely against putting
security design into doctrinal format, advocating against it for Israel as well
as any military seeking to codify design. Always an enigma, Naveh did draft
up some doctrinal products for the IDF and the U.S. Army in the 2004–
2005 period, perhaps in an early effort to gain momentum by appeasing the
institutional thirst for design in doctrine.[41] Yet in subsequent interviews and
discussions from 2011 through 2019, Naveh expressed hostility toward any
attempt to insert security design into codified military doctrine. SOD content would remain theoretical, and outside of doctrine for many years.

While the SOD community did produce research and theory, one might
expect that Naveh would lead the charge on this effort. Instead, he left the
bulk of professional writing to others. While Naveh's doctoral thesis exists
as a published book, it does not feature a clear, methodical explanation of
his theory on design. The 1996 book *In Pursuit of Military Excellence* is a difficult read that introduces many concepts, terms, and different disciplinary
sources for inspiration, but it is not a book about SOD. In Ryan's recollection
of his time with Naveh during the 2005–2009 period they worked together,
Ryan observed Naveh had "thus far refused to put his theory into writing
… he was reluctant to document a practice that was still evolving and defied
proceduralization".[42] In the research for writing this book, the author did
come across an unpublished, draft document Naveh did write for potential

doctrinal integration of SOD. Titled *Designing Campaigns and Operations to Disrupt Rival Systems*, it was written in 2005 for the American Training and Doctrine Command's "Future Warfare Studies" Division, yet this document only now exists in private collections and remains unpublished.[43] Naveh's work rarely followed the traditional model for military concept development and implementation. He would create like an artist, continuously drafting up new sketches while rarely keeping track, and often neglecting to move content from draft to published formalization. It would be up to his contemporaries, apostles, and advocates to help spread the word.

In the 2005–2010 period for SOD, Naveh's work was most accessible through student monographs largely produced at the U.S. Army School of Advanced Military Studies (SAMS) where multiple attempts were made to clarify Naveh's ideas, often for an American security audience.[44] Naveh did join with some collaborators for a doctrine-like monograph explaining design, yet this was done through a military contractor and had limited circulation.[45] He now relies on his design colleague Ofra Graicer to promote his latest ideas publicly through print media, videos, and conferences.[46]

Israeli design doctrinal products are usually classified and written in Hebrew. One exception to this was an English version of Israeli military design doctrine that was published in 2015 by the IDF doctrinal publishers. Neither Naveh nor Graicer were consulted during the writing of what would be a SOD variation on Naveh's ideas.[47] Naveh, when asked about this version of SOD, knew the authors as his former students, yet spoke disparagingly about the document and claimed it did not represent SOD or his ideas. This complicates assessing the value of the document, although it provides a limited example of IDF integration of SOD formally into their military decision-making and strategy beyond the classroom walls where Naveh teaches their generals.

Hirsch's autobiography published in 2016 would be a decade late in advocating the value of OTRI's new security design. Naveh's ideas and Hirsch's own direct applications of design against real-world security challenges would be showcased during Hirsch's combat experiences in the early 2000s, but publication of this book took years. This delay in part was caused by Israeli senior military leadership censorship, according to Hirsch. He would claim in the book that there was a lengthy character assassination tied to assigning operational blame in the fallout from the Second Lebanon War. This conflict resulted in opposing Palestinian forces taking serious tactical losses, yet it also created a firestorm of negative media coverage and political outrage. Strategic blame went to politicians, yet within the IDF, senior generals sought to pass operational blame to the SOD theories applied by select ground units. In SOD's first true battlefield experimentation, it would generate significant controversy and political infighting. Around 2002, Hirsch would initiate Operation Defensive Shield planning for the Israeli CENTCOM headquarters by conducting systemic operational design sessions:

At Kibbutz Ma'ale Hahamisha we held a systemic operation design (SOD) session, which is a form of situational assessment and knowledge-development meeting. It was chaired by J&S [Judea and Samaria] Division Commander Brigadier General Benny Gantz, who would later become the IDF's twentieth [IDF Chief of General Staff]. I led the development of the toolbox for realizing the principles of leveraging, and called this approach TSSBBE, standing for **tightening** (the grip around the Palestinian Authority territories), **shaping** (engineering projects intended to improve defenses, construct new roads, block routes, "re-obstacalize" the area, and encircle towns that served as launch sites for terror activities), **stings** (short-term raids with small – and low-signature forces), **butts** (heavy, high-signature raids, intended to grab and hold areas for longer periods of time), and **easing** (so as not to drag the entire Palestinian population into a full-scale popular struggle, there was a need to funnel pressure and make it easier on those members of the population who were not involved in terror.[48]

Hirsch would provide detailed examples of SOD set within an exceptionally challenging counterinsurgency campaign. Finally, a student of Naveh was demonstrating SOD in action. Unfortunately, it ultimately would contribute to the end Hirsch's career. His autobiography is one of outrage and retribution against character assassination, political in-fighting at senior levels of the IDF, and what Hirsch saw as a total misunderstanding of SOD. Hirsch saw great value in applying it to the operational and tactical challenges of fighting a complex counterinsurgency opponent, yet it was not a simplistic tool to be blunted into standardized patterns, nor simplified down so that the entire military might grasp it. Hirsch, along with Naveh, Graicer, and select others would form the minority voice that SOD was not to blame for Israel's strategic blunders in the Second Lebanon War. The opposing voices would prove louder, and today there exists across militaries a dominant narrative that SOD proved itself a failure in this conflict, although these assertions are made without any clarification beyond echoing the opinions of SOD critics within the IDF.

SOD would take much of the blame for how that conflict ended, and be purged from the Israeli Forces formally, despite opponents already successful in eliminating much of it on the eve of the conflict they would later deem SOD responsible for. Skeptics of SOD as an alternative to the modern military decision-making frame now had their banner example to suppress any further design experimentation in military affairs. Or at least, that is what they hoped. Despite these political maneuverings, Israeli design would within a generation rapidly spread across the western network of military partners. Today, it continues to spread, often in new variants that may not be recognizable to the parents of the original.

Despite such opposition and criticism, Naveh had lit the match for a different way to conceptualize, improvise, and create new war frames for innovation. Although SOD had many developmental problems as well as fierce institutional resistance, it was the very first complete offering of an alternative decision-making methodology. Indeed, it represented a new war philosophy to challenge the technologically rationalized, Newtonian styled, modern way of warfare. With Naveh's new design frame, one did not have to play by the rules, rigid terms, and methods of established practice; designers could create new words and new meaning to express unimagined ideas on warfare. Hirsch reflected on this unusual military education he received at OTRI in the late 1990s:

> The investigative method I learned to implement systemically analyzed the gaps between my system and that of my rival, in a relevant context. These gaps necessitated new definitions, sometimes revolutionary. Here the need for new conceptualization may be crucial. Many times, we find that the existing terminology is not suitable and cannot appropriately describe reality and the measures needed to face it, so new and updated terms are born. It is impossible to fight current campaigns using the terminology of past wars. New language creates new realities, and new realities create new terminology.[49]

SOD would undergo three major phases of reconfiguration, each representing a new period of design thinking for Naveh and his team of designers.[50] SOD, in all three evolutions still seeks to appreciate complexity and emergence in complex systems by freeing itself of all roots from traditional military theory and practice. It rejects formalized analytic and linear planning as wholly insufficient for militaries except in the simplest of contexts.[51] Although Naveh has advanced the framework over the years, the core concepts of SOD remain quite inspired from postmodern and multidisciplinary theories and models, with a strong emphasis on crafting unique and highly tailored, disposable methods for future conflict application. There are no "one-size-fits-all" approaches, and mass production through mechanization is furthest from the minds of those practicing Naveh's design concepts in security contexts. Each version would reflect Naveh's never-ending artistry, and his inability to permit SOD to become calcified into some doctrinal formalization.

Kilduff and Mehra provide insight into this iterative, recursive aspect of postmodern inspired activities: "We emphasize a postmodernism that is, above all, eclectic rather than exclusive. Thus, our postmodern perspective seeks to include and use techniques, insights, methods and approaches from a variety of traditions."[52] Postmodern concepts, despite losing popularity in late twentieth century academia, thrive in a minority of military circles that follow Naveh's lead on SOD. Weizman, in studying Naveh's unorthodox assimilation of postmodern ideas into modern military applications, reports:

military use of theory for ends other than those it was meant to fulfil is not dissimilar to the way in which progressive and transgressive theoretical ideas were applied in organizing post-modern management systems in business and as efficiency indicators in technological culture.[53]

The fluid, temporary, and improvisational manner in how SOD approaches complex warfare is oppositional to how modern militaries desire order, uniformity, predictability, and mechanistic processes to smooth out desired effects and behaviors.[54] Yet security designers mixing postmodern theory are also playing with fire. These difficult and controversial theories require careful consideration in how to explain, implement, and defend their application to institutional defenders of warfighting modernity.

SOD introduced to militaries exotic concepts from postmodern philosophy such as "rhizomes", "assemblages", "problematization", and "interiority/exteriority".[55] For most professionals, these concepts could only be located in the dense, abstract literature of postmodernists like Deleuze and Guattari, Derrida, Virilio, Baudrillard, Foucault, and others that regularly rallied against modern constructs such as the nation-state, capitalism, and the military instrument of state power. These difficult concepts rarely make any inroads into modern military education, with what Beaulieu-Brossard aptly coined a "minority of a minority of a minority" finding inspiration in these ideas.[56] Naveh would incorporate Deleuze's and Guattari's concepts of smooth versus striated spaces, Deleuzian folds, concepts of the nomad, and of conceptualizing through topology entirely different ways to consider war beyond the Newtonian styled limits. This would require a different social paradigm to fully understand the theories, conceptual models, methods, and language that contrasted starkly with the modern military logic toward war, one that Naveh sought to teach military designers to utilize through experiential learning.

These exotic concepts took root in early military designers, creating a double-edged sword that made SOD intellectually powerful yet obtuse, confusing, and outright intimidating for those seeking to remain in their preferred military mindset. Naveh, cited in Weizman's study of Israeli design thinking in warfare, stated that

> several of the concepts in *A Thousand Plateaus* became instrumental for us [in the IDF] … allowing us to explain contemporary situations in a way that we could not have otherwise explained … We wanted to confront the "striated" space of traditional, old-fashioned military practice with smoothness.[57]

Graicer, having taught this to Israeli senior leaders for decades, explains:

> For an operational commander, then, systemic thinking involves the ability to think and act in two modes: inwardly (the way the system is

organized, the relationships between its components, its ontology), and outwardly (the way one determines how the system relates to its ecology, what defines these relationships, and what determines its ontological manifestation.[58]

SOD's dense language and graduate-level educational approaches would prove to be strengths as well as weaknesses in Israel and elsewhere for thinking differently about war. Naveh's inspiration from postmodernists such as Derrida would have him adapt postmodern positions on the fluid nature of language, where "languages are *sown*. And they themselves pass from one season to another," as Derrida remarked.[59] SOD would not subscribe to any doctrinal military terms, it would not assume the conceptual models provided in military planning doctrines to be valid or even scientifically rational, and SOD would not permit any watering down of postmodern concepts so that they might "fit better" with existing, institutionalized military decision-making methods. This plays well with innovators and those frustrated with modern military processes and theories, but it creates havoc with stimulating the more conservative military professionals to critically reexamine previously unquestioned beliefs.

Naveh's SOD assemblage of various theories is potentially incommensurate with its own demand to use systems thinking that derives from the functionalist paradigm where scientific rationalism is paramount.[60] Naveh's unorthodox combination of postmodern concepts from other paradigms with that of complexity theory and systems thinking creates logical paradoxes that are not easily resolved. Hirsch again provides necessary context here with how he would create new language and terminologies for SOD concepts within the IDF, but also faces institutional resistance or rejection of these ideas:

> The more I delved into this [Systemic Operational Design] process, the more I found myself inventing new and exciting words and phrases to describe and define reality – both the threat and the operational response. With time, some of my terms became a huge success and led to the introduction of meaningful content into the new form ... There were, of course, failures. I coined the term "snailing" to describe an operational mode used by special forces, where the objective is not always to engage the enemy, but to disengage, dig in, and lie low – like a snail recoiling within its shell – and carry out a designated mission. This was not typical of "regular" military forces whose default mode of operations was to engage the enemy and reach a decisive victory. Though the tactic was extremely successful, the terminology was not: people want to be lions, not snails, and the term elicited a fair amount of objection.[61]

Skeptics of Naveh's first form of design as provided by OTRI in the late 1990s through 2005 declared the concepts not just too theoretical and difficult for the IDF to assimilate, but conceptually still far too experimental to implement into organizational practice in such a short period of time.[62] The ideas were so radical and disruptive that rational, pragmatic military professionals could not agree to such radical disruption, nor the questioning of institutionalized war beliefs.

In the period of the 1990s through the early 2000s in Israel, there are competing narratives on whether IDF was really eager for radical new ideas, or if they were shopping around for new concepts to bolt onto existing, legacy constructs in order to incrementally improve them. Graicer described this as a period where "the IDF at the time was not hearing drums of war and world attention laid elsewhere".[63] Other Israeli strategists found that same period to be one of internal frustration and fear of increased nation-state as well as surrogate regional threat due to post-1990 developments. Beaulieu-Brossard, a French-Canadian postmodern scholar who studied Naveh, offers that the second intifada period from 2000 to 2005 sustained an "emergency atmosphere that proved unfertile for theoretical nuances" such as dedicated investment of SOD into the IDF.[64] Hirsch also would reflect on how this period between the late 1990s and the 2006 termination of the Second Lebanon War was unstable, with the IDF less interested in radical new ideas and more preoccupied with incremental, institutionally palatable change. The defenders of these "tried and true concepts" lashed back at obtuse language and experimental war theories:

> There were many who did not recognize the fact that remaining relevant necessitated changing and reshaping the language. Not everyone understood the role language has so often played in revolutionizing a field. There was much ignorance and alienation, but I perceived this partly as my own failure. I should have led the revolution using different methods, which were more low key, quiet, sophisticated, and subversive. I should have led the train of change to its destination without sounding such a loud whistle and awakening those who prefer to sleep and wake up at the station.[65]

Despite critical opposition, SOD was soon ushered into battlefield experimentation, even if in localized and isolated applications. SOD would be implemented hastily, unevenly, with limited consideration to why one might not seek to change so much of a military organization overnight.[66] SOD would be employed in 2002 where Colonel Aviv Kochavi,[67] a Naveh student of SOD, would reframe his infantry brigade's urban operations when battling enemy forces in Balata in 2002.[68] These first battlefield experiments would validate some initial promises of military design advocates, but also increase the antibodies of those critics worried about whether such strange ideas were appropriate for a regimented, bureaucratically ordered, analytically minded warfighting organization.

Israeli adversaries prepared elaborate ambush areas along key avenues of approach in key neighborhoods in Balata. Deadly mines, boobytraps, rocket-propelled grenade equipped defenders, and skilled snipers prepared for the expected onslaught of Israeli infantry and armor into these areas. This was how the IDFs previously fought, how they trained, and what they declare is the one way to fight in their own military doctrine. In fact, Colonel Kochavi had been ordered to precisely do that and secure the hostile targets "by the book".[69] Instead, Kochavi used SOD to disrupt how his organization thought about urban warfare, and by designing an entirely different way to accomplish the mission, he would disrupt and confuse his enemies. They would refuse to maneuver down the neighborhood streets, they would ignore entering the deadly kill boxes, and yet the IDF would still move through the battlespace and capture or kill the enemy present.

In this operational example, Kochavi termed his design model "fractal geometry" and inverted the traditional battlespace so that his infantry would maneuver only through the buildings and avoid the treacherous, booby-trapped streets that the enemy expected them to maneuver down. Using the metaphor *of a worm eating its way through the apple*, Kochavi demonstrated the first published battlefield example of Israeli forces applying SOD in some recognizable way. His unit did not suffer a single enemy-inflicted casualty in what was expected to be a very costly operation in blood and sweat.[70] The original casualty estimates using the traditional techniques were expected to be much higher, yet still tolerable for the Israeli society demanding security action. Kochavi did not want to accept that framework, and thus designed another way. Between examples by Kochavi and Hirsch's subsequent recollections of using SOD in war, these IDF experimentations would become the first available case studies for military design, as well as the primary targets of criticism and misinformation by opponents of design thinking in war.

Why Is SOD so Different from Previous Developments in Military Thought?

SOD operates as a form of conceptual inquiry that attempts to transform fleeting, improvisational abstractions and experimentations into an iterative series of cognitive frames.[71] Each would progressively lead toward concrete understandings that converged into design deliverables that would be executed at the leadership of the commander who conducted the design. To accomplish this, these senior officers had to frame what their legacy system was and why it existed as so, and then deconstruct and rebuild as many conceptual and tangible aspects into a new system of becoming. In these designs, they might gain novel, unrealized advantage that previously could not be accessed using the legacy ways of warfare. This requires critical thinking about one's own military models, methods, and theory as it would to study one's enemy in this regard. SOD begins in philosophy, in that one cannot attempt to "think outside of the box" if the box itself is undefined.

Studying how SOD was performed early in its implementation, Weizman describes this iterative, reflective cycle of framing and deconstructing institutionalized military paradigms as follows:

> critical theory provides the military with a new language with which it can challenge existing military doctrines, break apart ossified *doxas* and invert institutional hierarchies, with their "monopoly" on knowledge ... the language of post-structuralist theory was used to articulate a critique of the existing system, to argue for transformations and call for further reorganizations.[72]

Using these new frames and operational concepts, the generals could move between cognitive frames and help transform their organizations as well as influence future states in the complexity of war.[73] This would invoke fierce resistance of the institution by those that feared removal of whatever concepts, models, forms, or functions that a SOD practitioner set in their sights, meaning that any potential application of SOD would create controversy, opposition, and even resentment. Military anti–intellectualism would be easily tacked onto these reactions, in that SOD itself required designers to learn entirely dissimilar, often unorthodox, and decidedly *non-military* ideas.

Naveh blended previously unrelated disciplines into a melting pot of concepts for military applications in ways that may initially appear counter-intuitive. Nesting systemic thinking with postmodern philosophy to generate novel military perspectives on strategy, organizational form, and creative action does appear in opposition with traditional military methodologies, doctrine, and epistemological norms. This is not to suggest that users of the modern military war frame have not utilized some (mostly post-positivist) complexity theory, or that architectural designers have not experimented with mixing systems thinking with sociology in various attempts to shift from the traditional Newtonian stylings of modern warfare.[74] Rather, Naveh represents the first of the military profession to formally incorporate multiple disciplines into an alternative sense-making and decision-making framework entirely independent of the legacy frame. SOD as a military design praxis *operates distinctly and independently of the legacy framework for planning military activities coherently.* One does not prepare to plan by executing a design effort first, or fit design within an existing strategic or operational planning framework. SOD exists independently, and likely will reconfigure, disrupt, or otherwise reconceptualize planning and strategy into novel, contextually customized, and "one-time-use-only" forms.

SOD methodology originally consisted of seven iterative stages. Designers first conduct systems framing, and then frame a "rival rationale" which later would be renamed "system of opposition".[75] They then follow a familiar design epistemological process of prototyping and experimentation. Naveh terms this as: "reframing operational heuristics and learning meta-theory ... reframing initial system frame ... problematization".[76] This combination of systems theory and the emphasis on postmodernism and alternative paradigms creates very difficult intellectual paradoxes for SOD theorists to navigate. The iterative nature combined with a "disposable unique composition" emphasis would advocate designers not to follow some design checklist or linear sequence of events as some commercial design methods are criticized over.[77] SOD, as a design praxis, was created to resist falling into some sequenced, formulaic, or doctrinally suitable methodology. SOD deliverables were to be analogous to works of art in the museum, not the mass-produced "paint-by-number" coloring books found in the museum gift shop. Yet this would perplex the uninitiated and the novice designers alike; how does one repeat success if one is expected to continuously create through non-repetitive, highly customized praxis and experimentation? How does a doctrine-dependent, engineering and formula-oriented military force move into such an artistic, improvisational, innovative mindset?

Designers would move to design their "strategy for transformation", and then conduct an operation framing as their last design stage, at least if they only ideate through a single iteration of SOD, which would be uncommon. SOD's objective is "to deconstruct all assumptions, and to design the problem, metaphorically, and literally through mapping the system of the rival".[78] Although this chapter is not intended to be a formal primer on how one exercises SOD, the concepts are summarized below. The emphasis on self-development, learning about learning (reflective practice, knowing-in-action), intuitiveness, and a high level of artistry toward warfare made for an extremely high intellectual bar. This would of course come back to haunt much of Naveh's reformative design vision, provide ample fuel for critics, and produce a small population of willing and able participants. Those seeking a design checklist with existing military terminology and "A plus B leads to C" systematic logics would become frustrated or otherwise decline to participate in SOD educational outreach efforts Naveh would pursue.[79]

Designers would need to configure "frames of understanding" first so that they were constructing a customized "systemic inquiry" that was specific to their context; thus, any sort of SOD doctrine would not really establish a design template. The template itself is crafted by the designers. SOD would require students to "construct frames of understanding about emerging ecologies and complex phenomena".[80] Then, students would propose the conceptual and aesthetic artifacts resulting from their systemic inquiry into the security challenge. This would involve an iterative "pattern of self-creating" where the design learning "regulated observation and observed systems in emerging strategic contexts".[81] As humans learn and creatively form models and frameworks differently and quite contextually, any attempt to create a uniform, convergent, or standardized design framework would prove illusive.

From there, Naveh's facilitators encouraged students to create a "regime of systemic adaptive learning" where cognitive movement between these temporary, improvisational frames would be developed through SOD's "operationalization" of design into a new plan of action. Each episode was supposed to be unique, in that Naveh rejected traditional educational forms and processes. He substituted postmodern alternatives such as Ranciere's *Ignorant Schoolmaster* approach that centered learning and discovery entirely upon the students, with no "master" pulling them along or demanding select paths to take.[82] Instead of forming the design praxis upon some universal war maxims or rules, SOD emphasizes an *interpretivist* paradigm positioning where context and socialized meaning were unavoidably systemic on framing complex warfare.[83] Considering military activities and complex warfare from anything other than a Newtonian styled, functionalist paradigm setting would upset the entire apple cart that comprises modern military methods, theories, language, and doctrine.[84] SOD praxis was difficult to explain even in person, and Naveh was always under pressure to simplify the design education into some written format for broader dissemination and study. Benson, the director at the U.S. Army School of Advanced Military Studies when Naveh would first start engaging with students, recalls this tension with:

> "[Shimon] Naveh, you've got to help me. I have to be able to teach [design]. You've got to write it down, it being SOD and the story of SOD and the path you took to it." And Shimon told me, "No, once it gets written down, it is carved in stone and all thinking stops. It cannot be written down; it must always be in motion. It always has to be thought about and applied as conditions change." I told him that wouldn't work. If we want to put this into our doctrine, we have to be able to teach it. Not only to a small group of officers, but to all of our folks and staff colleges. He could not accept that. [85]

Naveh's unpublished U.S. military doctrinal chapter he made for TRADOC does provide some assistance here in expanding how SOD functions. Each part of his SOD geometric graphic from Phase 1 gets several dense paragraphs of design prose, accompanied by lists of definitions explaining every major term in the corresponding paragraph. Despite this effort by Naveh to detail the inner workings of SOD, this document is entirely unclear on how one actually went about doing any of the things. An example sentence of: "The purpose or role of operation framing is to explain the space continuum of the positioning of forces by providing the meta-frame for key ideas on how the operation will unfold" is vague and without necessary context if not read with a SOD facilitator nearby.[86] Instead, the SOD draft TRADOC chapter acts as a point of clarification for a designer already familiar with how and why SOD is performed, and seems to be intended as a complimentary document to formal SOD facilitation. This would be a vulnerability in SOD, as novices would struggle unless they had access to Naveh or one of his experienced facilitators.

In an acknowledgement of this vulnerability, Naveh and Graicer admit that they approach design practice not as engineers, but as artists. In facilitating SOD, they continuously improvise, reframe, and generate different ways of pursuing SOD activities so that even classes of SOD students from one year to the next experience a tailored yet dissimilar SOD immersion. SOD was intended to be boutique, never mass-produced in factory-style military educational efforts. Without a SOD facilitator present to act as a sherpa on the design mountain, students would be lost and unable to reach any useful summit. Even Naveh's original graphical depiction of SOD below appears geometrical, but hardly explanatory to how one performs systemic operational design (Figure 3.1).

When Naveh and other SOD theorists coached design activities, they described the methodology as "praxis of discovery or the perpetual cognitive operation of spiraling from *description* through *problematization* to *synthesis*".[87] Again, Naveh appears to be combining theory on wicked problems and complexity from theorists such as Rittel[88] with concepts in postmodern philosophy,[89] creating some rather difficult intellectual demands for SOD theorists to navigate. "Problematization" comes from postmodernist Michel Foucault, but it also has expanded into sociology, and Naveh would combine it with experiential learning (reflective practice) to move designers away from analytical reductionism to that of a systems thinking "synthesis".[90] He offers a description of the designer journey toward "meta-cognitive awareness" where designers move between an emerging ontological awareness and the designer's reflection upon their institution's epistemological choices.[91] This is

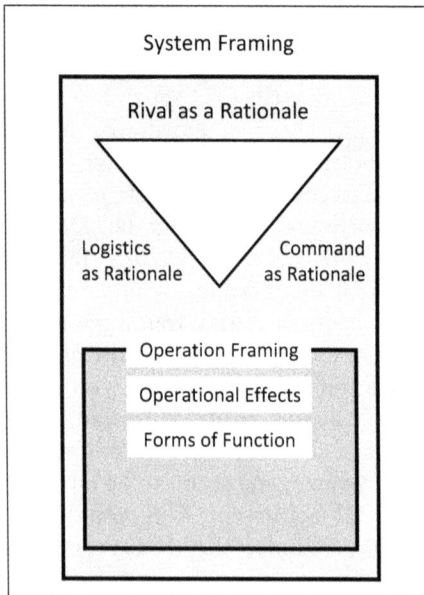

Figure 3.1 Systemic Operational Design 1995–2005 (Phase 1).
Source: Author's Creation.

entirely at the philosophical level of abstraction, and where SOD departs from the modern military paradigm that has no such philosophical equivalency.[92]

SOD emphasizes transformative qualities upon not only the ways of which an organization might craft new strategic options, but also deeply within the core institutional frame to challenge one's institutional outlook upon war. Naveh explains that the first step of self-transformation of one's own worldview represented the most demanding part of SOD, in that it requires philosophical introspection at things the military institution desires not to be questioned most of all.[93] SOD stresses "system framing" where the epistemic conditions generated by institutionalized belief systems must be put into a systemic context of tensions, overlaps, and interplay between what could be very different ways of considering complex reality and war. Naveh explains with: "conceptual tensions between prevailing paradigm and the context in emergence",[94] creating a demanding intellectual endeavor well beyond any sort of traditional military planning method. Militaries normally never question why their strategic or operational methodologies are as presented; staffs attempt to perform "center of gravity" analysis with greater efficiency and compliance to the doctrinal method. In planning, the planners will not question why they must pursue using the "center of gravity" model and not something entirely different. SOD encourages immediate divergence from the institutionalized frame, putting it in tension with how modern militaries desire decision-making activities to remain compliant and obedient to the ritualized processes upheld in doctrine and shared practices.

SOD begins with philosophical framing and the epistemic conditions that Naveh states are in tension with what is currently framed, with the future and novel system that is in an emergent process of becoming. Tension is not used in some engineering, quantitative, or mechanistic sense in SOD. Rather, Naveh takes a postmodern perspective of a fluid, dynamic reality where signs, symbols, artifacts, and codes interact within a plurality of forms and functions. Systemic tension occurs across the social construction of complex reality that humans generate, and effects of those tensions materialize into the physical reality where war unfolds. Jullien, a postmodernist Naveh credits with some of the sign/symbol theory found within SOD, provides in his study of Chinese and Western modes of thinking:

> We know that every culture is plural, as much as it is singular, and that it is ceaselessly mutating; that it is led to homogenize itself at the same time as well as re-identifying itself, that it conforms but also resists; imposes itself as the dominant culture while, at the same time, provoking dissent against it. Official and *underground:* the cultural is always being unfolded and moving faster, but only between these two movements.[95]

For military planners trained to approach war with a Newtonian styling and systematic, analytical frames, SOD's heavy use of postmodernism is likely off-putting. In Naveh's model for design education, some instructors estimated that ten to twenty percent of a typical SOD class would fail to achieve course objectives. Naveh, Graicer, and other SOD educators attributed this to those students being unable to let go of a "career planner" mindset or due to institutionalized barriers stemming from years of military factory-styled inculcation that promoted an engineering mindset as well as a curiously anti-intellectual reflex oppositional to anything without clearly defined objectives and goals. Some students insisted on viewing time, reality, and the linkages of events as linear, causal, able to be analyzed and objectified where war had the feel of a natural science. Naveh would challenge these beliefs by offering postmodern alternatives such as Jullien's disruption of time. Jullien's "silent transformations" of reality explained that emergent, systemic transformation is not dependent on individual observations that might be converted into formulas and data analysis. Complex reality's process of becoming was:

> infinitely gradual and not local but global, unlike action, and thus do not differentiate themselves, nor are they noticed in consequence ... transformations are silent in a still more insidious ... way, supported by our own very usage of language.[96]

SOD would explore things like time, space, causality, meaning, and purpose in ways that threatened the modern military preferred manner of treating complex warfare through engineering and hard science thinking.

Naveh used "narrative" in an entirely different manner than used in the modern analytic military frame. Modern military forces see a narrative as a story or some sequence of data provided by the military to support mission objectives, strategic messaging, public affairs efforts, disinformation campaigns, or propaganda activities. Data has a uniformity and stability in that military narratives work formulaically, whether one is collecting intelligence on enemy forces, or one is writing a speech for a general to give at a public event. Yet for SOD, the metaphoric devices underpinning the words revealed the core ontological and epistemological choices *and limitations* of how each paradigm seeks to interpret reality. Words must be interpreted in SOD, and never taken in some static, dictionary-styled manner of enforcement. Language is contextual, with shifting meanings that require an interpretivist approach rather than a functionalist, analytical one.

Hayden White explains: "the narrative is a vehicle rather than in the way that Morse code serves as the vehicle for the transmission of messages by a telegraphical apparatus".[97] SOD stipulated systemic logic where the content of the discourse "consists as much of its form as it does of whatever information might be extracted from a reading of it ... narrative utilizes other codes as well and produces a meaning quite different from that of an chronicle".[98] SOD encourages designers to disrupt all conventional assumptions of analytic objectivity concerning how the various stakeholders interact within war. This design model must be tailored, reconstructed each time with considerable introspection and critical realization of one's own mental framing of what the current security challenge or context is, *and what it is no longer.* This also frustrates traditional planners and strategists seeking identification of a familiar, definable "problem" that already has a corresponding "solution". Designers seek not to solve, but to dissolve, in that designing a future system where the original problem no longer is relevant creates a dissolution of the current frame.[99]

Despite many SOD novices falling away or rejecting the concepts, those that did would: "[see] through the crack in the wardrobe into Narnia", as SAMS faculty and design educator Butler-Smith described when certain students would open up to the challenging SOD materials, readings, and instruction.[100] Once a designer moved to mindfulness, they could "transform their learning … transform their institutional organization … transform the observed ecology or environment".[101] The emphasis in this military design rested not in some systematic "problem-solution" formula, but in the contextually unique learning that occurred in the journey that designers undergo.[102] There is a certain sort of intellectual initiation and painful personal self-examination and eventual liberation that is paired with Naveh's original security design praxis. While planners act to change their environment, designers exercising SOD will also change themselves.

To accomplish this, SOD emphasized a rigorous educational journey that later would be criticized as too abstract, dense, and elitist.[103] Naveh's original design praxis featured these seven phases that would together comprise a rigorous, self-paced learning system for students.[104] Students of SOD embraced a wide, unorthodox range of disciplines that demanded intense study and personal discipline. The intellectual demands of SOD were extensive and likely a shock to most military professionals previously conditioned to a more mechanistic, reductionist mode of decision-making.[105] One could not teach SOD with PowerPoint slides and index cards, as the U.S. Army would seek to do with their own version of design later in this story.

SOD requires the designer to "rigorously investigate the significance of different values, goals, and practices between the rival and oneself to identify connections and logical relationships within the system".[106] This not only required dedication, intellect, and time, but it also demanded an expansive vocabulary that would frequently sour the experience of those on the receiving end of SOD design deliverables. According to Graicer,

> SOD methodology is both the product and the process – it generates paradigm (guiding frameworks) that allow performance while constantly challenging these frameworks, through which new trends are perceived. By doing so, it allows the *creation* of the box as well as the process of *thinking outside the box.*[107]

Again, this emphasis on tailored, disposable, "one-time-application" products for novel warfare experimentation would move military practitioners well outside of normal comfort zones.

Military design skeptics would in turn accuse SOD practitioners of being overly intellectual, with SOD activities full of navel-gazing and "ivory tower perspectives" far removed from the gritty reality of the trenches where problems needed clear, repeatable solutions. A false dichotomy of "theory versus practice" often occurs in the SOD-planning divisions where institutional defenders claim SOD is impractical, overly intellectualized, and inferior to pure, proven planning processes.[108] It creates an impossible feedback cycle of demanding "proof" of SOD utility that must only be proven using the very same institutional frame that SOD seeks to disrupt. Despite these barriers, Naveh and SOD educators recognized and even expected student frustration and a steep failure rate. They suggested that some military professionals may even lack the intellectual rigor to survive the transformation, which reinforces the criticisms of intellectual elitism by SOD practitioners.[109]

Learning SOD was expected to be a painful journey where many would fall by the wayside. In Naveh's unpublished reflections on teaching SOD as well as select interviews, he frequently discusses where students experience a growing frustration, usually indicating to Naveh that they were "close to *cognitive culmination* ... [leading] to an increase in meta-cognitive awareness".[110] Design teams undergoing SOD education would appear to reach mental exhaustion, and often only after iteratively failing in the SOD process would they achieve some "moment of illumination" after experiencing a "state of deep frustration".[111] Again, this created an aura of intellectual elitism, one that would not go over well in many military circles.

In 2004–2006, many in the IDF would not bother with learning SOD, or quit early into the process. Many Israeli officers during this first period of SOD experimentation had little precious time to dedicate to learning such challenging concepts, while many others believing it to be "utter nonsense" due to the high intellectual bar to clear at the start.[112] SOD's simultaneous demands of abandoning much of one's previous military expertise and experience coupled with the highly sophisticated literature requirements appeared to limit the quantity of student interest as well as those able to make the intellectual pilgrimage. This pattern would continue in all variations of SOD, as well as even in some of the simplified military design hybrids explored later in this book.

For SOD educators applying the concepts in the classroom, it seemed that only after students consciously complete a cognitive breakout could they coherently engage in "the cognitive exploitation which concerned revisiting the conceptual products constructed in the course of the previous days".[113] This element of deep frustration and self-transformation likely created some significant barriers for how Naveh and his team of design educators intended to transform the larger IDF, as well as subsequent experimentation across western strategic partners and allies.[114] The experiential learning approach for SOD utilized a "thesis-anti-thesis-synthesis dialectic"[115] where designers iteratively play devil's advocate to challenge design biases and experiment in theorizing and then inquiring into the system while learning. This Hegelian philosophical construct drew from philosophy, again something modern militaries marginalized or ignored. To fully embrace SOD, one had to consider various social paradigms, deconstruct and challenge them using a postmodern lens, synthesis the design frames with systems thinking and complexity theory, and operationalize this all into a new military approach. Further, Naveh resisted codifying SOD into anything doctrinal, and as he and others facilitated SOD education they would continuously refine and reinterpret SOD itself.

Naveh would continue through the various incarnations of his SOD approach from the 1990s through 2019 to emphasize what can be described as a "postmodern discourse".[116] SOD fostered cognitive divergence where team members would challenge each other and themselves in ways that would be mentally exhausting. Yet the "thesis-anti-thesis-synthesis" dialectic would help designers break free of self-imposed mental limits, in that new knowledge construction is itself an intellectually excruciating activity.[117] Naveh is more direct in presenting this as follows:

> People think, "if you are my friend … you are my friend. You will support me! No, no … no. When there is a conflict of interest in a discussion, [pounds table] I will depose you! Because I am your friend, I know you."[118]

SOD would not permit any "sacred cows" to walk freely. When an organization requires game-changing innovation, everything must be open to deconstruction, in that the unrealized opportunity likely is obscured by instructional barriers to the design inquiry itself.

Despite SOD's illusive framework and lack of clear methodological structure as well as doctrine, we can attempt to articulate a pattern to the praxis. By considering the deeper "why, how, what" framework underpinning SOD's original seven steps to designing, we might present a framework to consider how SOD operates differently than commercial design praxis demonstrated in the first chapter. An epistemological interpretation of this first SOD methodology from 1995 to 2005 practice is depicted below based on the author's synthesis of SOD research and study (Figure 3.2).

The illustration above frames SOD in an epistemological map created by the author and is but one of many ways to conceptualize why SOD exercises as it does. The SOD methodological graphic is repeated here, and remains the most recognized of all the SOD variations outside of Israel. SOD starts with a directive to construct frames of understanding about the emerging ecologies and complex phenomena involving both the designers, their institution, rivals in the system, and other significant social, cultural, as well as physical and synthetic phenomenon comprising the broader picture. This is a "system framing" as depicted on the right in the illustration. The numbering relates the epistemological constructs to how the SOD methodology functions.

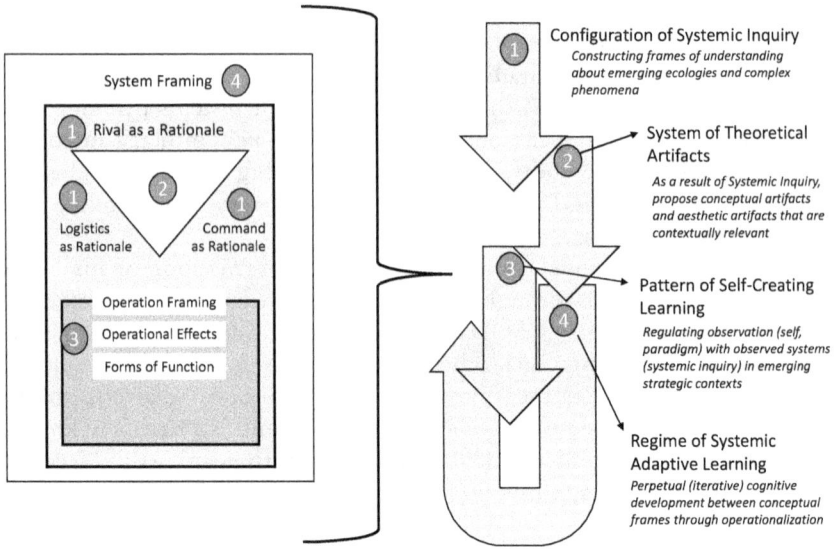

Figure 3.2 Epistemological Framing of SOD Phase 1.
Source: Author's Creation.

Systemic framing requires designers to become reflective practitioners and commence at the philosophical level of conceptualization. This "thinking about one's thinking" requires one to frame the "self" which is individualized, but also must be representational of the larger institutions and group identities that shape each member. Systemic is not to be confused with systematic logic. *Systematic* thinking seeks to break down complexity into reduced, simplified parts to link inputs to outputs. Systematic thinking is an engineering mindset where one can subsequently apply mathematical or classical mechanics themed models. Systemic thinking moves in the opposite direction in that complex systems are composed of more than just their parts; synthesis is only illuminated by moving into abstraction where systems interact with larger ones, and SOD advocates for systemic over systematic in much of how the design inquiry unfolds.

Once the SOD practitioner can sufficiently build this self-reflective configuration of systemic inquiry, they transition to the second major arrow illustrated above. It is important to mention that jumping from arrow to arrow in an epistemological framework is not sequential or linear; these constructs are abstractions and in SOD one is free to move irregularly and often circle back and forth. Indeed, SOD is expected to be nonlinear, iterative but perpetually in drift, where a process of *becoming* is never in some checklist or repeatable recipe for designing. Experienced SOD practitioners might move in wildly unexpected cognitive paths that make sense to them at an epistemological level, but for observers attempting to map some SOD methodology it could appear random or erratic. Advanced SOD designers may construct entirely novel designs, and also along the way follow an emergent, novel design process itself.

The second epistemological arrow shown above is where designers propose theoretical artifacts that can be conceptual (abstract, tacit) or in other configurations that may be aesthetic (symbolic, explicit, rendered). This is done for addressing tensions, vulnerabilities, risks, as well as emerging opportunities the designers realize when framing rivals, logistics and command. Next, these frameworks are depicted in elaborate drawings, topological conceptualizations, poetic narratives, or some hybrid creation therein. Rarely are they constructed in some conventional or institutionally preferred form. Then, designers attempt to generate a pattern of "self-creating learning" to prototype, test, reform, and develop the designs through ever-increasing engagements within the design team and any additional experts or leaders. Students of SOD "learn about learning" to design in an intimate praxis where theory, model and method affect one another in robustly informed and intuitive experiments. The "thesis-antithesis-synthesis" Hegelian dialectic is an epistemological underpinning to this activity.

In the fourth arrow shown above, the epistemological representation twists back upon itself indicating an iterative, self-reflective activity that is perpetual and potentially self-destructive. Designers circle back to earlier activities and ideas to erase them and start over; multiple iterations smash earlier concepts to build upon those failings with valuable insights previously unavailable. New ideas test one another, and in this emergent process, previously unimaginable, unrealized ideas come into light. Often, the final design concept a SOD team develops comes from many unexpected connections and developments that could never have been anticipated at the start of the design journey. This "regime of systemic adaptive learning" is where SOD praxis is beholden to no institutional form or function.

Naveh originally built SOD for Israeli military contexts, not for other nations that would soon have interest. He didn't originally consider how SOD might work around the world for other military forces facing different strategic and geopolitical challenges. He built a unique security design for Israelis, by Israelis, to apply toward Israeli threats in wars with Israel. Naveh envisioned it for enabling Israeli military forces to out-think, out-imagine, out-adapt, out-innovate, and out-perform the many clever and cunning enemies of the Israeli state. Naveh reflected on his own struggles as an infantry commander and later a General Officer and sought to create something new. SOD was created to offer Israeli Generals a novel way to design the application of organized violence that had little obligation to the legacy system of established IDF decision-making, doctrine, and training.

Naveh has an insatiable thirst for knowledge, making his work intellectually demanding for others to understand and implement. His lack of patience for conservative thinking in war and his utter contempt of intellectual laziness make for a difficult selling point for SOD. These would all weigh upon the perceived value of SOD as well as a building animosity for both these radical ideas and the combative personality inspiring this military insurgency of thought.

Banishment: The Removal of Naveh and SOD Praxis from Israel

Despite Naveh being the father of the military design revolution and a fierce advocate of what SOD should and should not become, he soon would lose control of it. "At some point it went rogue; whereby the publishing of design manuals was to be the sign of its reduction *ad absurdum*, until its own father could not recognize it".[119] Graicer, Naveh's closest design facilitator and collaborator for most of his work in the IDF, would also admit that doctrine could be relevant in some contexts, depending on the design challenge. She states, "SOD takes doctrine to be the apotheosis of the organization's intellectual journey, as much as the point of departure for the next journey".[120] Using an architectural term for a church's southern library area, Graicer softens Naveh's anti-doctrine position by framing most military doctrine as useful *if only to help designers frame where they are starting from* so that they can cleanly leap further away. Yet more often than not, militaries would fix design into their doctrine in order to tame design to obey deeper institutional beliefs on warfare.

As quickly as SOD emerged in the late 1990s within the previously traditional Israeli military landscape for decision-making in complex warfare, it was extinguished by 2006. After the transition of leadership from Lieutenant General Ya'alon to Lieutenant General Halutz in 2006, the new Israeli Chief of Staff shut down OTRI and began to erase SOD and Naveh's group from any position of influence upon the IDF.[121] By 2007 the IDF as an organization collectively rejected military design. Naveh as well as his SOD proponents were no longer welcome in IDF training or education environments. Officers associated with SOD were marginalized, or forced into retirement. This would remain unchanged in Israel for the next six years. From 2006[122] through 2013, Naveh and Graicer along with a small group of SOD theorists and educators would teach SOD outside of Israel, essentially banished. The maxim of "no man is a prophet in his own land" applies in that Naveh spent nearly a decade fostering an international design movement away from his home country.

There remains a great deal of political and interpersonal arguments over what precisely happened during the 2005–2006 period in the top leadership levels of the IDF, to include why SOD was suddenly abandoned. Both sides of this SOD debate will be summarized to help frame the expansion of military design out of Israel and into other militaries. To this day, camps are fiercely divided on this point, and the whole truth may never fully be appreciated.

Was SOD the catalyst for necessary change in the IDF, or did it hamstring senior military leaders and confuse the force? Either way, Naveh's design ideas gained enough attention internationally to create new demands from outside the country. The U.S. Training and Doctrine Command (TRADOC) would even entertain some former OTRI design products produced by Naveh. These would be translated into English and considered in draft form, if only internally to their Future Warfare Studies Division.[123]

By June 2006 and just a few weeks before the Lebanon War broke out, the gradual expansion of military design methodology into the IDF was halted. All SOD education as well as unit implementation was terminated.[124] Advocates such as Naveh were dismissed from the OTRI facility and no longer welcome in IDF activities.[125] OTRI was disbanded and any SOD activities ceased by order of a new Chief of General Staff. This career aviator, according to Naveh and other SOD advocates, rejected the main cognitive stances that SOD advocated, instead encouraging concepts such as "effects-based operations" (EBO) that play to aviation reasoning and precise targeting.[126] It is understandable that a military aviator and senior leader in the mid-2000s would be drawn to EBO, as that style of reductionist, systematic conceptualization had spread from the United States Air Force in the 1980s–1990s across most partners and allies.[127] The clean, clear, kinetic reasoning of EBO made sense to warfighters that positioned cause to effect, where a bomb dropped onto a target could provide quantified, objective, and observable results. Whether such reasoning could extend to all things in war is another matter that SOD proponents would certainly challenge.

History does not repeat itself, but it does rhyme, as the saying goes. This tension in militaries between using EBO or SOD would not just occur in Israel. Marine General James Mattis, around the same time SOD was being purged in Israel, would embrace Naveh's design ideas in the United States and attempt to introduce them into the U.S. Marine Corps. Mattis has mixed results in the Marine Corps, but upon moving to command U.S. Joint Forces Command in 2008, Mattis banished EBO in a memorandum and began efforts to replace it with some evolution of Naveh's design praxis.[128] The EBO mindset could not tolerate SOD within it, whereas SOD could position EBO as a kinetic, precise model that could be applied in select, tailored contexts. As EBO advocates sought a near total takeover of military decision-making to become adherent to EBO logic, SOD became a clear threat.

Mattis would later recall:

> While I agreed then (and now) with the U.S. Air Force's application of EBO in their targeting process, [Joint Forces Command]'s misapplication of USAF's targeting doctrine was the fatal flaw. Basically, Joint Forces Command's misuse of EBO occurred because the Air Forces tried to introduce an approach to warfighting at odds with war's fundamental nature.[129]

At this particular period in the early 2000s, militaries would struggle between seeking a systematic, technically rationalized manner of conceptualizing thought and action in war, and that of a systemic, multi-paradigmatic, design process that rejected standardizations and reverse-engineered, formulaic analysis. IDFs would move in one direction, while the American militaries seemed to explore a little of both.

Aside from Naveh and his SOD proponents, there is scant evidence Halutz advocated for EBO specifically. While it is clear that he rejected SOD, it is feasible that Halutz, as a career aviation officer, had some exposure to EBO. More likely, his own staff and military organizations were already steeped in EBO culture, and as they produced strategic and operational deliverables for him to act upon, they may have been developed through EBO lines of thinking. It is a fair assumption that no SOD materials or activities were provided to Halutz other than the ones that caused his decision to purge the IDF of design.

The 1990s and early 2000s for the IDF appear to be in retrospect a difficult period of experimentation and almost a tribal environment of various military groups advocating for various reforms, radical ideas, or legacy concepts. The subsequent Lebanon war was executed with a mix of some IDF applying SOD in tactical and operational missions primarily within ground forces, and the bulk of aviation and naval forces relying entirely upon traditional linear planning. EBO was popular in the air forces, and likely a prominent construct under Halutz's command of the organization. The IDF was institutionally confused, with two different factions of pro and anti-SOD creating deep institutional tensions during the outbreak of an actual war. EBO had not, at that point, been used outside of clearly kinetic, largely aviation-oriented endeavors, while SOD was a novel enterprise consisting of small groups of converts.

Gracier, in recalling the overall approach to the Hezbollah threat, characterized the bias of the new Chief of General Staff, Dan Halutz as

> shallow mindedness or one-dimensionality, as things seem from a cockpit … the IDF was headed by an aviator. Trained in the Israeli Air Force, he never participated in a command course at any level … and was totally ignorant when it came to Land or Naval force employment.[130]

Unfortunately, there are no available rebuttals from Halutz available on the topic of SOD and intellectual development within the IDF for that period. SOD and EBO perspectives on how the IDFs ought to construct war plans and strategy against the Lebanon threat were strikingly opposite with little middle ground.

Naveh was far more critical in his assessment of Halutz,

> he might have been a good fighter pilot … but he's totally innocent of any education that could have prepared him for the challenge that

awaited him as a general. Being both arrogant and ignorant, he never bothered, like so many generals, to really study. He's the kind of man that, if you can't really comprise your words into two lines, he'll never go through it.[131]

Naveh does have a pattern of *ad hominin* attacks on critics and those he feels have wronged him or the SOD movement, and he has also been accused of being a boastful, even arrogant individual. However, being pushed into retirement and seeing your think tank and design efforts erased provides necessary context for hurt feelings and grudges.

After Halutz disbanded OTRI and ejected SOD as a concept, "SOD language became heretical. Those who were using SOD language ... just fell back on conventional tactical language".[132] The postmodern content was erased, and the military establishment continued a long march toward codifying warfare exclusively through pseudo-scientific terminology, models, and reasoning. Traditional military historians such as Yagil Henkin, an educator at the IDFs Command and General Staff College, would offer a common skepticism on the postmodern language in SOD as follows: "the same ideas can be expressed without using interpretive language".[133] Henkin echoes what most SOD opponents harbor, in that military forces should clearly understand what they are doing, and such conceptualization ought to be accomplished in simple, clear, universal terms. Design, as an enterprise oriented toward innovation and change, struggles to operate when forced to use institutionalized processes or adhere to set rules. Naveh's SOD did draw from perhaps the most antagonistic discipline (postmodern philosophy) to choose from, but most all design rejects the oversimplified military frame that Henkin extends. Dorst, a commercial design theorist, explains this as follows:

> Conventional problem-solving requires us to stop the world, isolate the problem, and come up with a one-off solution ... this approach is curiously nonexperimental, and underlying it is the apparent need to attain complete closure before the solution is put into action.[134]

By 2005, only Israeli colonels and first-star general officers had taken SOD courses, but the total numbers reflected a tiny minority of the overall IDF senior leadership. Of those that attended a SOD course, "some dropped right after the first course, and others regularly abstained. Some students felt this teaching as imposed learning and the language as far too unfamiliar for what they were used to".[135] Statistically, few officers across the IDF's centralized military hierarchy ever personally experienced or even understood SOD. This small band of SOD practitioners would face exceptional institutional barriers to quickly expand SOD thinking in ways that could be readily understood. That this all was supposed to happen during the onramp for a major military conflict would only add to the tensions in play. Along the way, SOD had also accumulated many anti-designers that would see any setbacks or failures in war as "proof" of SOD being a worthless enterprise.

By the eve of the war in 2005 only fourteen graduates of any previous SOD course remained on active duty in the IDF, with the vast majority of senior leaders unfamiliar or skeptical of SOD's value in combat.[136] This lack of SOD-educated leaders meant that Naveh's intent for "the wisdom of SOD [to] trickle down inside the units of all corps" did not occur in time for war, and possibly hastened IDF formal rejection of it.[137] Few senior leaders understood the radical new concepts, and many more saw the deliberate purging of SOD advocates on the eve of the conflict.[138] Hirsch reflected on this deliberate removal of design theory, OTRI influence, and SOD advocates within the IDF in the months leading up to the Second Lebanon War:

> A few months before the start of the 2006 Second Lebanon War, the [OTRI] institute's prominence began to fade, for there were those who systematically acted to reduce its influence. It is true that Dr. Naveh and his people provoked substantial objection and even resentment by a senior group within the IDF. It was no longer an academic, idea-driven debate. Because of this heightened cultural war, there were those who found ways to oust this group of scholars from within the ranks of the IDF, using various excuses. It was a sad and troubling move, and I believe also negative for the future of the IDF … Many senior officers and prominent institutions from other countries value and adopt the unique ideas that emanated from OTRI, yet here in Israel we didn't seem to find room for them and their ideas. A few weeks before the Second Lebanon War, I received a letter from one of OTRI's senior leaders, warning me that I was to be next, as the IDF senior leadership was tired of critical and transformational officers, representing the approach of OTRI. He warned that they would soon find a way to get rid of me and were only waiting for an opportunity.[139]

As with all wars, the aftermath has many fingers pointing to what is to blame. The IDF experienced nothing different, and in many later accounts, the critics of SOD took advantage of this opportunity to pin responsibility upon pre-conflict SOD experimentation. Faultfinders would claim the SOD infatuation with esoteric ideas and theories muddled the waters for the military,[140] while critics of EBO found opportunity to challenge the utility of that concept as well.[141] There were accusations that the IDF lusted after the latest technology, doctrine, or contemporary foreign policy positions while in some sort of identity crisis.[142] The Second Lebanon War remains a controversial topic, at the political level and deep within the IDF on what went wrong. One particular narrative that proves enduring is that Naveh and his SOD ideas were to blame, despite the contradictory evidence that hardly enough IDF leaders invested in learning it before the fighting started.

Naveh's strategy for inculcating SOD across the military consisted of seeding the senior officers with these radical concepts so that they could use their position and power to further drive SOD down to subordinate commanders and into tactical units. Yet few leaders took on the high intellectual demands of SOD, and many of those that did would be eliminated from the IDF before war started. Prior to the Battle of Bint-Jebel, the commander of the Golani Brigade was asked if he was aware of SOD concepts. Specifically, he was told that Colonel Gal Hirsch was promoting the idea of "a swarm of hornets" as a metaphor both representative of SOD, possibly combined with components of swarm theory.[143] Hirsch was the Israeli top operations officer (Joint Operations Officer or "J3", Israeli Central Command) during the earlier 2002 demonstration of successful design by Colonel Aviv Kochavi. What transpired in this exchange is indicative of the tensions gripping the IDF at the time.

Colonel Tamir Yeda'i, the tactical-level commander of the Golani Brigade replied to the question of whether he was familiar with this new "swarm of hornets" concept as follows: "I am not familiar with the chirping of crickets".[144] Yeda'i was not alone in skepticism of SOD's creation of new terms and assimilation of previously unconsidered theories into warfare such as swarming or other ideas that clashed with the traditional, mechanistic, and highly engineering-centered language and doctrine of modern warfare. In response to these skeptics, Hirsch remarks on the introduction of maneuvering through design-induced "swarm" with:

> I valued this mode of operation … I proved its suitability during training and exercises throughout the preparation and learning period before the war. I viewed criticism and mocking of terminology such as the "cloud of wasps" and "humming effect", and the "swarms" as a lack of creativity and imagination.[145]

Hirsch, as a SOD advocate, would become a target of criticism with how he attempted to introduce design thinking into the IDF, with Yeda'i's biting sarcasm a mild example compared with more vocal antagonists.

SOD would accentuate what was not just a tension in the IDF but across most western modern militaries over innovation, intellectually safe areas to apply in warfare versus those deemed dangerous, and battles between military intellectuals and those that prefer violence of action over thinking. Hirsch reflected on this in his autobiography as follows:

> Intelligence turned overnight into a handicap, and intellectualism became "lack of combativeness …" the complaints on [SOD's] sophisticated language, detachment from the field, and nicknames such as "astronaut" and "daydreamer" had acquired a foothold … I became known as a man of the book … not the sword. My propensity to write articles and brief senior foreign generals in English was considered by some a disadvantage.[146]

Despite this open hostility to SOD concepts in the IDF at senior levels, Hirsch in the 2005–2006 Lebanon War period would command the Galilee Regional Division and apply SOD concepts in multiple major operations. Later, he would be criticized for what was critiqued as a confusing application of design, and even faced disciplinary action by the IDF. However, an investigation by the Israeli national committee on his division's actions would refute this early criticism and largely clear Hirsch's name. Today, while Hirsch is retired yet politically active in Israeli security and politics, Yeda'i is a Major General in the IDF, commanding the Homefront Command, and has graduated Naveh and Graicer's design program in Tel Aviv. An interesting anecdote is that in a 2019 interview, Graicer stated that Yeda'i now appears to find value in SOD, and has developed a positive stance on using it.[147]

Naveh in 2006 witnessed the rise and sudden fall of military design in his own nation, exercised and rejected by his own military. Most SOD graduates were forced out, alienated, or no longer able to safely use SOD concepts in the fall-out of the Second Lebanon War. OTRI was disbanded, and the grand intellectual experiment by the IDF seemed doomed. Yet as certain senior Israeli military leaders conducted their purge of design, Naveh's ideas took hold internationally. With what started with some occasional lectures and visits to American training centers and exercises soon blossomed. By 2006, Naveh and select SOD practitioners set off on a new voyage, to teach military design in the largest, most technologically advanced, and heavily resourced military on the planet. American leaders in the Army and Marine Corps were already alarmed about security quagmires in Iraq and Afghanistan, and this strange Israeli intellectual became a focus of intrigue, investment, and also outrage.

Notes

1 Carl Builder, *The Masks of War: American Military Styles in Strategy and Analysis* (Baltimore: John Hopkins University Press, 1989).
2 Simon Critchley, *On Humour, Thinking in Action* (New York: Routledge, 2002), 16–22.
3 Protzen and Harris, *The Universe of Design: Horst Rittel's Theories of Design and Planning*, 140.
4 Rika Preiser, Paul Cilliers, and Oliver Human, "Deconstruction and Complexity: A Critical Economy," *South African Journal of Philosophy* 32, no. 3 (2013): 261–264.
5 Gal Hirsch, *Defensive Shield: An Israeli Special Forces Commander on the Front Line of Counterterrorism, the Inspirational Story of Brigadier General Gal Hirsch* (Jerusalem: Gefen Publishing House, Ltd, 2016), 430–431.
6 Ofra Graicer, "Between Teaching and Learning: What Lessons Could the Israeli Doctrine Learn from the 2006 Lebanon War?," Experticia Militar, October 2017, 22–29; Ofra Graicer, "Self Disruption: Seizing the High Ground of Systemic Operational Design (SOD)," *Journal of Military and Strategic Studies* 17, no. 4 (June 2017): 21–37.

7 Naveh's contemporaries included Zvi Lanir, Dovik Tamari, Yosef Kuperwaser and others. This group featured a blend of intellectuals, academics, military insiders, and also experts in dissimilar fields far removed from the Israel Defense Forces.

8 Markus Mäder, *In Pursuit of Conceptual Excellence: The Evolution of British Military-Strategic Doctrine in the Post-Cold War Era. 1989–2002* (Bern, Germany: Peter Lang AG, 2004), 48–72.

9 Charles Moskos, John Williams, and David Segal, "Armed Forces after the Cold War," in *The Postmodern Military: Armed Forces after the Cold War*, ed. Charles Moskos, John Williams, and David Segal (New York: Oxford University Press, 2000), 1–13.

10 Mäder, *In Pursuit of Conceptual Excellence: The Evolution of British Military-Strategic Doctrine in the Post-Cold War Era. 1989–2002*, 61–65; Andrew Nocks, "The Mumbo-Jumbo of Design: Is This the Army's EBO?," *Small Wars Journal*, September 20, 2010.

11 Yagil Henkin, "On Swarming: Success and Failure in Multidirectional Warfare, from Normandy to the Second Lebanon War," *Defence Studies*, 14, no. 3 (2014): 310–332.

12 Shimon Naveh, *In Pursuit of Military Excellence: The Evolution of Operational Theory* (New York: Frank Cass, 1997); Aleksandr Svechin, *Strategy*, ed. Kent Lee (Minneapolis, Minnesota: East View Publications, 1991).

13 Hirsch, *Defensive Shield: An Israeli Special Forces Commander on the Front Line of Counterterrorism, the Inspirational Story of Brigadier General Gal Hirsch*, 126–127.

14 Paul Van Riper to Ben Zweibelson, "Re: Design for Defense Book PDF Manuscript," September 20, 2021.

15 Shimon Naveh, "Between the Striated and the Smooth: Asymmetric Warfare, Operational Art and Alternative Learning Strategies" (Unpublished manuscript, undated), 2–3; Dima Adamsky, "The Israeli Revolution in Military Affairs," in *The Culture of Military Innovation: The Impact of Cultural Factors on the Revolution in Military Affairs in Russia, the US, and Israel, First* (Stanford, California: Stanford University Press, 2010), 99–101.

16 Alex Fishman, "How the Tides Have Turned for Gal Hirsch," Y-Net News.Com, August 29, 2015, online edition, www.ynetnews.com/articles/o.7340.L-4695 324.html

17 Weizman, *Hollow Land: Israel's Architecture of Occupation*, 210–212.

18 Shimon Naveh and Ofra Graicer, Naveh audio 15OCT2019 34min22sec Day 2 morning.MP3, interview by Ben Zweibelson and Nathan Schwagler, personal interview, October 15, 2019, 6:58.

19 Shimon Naveh, *Systemic Operational Design: Designing Campaigns and Operations to Disrupt Rival Systems (Draft Unpublished)*, Version 3.0, unpublished draft (Fort Monroe, Virginia: Concept Development & Experimentation Directorate, Future Warfare Studies Division, US Army Training and Doctrine Command, 2005), 2.

20 Chris Argyris and Donald Schön, *Theory in Practice: Increasing Professional Effectiveness* (San Francisco, California: Jossey-Bass Publishers, 1974); Donald Schön, *The Reflective Practitioner: How Professionals Think in Action*, 1st Edition (New York: Basic Books, 1984); Donald Schön, "Knowing-in-Action: The New Scholarship Requires a New Epistemology," *Change: The Magazine of Higher Learning* 27, no. 6 (December 1995): 27–34; Chris Argyris, "Double Loop

Learning in Organizations," *Harvard Business Review*, September 1977, https:// hbr.org/1977/09/double-loop-learning-in-organizations

21 Karl Weick, "Organizational Communication: Towards a Research Agenda," in *Communication and Organizations: An Interpretive Approach*, ed. Linda Putnam and Michael Pacanowsky (Beverly Hills: Sage Publications, 1983); Richard Daft and Karl Weick, "Toward a Model of Organizations as Interpretation Systems," *The Academy of Management Review* 9, no. 2 (April 1984): 284–295; Karl Weick, *Sensemaking in Organizations*, Foundations for Organizational Science (Book 3) (California: Sage Publications, 1995).

22 U.S. Joint Chiefs of Staff, *Joint Publication 3–0; Joint Operations*, Incorporating Change 1 (Suffolk, Virginia: U.S. Department of Defense, 2018), II–3, https:// irp.fas.org/doddir/dod/jp3_0.pdf

23 *Joint Publication 5–0: Joint Operation Planning* (Washington, DC: US Department of Defense, 2006), IV–2.

24 *Joint Publication 5–0: Joint Operation Planning*, IV–2.

25 Christopher Paparone, "Designing Meaning in the Reflective Practice of National Security: Frame Awareness and Frame Innovation," in *Design Thinking: Applications for the Australian Defence Force*, ed. Aaron Jackson and Fiona Mackrell, editor's manuscript pre-publication version, Joint Studies Paper Series 3 (Canberra, Australia: Defence Publishing Service, 2019), 1–18; Christopher Paparone, "How We Fight: A Critical Exploration of US Military Doctrine," *Organization* 24, no. 4 (2017): 516–533, https://doi.org/o0r. g1/107.171/1773/5103505804814716769933853

26 Larry Kay, " 'A New Postmodern Condition': Why Disinformation Has Become So Effective," *Small Wars Journal*, February 27, 2020, 4.

27 Schön, *The Reflective Practitioner: How Professionals Think in Action*; Paparone, "Designing Meaning in the Reflective Practice of National Security: Frame Awareness and Frame Innovation"; Philippe Beaulieu-Brossard and Philippe Dufort, "Introduction to the Conference: The Rise of Reflective Military Practitioners" (Hybrid Warfare: New Ontologies and Epistemologies in Armed Forces, Canadian Forces College, Toronto, Canada: University of Ottawa and the Canadian Forces College, 2016).

28 Thomas Graves and Bruce Stanley, "Design and Operational Art: A Practical Approach to Teaching the Army Design Methodology," *Military Review*, August 2013, 53–59; Wayne Grigsby et al., "Integrated Planning: The Operations Process, Design, and the Military Decision Making Process," *Military Review* XCI, no. 1 (February 2011): 28–35.

29 In his first period of design education from the 1990s through 2005, Naveh largely educated Israeli colonels with sporadic access to general officers. Later, he would teach to field grade level officers in the American and Australian militaries from 2005 to 2013. Later still, he would teach exclusively to senior colonels and general officers from 2013 onward.

30 Graicer, *Two Steps Ahead: From Deep Operations to Special Operations – Wingate the General*, 33.

31 Excluding required courses on command or various administrative briefings after promotion to higher ranks, military "flag officers" do not continue any formal schooling in most nations.

32 Seth Johnson, ed., *IBM Design Thinking Field Guide Version 3.1* (IBM Corporation, 2015), ibm.biz/idt_fieldguide

33 Shimon Naveh and Ofra Graicer, Naveh audio 14OCT2019 9min19sec Day 1 afternoon.MP3, interview by Ben Zweibelson and Nathan Schwagler, personal interview, October 14, 2019.

34 Naveh in personal correspondence with the author in 2010 stated that mid-grade officers such as Majors and Lieutenant Colonels might be the better audience for military design; however, in 2019 during recorded interviews in Tel Aviv with the author, Naveh reflected back on this and shifted his position back to one that the proper target audience for systemic operational design were General Officers. In 2006–2010 Naveh was shunned by the Israeli War College but accepted as a mentor and educator at the U.S. Army School of Advanced Military Studies, yet by 2019 this relationship reversed and Naveh was again working at the Tel Aviv Israeli War College while essentially banished from Leavenworth.

35 Barry Posen, *The Sources of Military Doctrine: France, Britain, and Germany Between the World Wars* (Ithaca, New York: Cornell University Press, 1984), 44–57; John Nagl, *Learning to Eat Soup with a Knife: Counterinsurgency Lessons from Malaya and Vietnam* (Chicago: University of Chicago Press, 2002), 8; Daniel Bolger, *Why We Lost: A General's Inside Account of the Iraq and Afghanistan Wars* (New York: Eamon Dolan Books, 2014), 433.

36 James Gibson, *The Perfect War: Technowar in Vietnam*, First Edition (Boston: The Atlantic Monthly Press, 1986), 463–467; Siniša Malešević, *The Sociology of War and Violence* (Cambridge, United Kingdom: Cambridge University Press, 2010), 28.

37 Shimon Naveh, "Between the Striated and the Smooth: Asymmetric Warfare, Operational Art and Alternative Learning Strategies" (Unpublished manuscript, undated); Shimon Naveh, *Systemic Operational Design: Designing Campaigns and Operations to Disrupt Rival Systems (Draft Unpublished)*, Version 3.0, unpublished draft (Fort Monroe, Virginia: Concept Development & Experimentation Directorate, Future Warfare Studies Division, US Army Training and Doctrine Command, 2005); Shimon Naveh, "The Australian SOD Expedition: A Report on Operational Learning" (Unpublished manuscript, December 2010).

  Alex Ryan, "A Personal Reflection on Introducing Design to the U.S. Army." The Medium (blog), November 4, 2016, https://medium.com/the-overlap/a-personal-reflection-on-introducing-design-to-the-u-s-army-3f8bd76adcb2

38 Hirsch, *Defensive Shield: An Israeli Special Forces Commander on the Front Line of Counterterrorism, the Inspirational Story of Brigadier General Gal Hirsch*, 128.

39 Weizman, *Hollow Land: Israel's Architecture of Occupation*; Weizman, "Walking through Walls: Soldiers as Architects in the Israeli/Palestinian Conflict."

40 One example of Naveh's dense, cryptic presentation slides can be found here: Weizman, "Walking through Walls: Soldiers as Architects in the Israeli/Palestinian Conflict," 199.

41 Naveh, *Systemic Operational Design: Designing Campaigns and Operations to Disrupt Rival Systems (Draft Unpublished)*.

42 Ryan, "A Personal Reflection on Introducing Design to the U.S. Army."

43 Graicer provided this document to the author in 2020 after Naveh discovered it in his own personal storage.

44 One of the first examples was written by multiple SAMS faculty in 2005. See: William Sorrells et al., "Systemic Operational Design: An Introduction" (US Army School of Advanced Military Studies Monograph, Fort Leavenworth, Kansas, May 26, 2005), ATZL-SWV, OMB No. 0704–0188.

45 Naveh disseminated these books to the few students willing to meet with him off campus. See: Shimon Naveh, Jim Schneider, and Timothy Challans, *The Structure of Operational Revolution: A Prolegomena, A Product of the Center for the Application of Design* (internally produced publication: Booz Allen Hamilton, 2009).

46 Graicer, "Self Disruption: Seizing the High Ground of Systemic Operational Design (SOD)"; Graicer, "Between Teaching and Learning: What Lessons Could the Israeli Doctrine Learn from the 2006 Lebanon War?"; Graicer, "Beware of the Power of the Dark Side: The Inevitable Coupling of Doctrine and Design."

47 IDF Doctrine and Training Division, "Design: Learning and Knowledge Development Processes for the Development of Concepts at the General Staff Headquarters and the Major HQ Levels" (Israeli Defense Force Training and Doctrine Division, November 2015).

48 Hirsch, *Defensive Shield: An Israeli Special Forces Commander on the Front Line of Counterterrorism, the Inspirational Story of Brigadier General Gal Hirsch*, 163.

49 Hirsch, 128.

50 Graicer, "Self Disruption: Seizing the High Ground of Systemic Operational Design (SOD)."

51 Naveh, Schneider, and Challans, *The Structure of Operational Revolution: A Prolegomena*, 78–79.

52 Martin Kilduff and Ajay Mehra, "Postmodernism and Organizational Research," *Academy of Management Review* 22, no. 2 (1997): 457.

53 Weizman, "Walking through Walls: Soldiers as Architects in the Israeli/ Palestinian Conflict," 211.

54 Antoine Bousquet, "Cyberneticizing the American War Machine: Science and Computers in the Cold War," *Cold War History* 8, no. 1 (February 2008): 77–102; Gibson, *The Perfect War: Technowar in Vietnam*, 20–27, 464–467; Malešević, *The Sociology of War and Violence*, 79–85.

55 Graicer, *Two Steps Ahead: From Deep Operations to Special Operations – Wingate the General*, 38–42; Naveh, "Northern Storm: A Narrative of Reflective Command, Systemic Learning, and Operational Design 2002–2005"; Ben Zweibelson, "Three Design Concepts Introduced for Strategic and Operational Applications," National Defense University PRISM 4, no. 2 (2013): 87–104.

56 Beaulieu-Brossard, "Encountering Nomads in Israel Defense Forces and Beyond," 2020, 4.

57 Weizman, *Hollow Land: Israel's Architecture of Occupation*, 200–201.

58 Graicer, *Two Steps Ahead: From Deep Operations to Special Operations – Wingate the General*, 37.

59 Jacques Derrida, *Of Grammatology*, trans. Gayatri Spivak, Fortieth Anniversary Edition (Baltimore: John Hopkins University Press, 2016), 235.

60 Linda Putnam, "The Interpretive Perspective: An Alternative to Functionalism," in *Communication and Organizations: An Interpretive Approach*, ed. Linda Putnam and Michael Pacanowsky (Beverly Hills: Sage Publications, 1983), 40–45.

61 Hirsch, *Defensive Shield: An Israeli Special Forces Commander on the Front Line of Counterterrorism, the Inspirational Story of Brigadier General Gal Hirsch*, 129.

62 Adamsky, "The Israeli Revolution in Military Affairs," 99–101.

63 Graicer, "Self Disruption: Seizing the High Ground of Systemic Operational Design (SOD)," 24.

64 Beaulieu-Brossard, "Encountering Nomads in Israel Defense Forces and Beyond," 2020, 13.

65 Hirsch, *Defensive Shield: An Israeli Special Forces Commander on the Front Line of Counterterrorism, the Inspirational Story of Brigadier General Gal Hirsch*, 129–130.
66 Adamsky, "The Israeli Revolution in Military Affairs," 99–100; Weizman, *Hollow Land: Israel's Architecture of Occupation*, 212–215.
67 English variations of his name spelling also include "Kokhavi."
68 Eyal Weizman, "Lethal Theory," Log 7 (Winter/Spring 2006): 53–77; Weizman, "Walking through Walls: Soldiers as Architects in the Israeli/Palestinian Conflict."
69 Yagil Henkin, "On Swarming: Success and Failure in Multidirectional Warfare, from Normandy to the Second Lebanon War," *Defence Studies* 14, no. 3 (2014): 310–332.
70 Henkin, "On Swarming: Success and Failure in Multidirectional Warfare, from Normandy to the Second Lebanon War," 322. The one Israeli casualty occurred through friendly fire with apparent sniper misidentification.
71 Graicer, "Self Disruption: Seizing the High Ground of Systemic Operational Design (SOD)," 25–26.
72 Weizman, "Walking through Walls: Soldiers as Architects in the Israeli/Palestinian Conflict," 215.
73 Naveh, "Between the Striated and the Smooth: Asymmetric Warfare, Operational Art and Alternative Learning Strategies," 4.
74 Christopher Paparone and George Topic, Jr., "From the 'Swamp' to the 'High-Ground' and Back – Educating Logisticians to Operate in Complexity: Part One," *Logistics in War: Military Logistics and Its Impact on Modern Warfare*, May 13, 2017, https://logisticsinwar.com/2017/05/13/from-the-swamp-to-the-high-ground-and-back-educating-logisticians-to-operate-in-complexity-part-one/; Antoine Bousquet and Simon Curtis, "Beyond Models and Metaphors: Complexity Theory, Systems Thinking and International Relations," *Cambridge Review of International Affairs* 24, no. 1 (2011): 43–62.
75 In personal correspondence, Graicer recalled how Naveh originally used the term "rival rational" for 1995–2006, and when they were working with the Australians in the 2006–2009 period Graicer would propose this new term instead.
76 Shimon Naveh, "The Australian SOD Expedition: A Report on Operational Learning" (Unpublished manuscript, December 10, 2010), 7.
77 Lee Vinsel, "Design Thinking Is Kind of Like Syphilis – It's Contagious and Rots Your Brains," *Noteworthy – The Journal Blog* (blog), December 6, 2017, https://blog.usejournal.com/design-thinking-is-kind-of-like-syphilis-its-contagious-and-rots-your-brains-842ed078af29
78 Beaulieu-Brossard, "Systemic Operational Design or How I Began to Worry about the Dual Use of Critical Concepts," 6.
79 Ofra Graicer, "Self Disruption: Seizing the High Ground of Systemic Operational Design (SOD)," *Journal of Military and Strategic Studies* 17, no. 4 (June 2017): 21–37; Ofra Graicer, "Beware of the Power of the Dark Side: The Inevitable Coupling of Doctrine and Design," *Expertica Militar*, October 2017, 30–37.
80 Shimon Naveh, "The Australian SOD Expedition: A Report on Operational Learning" (Unpublished manuscript, December 2010), 6.
81 Naveh, 6.
82 Jacques Ranciere, *The Ignorant Schoolmaster: Five Lessons in Intellectual Emancipation*, trans. Kristin Ross (Stanford, California: Stanford University Press, 1991), 14–15.

83 Gibson Burrell and Gareth Morgan, *Sociological Paradigms and Organisational Analysis: Elements of the Sociology of Corporate Life* (Portsmouth, New Hampshire: Heinemann, 1979); Marianne Lewis and Andrew Grimes, "Metatriangulation: Building Theory From Multiple Paradigms," *Academy of Management Review* 24, no. 4 (1999): 672–690; Majken Schultz and Mary Jo Hatch, "Living with Multiple Paradigms: The Case of Paradigm Interplay in Organizational Culture Studies," *Academy of Management Review* 21, no. 2 (1996): 529–557; Linda Putnam, "The Interpretive Perspective: An Alternative to Functionalism," in *Communication and Organizations: An Interpretive Approach*, ed. Linda Putnam and Michael Pacanowsky (Beverly Hills, CA: Sage Publications, 1983), 31–54.

84 Christopher Paparone, *The Sociology of Military Science: Prospects for Postinstitutional Military Design* (New York: Bloomsbury Academic Publishing, 2013); Ben Zweibelson, "Professional Reading Lists: Thinking Beyond the Books and into Military Paradigmatic Biases," *Air and Space Power Journal* 30, no. 2 (Summer 2016): 15–37.

85 Kevin Benson, Interview of Colonel (retired) Kevin Benson, former SAMS Director, 04 AUG 2021 mp3 audio file, interview by Ben Zweibelson, mp3 Audio File, August 4, 2021, 23:36 to 24:15.

86 Naveh, *Systemic Operational Design: Designing Campaigns and Operations to Disrupt Rival Systems (Draft Unpublished)*, 20.

87 Naveh, "The Australian SOD Expedition: A Report on Operational Learning," 9; Graicer, *Two Steps Ahead: From Deep Operations to Special Operations – Wingate the General*, 33.

88 Protzen and Harris, *The Universe of Design: Horst Rittel's Theories of Design and Planning*, 110–117.

89 Deleuze and Guattari, A Thousand Plateaus; Michel Foucault, "Discourse and Truth: The Problematization of Parrhesia" (lecture, University of California at Berkeley, November 1983), http://foucault.info//system/files/pdf/DiscourseAndTruth_MichelFoucault_1983_0.pdf

90 Michel Foucault, "Discourse and Truth: The Problematization of Parrhesia" (lecture, University of California at Berkeley, November 1983), http://foucault.info//system/files/pdf/DiscourseAndTruth_MichelFoucault_1983_0.pdf

91 Naveh, "The Australian SOD Expedition: A Report on Operational Learning," 10.

92 Few PME programs offer any philosophy education, and those that do only promote select institutionalized war philosophies. Attempts to challenge the Westphalian ordered, Clausewitzian framed modern war frame are met with concern or outrage. SOD begins with the position that this modern frame is flawed.

93 Naveh, 11.

94 Naveh, *Systemic Operational Design: Designing Campaigns and Operations to Disrupt Rival Systems (Draft Unpublished)*, 4–7.

95 Francois Jullien, *The Silent Transformations*, trans. Krzysztof Fijalkowski and Michael Richardson (New York: Seagull Books, 2011), 25–26.

96 Jullien, 98.

97 Hayden White, *The Content of the Form: Narrative Discourse and Historical Representation*, paperback edition (Baltimore: The John Hopkins University Press, 1990), 41.

98 White, 42.

99 Russell Ackoff, "On the Use of Models in Corporate Planning," *Strategic Management Journal* 2, no. 4 (December 1981): 353–359.

100 Alice Butler-Smith, "Operational Art to Systemic Thought: Unity of Military Thought," in *Cluster 3* (Hybrid Warfare: New Ontologies and Epistemologies in Armed Forces, Canadian Forces College, Toronto, Canada: unpublished – Canadian Forces College internal document, 2016), 1–5.

101 Naveh, "The Australian SOD Expedition: A Report on Operational Learning," 11.

102 Naveh and Graicer, Naveh audio 15OCT2019 34min22sec Day 2 morning. MP3, 15:19.

103 Vego, "A Case Against Systemic Operational Design," 69–71; Weizman, *Hollow Land: Israel's Architecture of Occupation*, 214.

104 Graicer, *Two Steps Ahead: From Deep Operations to Special Operations – Wingate the General*, 33–38.

105 Henry Mintzberg, Duru Raisinghani, and Andre Theoret, "The Structure of 'Unstructured' Decision Processes," *Readings in Decision Support Systems*, 1976, 134.

106 Naveh, *Systemic Operational Design: Designing Campaigns and Operations to Disrupt Rival Systems (Draft Unpublished)*, 10.

107 Graicer, 33.

108 Milan Vego, "A Case Against Systemic Operational Design," *Joint Forces Quarterly* 53 (quarter 2009): 70–75.

109 Naveh, Interview with BG (Ret.) Shimon Naveh, 4.

110 Naveh, "The Australian SOD Expedition: A Report on Operational Learning," 13.

111 Naveh, 17.

112 Matt Matthews, "We Were Caught Unprepared: The 2006 Hezbollah-Israeli War," The Long War Series (Fort Leavenworth, Kansas: U.S. Army Combined Arms Center Combat Studies Institute Press, June 29, 2012), 28.

113 Naveh, "The Australian SOD Expedition: A Report on Operational Learning," 19.

114 Weizman, *Hollow Land: Israel's Architecture of Occupation*, 214–215.

115 Beaulieu-Brossard, "Systemic Operational Design or How I Began to Worry about the Dual Use of Critical Concepts," 7; Graicer, *Two Steps Ahead: From Deep Operations to Special Operations – Wingate the General*, 33.

116 Hatch, *Organization Theory: Modern, Symbolic, and Postmodern Perspectives*, 43.

117 Haridimos Tsoukas, "A Dialogical Approach to the Creation of New Knowledge in Organizations," *Organization Science* 20, no. 6 (December 2009): 943–945.

118 Shimon Naveh and Ofra Graicer, Naveh audio 14OCT2019 2hr28min Day 1 patio.MP3, interview by Ben Zweibelson and Nathan Schwagler, personal interview, October 14, 2019.

119 Graicer, "Self Disruption: Seizing the High Ground of Systemic Operational Design (SOD)," 21–22.

120 Graicer, "Between Teaching and Learning: What Lessons Could the Israeli Doctrine Learn from the 2006 Lebanon War?," 26.

121 Graicer, "Self Disruption: Seizing the High Ground of Systemic Operational Design (SOD)," 26.

122 Naveh first began lecturing on SOD internationally in 2003–2005, gaining interest in the United States.
123 Naveh, *Systemic Operational Design: Designing Campaigns and Operations to Disrupt Rival Systems (Draft Unpublished).*
124 Naveh, Interview with BG (Ret.) Shimon Naveh.
125 Naveh retired from active duty in the IDF as a Brigadier General in 1993, but as part of his retirement plan he was commissioned by the IDF as a civilian academic to write a doctoral dissertation on operational art, return to the IDF and establish the military think tank known as OTRI.
126 Graicer, "Between Teaching and Learning: What Lessons Could the Israeli Doctrine Learn from the 2006 Lebanon War?," 25–28; Graicer, "Self Disruption: Seizing the High Ground of Systemic Operational Design (SOD)"; Naveh, Interview with BG (Ret.) Shimon Naveh.
127 Paul Davis, "Effects-Based Operations (EBO): A Grand Challenge for the Analytical Community" (RAND Corporation, 2001), www.rand.org/ pubs/monograph_reports/MR1477.html; John Harris, "Effects-Based Operations: Tactical Utility" (U.S. Army Command and General Staff College, 2004), https://apps.dtic.mil/sti/pdfs/ADA428961.pdf
128 James Mattis, "USJFCOM Commander's Guidance for Effects-Based Operations," *Joint Forces Quarterly* 4th Quarter, 2008, no. 51 (2008): 106–108.
129 James Mattis to Ben Zweibelson, "RE: Design for Defense Book PDF Manuscript," September 16, 2021.
130 Graicer, "Between Teaching and Learning: What Lessons Could the Israeli Doctrine Learn from the 2006 Lebanon War?," 28.
131 Naveh, Interview with BG (Ret.) Shimon Naveh.
132 Beaulieu-Brossard, "Systemic Operational Design or How I Began to Worry about the Dual Use of Critical Concepts," 8.
133 Henkin, "On Swarming: Success and Failure in Multidirectional Warfare, from Normandy to the Second Lebanon War," 321.
134 Kees Dorst, *Frame Innovation: Creating New Thinking by Design*, Design Thinking, Design Theory (Cambridge, Massachusetts: MIT Press, 2015), 15.
135 Beaulieu-Brossard, "Systemic Operational Design or How I Began to Worry about the Dual Use of Critical Concepts," 8.
136 Beaulieu-Brossard, 8.
137 Beaulieu-Brossard, 8.
138 Naveh, Interview with BG (Ret.) Shimon Naveh.
139 Hirsch, *Defensive Shield: An Israeli Special Forces Commander on the Front Line of Counterterrorism, the Inspirational Story of Brigadier General Gal Hirsch*, 139.
140 Henkin, "On Swarming: Success and Failure in Multidirectional Warfare, from Normandy to the Second Lebanon War."
141 Matthews, "We Were Caught Unprepared: The 2006 Hezbollah-Israeli War," 64.
142 David Johnson, "The Second Lebanon War," in *Hard Fighting: Israel in Lebanon and Gaza* (Santa Monica, California: Rand Corporation, 2011), 18, www.jstor. org/stable/10.7249/mg1085a-af.10
143 Henkin, "On Swarming: Success and Failure in Multidirectional Warfare, from Normandy to the Second Lebanon War," 312.

144 Henkin, 323. Henkin cites the original source for this quotation as Prisoners in Lebanon: The Truth about the Second Lebanon War by Ofra Shelah and Yoav Limor, p. 197 which is available only in Hebrew. This research relies on Henkin's translation to English in his own work.

145 Hirsch, *Defensive Shield: An Israeli Special Forces Commander on the Front Line of Counterterrorism, the Inspirational Story of Brigadier General Gal Hirsch*, 318.

146 Hirsch, 172.

147 Personal correspondence with Graicer on February 23, 2020.

4   Design Comes to America

The Army Assimilation of SOD

The U.S. Army is the most technologically advanced, resourced, and most specialized land force created in human history. Yet for an institution so large and sophisticated, its ability to implement change and innovation is often inhibited by the scale, scope, and institutional self-preservation behaviors found in any similar entity. Over the last twenty-five years, we have witnessed significant experimentation and growth in design practice that also coincides with challenging "postconventional" warfare contexts, boiling counterinsurgencies, and a complex new competitive security environment that potentially defies previous description.[1] The U.S. Army has been a central actor and primary instrument of national power in the perpetual application of organized violence for national security designs.

Since at least the late 1990s, an expansive growth and development of non-state actors has clouded what previously adhered to Westphalian, nation-state oriented definitions of war and diplomacy. With the rise of social media and cyberspace, the arrival of a second quantum revolution, expanded exploration and militarization of space, humans now confront wickedly complex and novel combinations of war, terror, and security challenges unlike any previous generation. Understandably, this also creates fear and apprehension that existing ideas, methods, and organizational structures are no longer relevant. All militaries must undergo transformation and development, with some taking this proactively and others being dragged there through defeat, failure, and frustration on the battlefield. The American military would gradually experience all of these things over the last two decades with a "Global War on Terror", attempts at nation-building and creating western-oriented, "mini-America" styled democracies out of adversaries, along with an ever-expanding "Great Power Competition" between traditional rivals and a growing China threat to American unilateral power.

The Israeli initial experimentation in this new security design praxis toward a novel way of sense-making in war would inspire the attention, focus, as well as apprehension for implementation by the U.S. Army. Struggling in multiple counterinsurgency quagmires across Africa, the Middle East and elsewhere, the American Army would move rapidly toward exploring new ideas and how to implement them to somehow restore control and stability

DOI: 10.4324/9781003387763-5

in what seemed chaotic and ever-changing misfortunes in Iraq, Afghanistan, and beyond.[2] The Army would also defend and preserve ideas and things considered sacred or essential while paradoxically grasping toward emerging and unfamiliar concepts and novel war tools. These new ideas promised some sort of valuable investment, but to implement them would often require change and the retirement of existing (and favorite) tools for war; often the institution refuses to let go of favorite tools until it is too late.[3]

Innovating is hard, and institutions tend to throw up barriers to it. Innovation as it emerges has no predecessor or "proof" if it indeed is the thing or idea that the organization needs to move toward while also accepting risk that this novelty cannot feature robust evidence of functionality. Only legacy constructs have those, and indeed they often are the things ritualized into institutional belief systems while gradually becoming irrelevant to emerging contexts.[4] Old tools fade, but our memory that they worked well is projected upon tomorrow's challenge that requires an entirely new tool not in our inventory. Returning to Ackoff's explanation that designers dissolve legacy systems by designing future ones that eliminate current problems while ushering in entirely unanticipated ones, designing toward innovation means that the new concept *will be unfamiliar* in any legacy assessment.

In the mid-2000s in multiple combat zones around the globe, the U.S. Army faced this paradox and would look for answers to seemingly impossible new problems. Colonel (retired) Steven Leonard, heavily involved with design doctrinal development between the U.S. Army and Marine Corps, offers a useful summary:

> Doctrinally, we had come out of the Gulf War period drawing a lot of bad lessons … we moved into this period of about twelve or thirteen years where we really felt that "war is really easy" as it is all technologically based, we moved into "Revolutions in Military Affairs" and "hey, what is the next computer chip that will be the next revolution?" We got so wrapped up in that where we lost the cognitive things that had driven us and we had moved to this mindset where "all you need is a good battle rhythm and you can beat anybody" and "victory is with the staff … the staff knows how to win", so when you hit the 2004–2008 timeframe, design hit at exactly the same time we were trying to un-[expletive] our doctrine.[5]

The timing would be perfect for something controversial and disruptive to gain the attention of the intellectual core of the U.S. Army. The Army would quickly discover Shimon Naveh and this new Israeli concept of security design praxis; the Army would invite these Israeli designers into the heart of U.S. Army intellectualism in Fort Leavenworth, Kansas, and attempt to extract these radical ideas into something malleable for American military utilization. In these pressure-cooker security contexts where one's enemies refused to play by the traditional rules of war, new ideas that are radical, disruptive require significant momentum and support to break through institutional barriers.[6] Naveh's design would come to America, but the American Army would determine that form, function, and experience. This chapter explains that story.

The American Army sent strategic thinkers and intellectuals to Israel in 2004–2005 to learn about Naveh's systemic operational design and would later invite him back to Fort Leavenworth to begin educational exchanges with the SAMS students and faculty. The Army's "Training and Doctrine Command" (TRADOC) was already on the hunt for something new, and design as a strange new concept appeared to have potential. TRADOC wanted to implement new ways to think and act in warfare into doctrine and immediate practice in Iraq and Afghanistan to help turn the tide. Partially due to the similar ground missions in the boiling counterinsurgencies, the American Army partnered with the U.S. Marines as well as international military intellectuals to reconsider how they would conduct counterinsurgency operations. This initial collaboration resulted in a striking revision of *Field Manual 3–24: Counterinsurgency* (*FM 3–24/MCWP 3–33.5*) in 2006. It is in this revised doctrine where the American military first mentions "design" in the Israeli sense. However, this would not be a smooth journey, nor would Naveh be pleased with how design would transform once in American hands.

Early Interest in Thinking Differently about Warfare: The Long View of Iraq

In late 2003, the new Director of the SAMS program was Colonel Kevin Benson, who had just returned from combat in the Iraqi Invasion. He was concerned that the premiere intellectual center for the Army at Fort Leavenworth was still too focused on teaching much of the older classics, where studying Napoleon, the World Wars, and the ancient Greek accounts of the Peloponnesian War formed the basis of what was an operational planning school. Benson felt differently.

> I had just come back from Iraq, and I knew in my heart and in my gut that this was going to be a long war, and I didn't want to waste our time … I knew we had to get SAMS thinking about the long war and how to think our way through it.[7]

Timing would be fortuitous in that his faculty felt reforms were needed as well.

Dr. Peter Schifferle, who oversaw the SAMS Senior Fellows program, had recommended to Benson that the SAMS program pull out of a major military academic exercise called "Unified Quest" planned for 2004.[8] This was an inter-service and joint military futures exercise that occurred annually hosted by the U.S. Army. According to Benson, Schifferle found the exercise to be lacking in value for their program, and that reforms were necessary in light of institutional frustration brewing overseas in Iraq and Afghanistan. Yet pulling out of this major exercise would create serious political fallout within the Department of Defense, as SAMS holds a premier status of sorts as the Army's "intellectual tip of the spear" where select officers gain exceptional education and experience to take on the toughest operational staff roles around. SAMS skipping the exercise would cause a stir.

Army Colonel Bob Johnson (assigned at TRADOC) would catch wind of this and contact Benson. Johnson, the Unified Quest project manager, would plead for SAMS to come to the exercise. There was going to be something new and radical presented. Unbeknownst to Benson and SAMS faculty, Johnson, and several at TRADOC had stumbled across design in the Israeli Defense Forces decided to introduce them to the American land forces at this exercise. SAMS needed to come see this strange Israeli retired general and his team that spoke of strange concepts, used only white boards, and used processes unlike anything in existing planning doctrine to create new warfighting concepts.

Johnson decided to fly out to Fort Leavenworth in January 2004 and bring with him the founder of SAMS, Hubba Wass de Czege along with Shimon Naveh to provide the SAMS faculty and students with a military design overview. Naveh would lecture SOD concepts in the United States and in Israel at OTRI until the think tank was eliminated. Yet in a brief year of overlap of 2003–2004, Naveh would manage to intrigue enough American military thinkers to shift the movement into an entirely new defense organization. Naveh's concepts were still considered rather dense, and Brigadier General David Fastabend, the U.S. Army Deputy Director for Futures, instructed Johnson and others to travel back to Israel and spend more time with Naveh at OTRI. The American Army had a taste of SOD and was intrigued.

After Johnson returned from another week in Israel studying SOD, the U.S. Army started formally experimenting with SOD. It started at the TRADOC staff level, with a small population of experienced planners and in coordination with SAMS. Few people in the U.S. Army had heard of military design, and there were many skeptics including those within this first formation of design experimenters. Benson felt that TRADOC was "sandbagging him and SAMS a bit" over whether to invest valuable student time into this obscure theory and methodology, but Johnson convinced him to allow SAMS to become the laboratory for Army experimentation and evaluation of the foreign and dense ideas.[9] Johnson approached the U.S. Army War College in Carlisle at the same time as SAMS, but only Benson agreed to open the door to military design experimentation.[10]

Benson remembers that Naveh's first critical demonstration of SOD to the U.S. military occurred in "Unified Quest". For the 2004 iteration, Naveh was asked to wargame SOD against both the traditional military planning process and that of EBO, which was advocated by proponents of air power and strategic strike capabilities. Johnson recalls: "it was a side-by-side, qualitative comparison of SOD along with classical elements of operational [art and planning] ... the stuff that is in doctrine".[11] They would compare them with EBO and

> the overwhelming consensus was that the classical elements were useful up to a point, that EBO was nothing more than a targeting methodology that [the senior leaders] felt had been over-promoted beyond what it was intended for, and that SOD had real potential.[12]

That the SOD performance went so well was even more curious, in that the SAMS students demonstrating the design methods had only been taught SOD for a few sessions prior to the team heading to the exercise. In the span of a few short months, the Army went from a small experimentation to taking on the institutionalized preferred decision-making apparatus in a wargame.

Once SAMS agreed to attend Unified Quest, they put together a small design team. Benson and Johnson assembled an ad hoc group of SAMS students along with Tim Challans and Jim Schneider who were on SAMS faculty and would later join Naveh in developing an Americanized version. After a few weeks of SOD education, their design team then performed at the 2004 Unified Quest exercise. Benson recalls:

> It was rock and roll ... God, it was so exciting. I was suspect at the time of "Effects Based Operations" that was just starting to come about ... [EBO] struck me as such a linear concept. So, we had five or six sessions as I recall with these three guys. They were fantastic. I had Naveh talk to everyone in SAMS ... it was 2004, and it was such an exciting concept ... the purpose of SOD was to think through the challenges of warfare and reduce the number of enemy options ... if we could observe and learn faster, we could reduce the enemy's range of options ... this small group and I went to Unified Quest in 2004. There, one group did [the Military Decision-Making Process], another group did EBO. And they said: "oh, you SAMS guys, you can do that SOD stuff." We were a bolt-on to the exercise. I remember the Red Team [opposing force] commander in that exercise coming up to our group and saying that he did not understand what the SAMS team was doing to his enemy forces and it was confusing him. We felt were frustrating the enemy, and while the other teams were getting eviscerated, but our SOD team was staying ahead of him and keeping him completely confused. [13]

SOD's initial success at the 2004 Unified Quest war games would set into motion a long-term relationship for Naveh's designers and the American Army. By the 2005 iteration of Unified Quest, the exercise was much larger, and more SAMS students participated. The SAMS students under Naveh's tutelage as well as Wass de Czege would continue to excel and confound the other forces at the exercise. Benson realized in 2005 that "what OTRI and SOD was producing would have a profound effect on all of the western militaries because of the SOD work in these seminars."[14] During that Unified Quest exercise, Naveh gave custom SOD engagements to numerous senior military leaders showing up after hearing about this new military design method. Students began reading extensively on systems theory, complexity theory, chaos theory, architecture, and postmodern literature as Naveh's SOD infiltrated the SAMS curriculum.[15] This would transform SAMS into a military design laboratory for 2005 through 2009. Much of this would occur under a new SAMS director, Colonel Stefan Banach. It would also generate controversy, frustration, and political infighting within the Army.

Naveh's SOD deconstructed not just the traditional decision-making methodologies found in doctrine but would also tear down the epistemo-logical foundations supporting the continued relevance of many doctrinal beliefs, traditional military strategy, and especially operational art and oper-ational planning for warfare. While Naveh would open the American mili-tary up to unorthodox postmodern concepts, dense theory, and a challenging new way to design complex warfare through a reconceptualization of war itself, competitors would argue for less exotic transformations.

Advocates of a "systems thinking in warfare" envisioned a careful fusion of complexity theory, general systems theory, and several other adaptations from other non-military *scientific* fields would offer minor adjustments to the existing military decision-making methodology. Naveh's insistence on postmodern philosophy, multiple social paradigms, and other dense theories would be resisted, as would any serious disruption to the established mili-tary decision-making methodology and contemporary doctrine. Two camps would emerge on military design, with one being the original Naveh "purists" insisting on radical transformation and the slaughter of institutionally sacred cows concerning war, warring with the pragmatic reformers who wanted incremental improvements that were readily accessible to the entire force.

Dan Cox, on SAMS faculty during part of this period, reflected on the SOD experimentation spanning 2004–2010 as follows:

> the SAMS department moved away from postmodernism and Naveh's SOD in 2010 prior to the publication [of the first Army formal design chapter in *Field Manual 5–0*] … but we were all disappointed in the final doctrinal output which seemed linear and somewhat like a checklist.[16]

Over a six-year period of intense experimentation and academic research into military design for the U.S. Army, these two camps on design would become increasingly antagonistic, with egos and reputations on the line for many. All of this occurred while a surge of American military forces continued to face frustration, confusion, and a failure of strategic and oper-ational endeavors in Iraq, Afghanistan, and elsewhere.

Ryan, in his retrospective blog on being part of the Army's effort to intro-duce design thinking to the force, highlights one of these design camps as "the home team" where they wanted to follow design in Naveh's deeply intellectualized, self-disruptive mode.[17] The "away team", in fierce opposition to Ryan's home team, wanted all design to operate within the existing institutional framework in a pragmatic manner. The home team rallied around the complete irrelevance of legacy systems and outdated modes of warfighting, while the away team demanded that design would only be useful if it is digestible to the masses. Ryan neatly frames this paradoxical tension as follows:

> how do you introduce a new paradigm when you cannot have a discussion at the level of paradigms? The Home team was mostly ignored or derided by Army leaders. For every 100 students, they would convert one or two devoted acolytes … the Away Team was better received by students. But because not of these students were required to challenge their fundamental beliefs, they were never able to really reframe.[18]

The Away Team would, particularly in the 2009–2022 period, win this war and firmly entrench a design methodology for the U.S. Army that would not be able to rock the boat on anything dear to the institution.

There exist two Army design documents that highlight these institutional tensions between two different design advocates. The first is the June 16, 2006, final version of *Field Manual 3–24: Counterinsurgency* draft document circulated within the army before final edits and publication. The second is the final published version of that same manual, *FM 3–24/MCWP 3–33.5,* published on December 15, 2006. Both featured a military design chapter, but only the final published version would present design officially to both the U.S. Army and the U.S. Marine Corps. The distinctions between the draft design chapter and the final version provide compelling evidence of how the U.S. military struggled to incorporate a military design method into their warfighting frame, and why Naveh's original SOD would not simply be adapted as it was.

Naveh's SOD influences are quite apparent in the draft version of *FM 3–24's* design chapter. While some of the stylistic changes and substituted terminology are subtle between the two, several major differences and heavy editing of the draft design chapter indicate how and why the Israeli design theory would be removed from the final version. Further, the systems theory would be incorporated, but only within an assimilated manner where the dominant functionalist military war frame remained overriding across the military institution. Essentially if we examine earlier draft Army design doctrine and concepts in the 2004–2007 period, we find more direct influence of SOD praxis. Later in the 2009–2010 period when formal Army doctrine is published on design, that influence had weakened, and the outputs would no longer feature much of the SOD theory, models, language, or methodology.

The opening paragraph of the final published *FM 3–24/MCWP 3–33* design chapter would stipulate: "For Army forces, this chapter applies aspects of command-and-control doctrine and planning doctrine to counterinsurgency campaign planning".[19] However, the draft version originally led with: "This chapter describes conditions for designing counterinsurgency campaigns and the associated planning and operations,"[20] placing design activities in a superior status above military planning itself. Somewhere in the draft editorial discussions, a battle was waging on whether "design" captured planning in warfare, or if it was merely an optional step within a broader planning requirement that remained universal. Epistemologically, this reflects a tension on when, how, and why change occurs, and whether humans ought to approach complexity through a standardized or a custom, ever-changing sort of approach. In yet another example of this tension, the draft 2009 version of *Design: Field Manual Interim 5–2* (an unofficial document provided by TRADOC) also placed "design" at the precipice above subsequent planning activities.

The original epistemological stance that "design" would operate abstractly, above, and externally to operational planning logic would exist in the draft *FM 3–24* chapter as evidence of Naveh's significant and early influences on American Army intellectuals in 2004–2006. Naveh's influence appears in how the draft *FM 3–24/MCWP 3–33* version positioned military planning as a subordinate and entirely malleable, customized element within military design. SOD ideas abound in the draft version, while most of these would be removed or severely marginalized in the final publication. In another example, the draft publication also stated: "In this sense, design guides and informs planning",[21] the final version would be reworded to: "Design provides a means to gain understanding of a complex problem and insights towards achieving a workable solution".[22] Individually, multiple Army officers exposed to SOD would start to see design as a conceptual vehicle to challenging the entire military paradigm for what war is, and how one might re-imagine how to fight differently. Collectively, large military institutions such as the U.S. Army and the U.S. Marine Corps had no interest in disrupting a well-oiled, traditional bureaucracy with vast tombs of planning literature now under threat of total revision or elimination. Big military would resist Naveh's heretical ideas in America, just as did in Israel.

Another major difference between the draft and final versions of *FM 3–24/MCWP 3–33* involves how SOD went about creating the conditions for innovation in war. Naveh and his designers sought an initial self-reflection, where designers concentrated on their "self" to reflectively consider what their frame was, why it became this way, and how such a conceptual frame may limit one's ability to imagine or experiment beyond what the institution has already conditioned.[23] In other words, SOD required designers to design upon themselves first, before moving to consider adversaries, external environments, or determining some strategic goal or operational objective to pursue. The idea of an objective observer in design is not possible in that any attempt to sense-make a social reality (including war) occurs through a conceptual frame.[24] This breaks with the modern military paradigm covered in the second chapter, where a reverse-engineered goal is nested within systematic, technological, and what is best described as a Newtonian styled way to conceptualize all warfare activities. The draft version attempted to address this by summarizing the dense postmodern concepts in Naveh's original SOD. Termed "the nature of design", the draft *FM 3–24/MCWP 3–33* described this with:

> The design team engages in constructing and continuously modifying two complementary logics, or mental models. The first is the governing logic of the problem. The aim here is to rationalize the problem situation – to construct a logical explanation, in the form of an abstract model that unravels the problem logic. The essence of this counterlogic [sic] is the success mechanism or sequence of interactions envisioned to achieve a desired end state.[25]

Yet the final version of the same chapter would eliminate the entire counter-logic construct, rendering much of Naveh's self-disruption moot in what the American military sought to indoctrinate to the total force. This illustrates a single-paradigm dominance where military theorists and doctrine writers could assimilate select design methods but misunderstand and eventually discard the deeper theoretical content. SOD's original emphasis on exploring "conceptual tensions between prevailing paradigm and the context in emergence" would be abandoned for no consideration of paradigms at all aside from not criticizing (or even acknowledging) the one upheld institutionally.[26] Gharajedaghi provides a useful summary of paradigm shifting in how organizations abandon obsolete ways of thinking toward novel ones that provide greater utility and opportunity:

> A shift of paradigm can happen purposefully by an active process of learning and unlearning. It is more common that it is a reaction to frustration produced by a march of events that nullify conventional wisdom. Faced with a series of contradictions that can no longer be ignored or denied and/or an increasing number of dilemmas for which prevailing mental models can no longer provide convincing explanations, most people accept that the prevailing paradigm has ceased to be valid and that it has exhausted its potential capacity.[27]

Army designers, if only reading their design doctrine, remain trapped within the dominant institutional paradigm, unable to design beyond those epistemological and ontological structures. This is in keeping with how modern military doctrine is published to maintain tight, hierarchical control and order on what is thought about in war, and what must not be thought about. Thus, while Naveh's SOD provided the logical breadcrumbs to motivate military designers to challenge their institutional paradigms concerning war, Army Design Methodology (ADM) would impose rigid conformity to the existing army way of doing business. One design created heretics, and the other would generate practical planning options potentially unrealized if a staff merely executed detailed planning without ADM.

While the U.S. Army co-authored *FM 3–24/MCWP 3–33* with the U.S. Marine Corps, it appears the Marines wrote nearly all the draft and then final design chapters, with little to no input from the Army. As this doctrine was in production around the start of Naveh's involvement with SAMS, it creates an important nuance that will be elaborated upon in the next chapter on how the Marines were introduced to Naveh's design. Essentially, as Naveh was invested into teaching original SOD to SAMS in the 2004–2009 period, TRADOC and the broader army community was detached from that experience. The Marines would learn of design from Naveh as well, and design took a curious path from conceptualization to the final, official publication *FM 3–24/MCWP 3–33* in 2006.

Marine General James Mattis recollects on how *FM 3–24/MCWP 3–33* was divided between the two organizations. "When General Petraeus and I broke up FM 3–24's chapter writing assignments between Leavenworth and Quantico, the only chapter we Marines specifically requested was the fourth chapter, *Design*. So, the initial draft was done at Quantico."[28] He would go on to state that while the Army provided review and input of the drafts, the design chapter was of Marine authorship exclusively. There were also efforts to construct a design instructional pamphlet or publication for Joint Forces Command in this period, indicating a broader interest in adapting Naveh's concepts into some tangible methodology and doctrinal language for U.S. services and components to begin to use. Yet many factions held their own views and agendas, and soon enough they would wage a war of competing design narratives.

Johnson believed that while TRADOC was not yet convinced at that time of design's potential, the Marines were completely on-board with many of Naveh's ideas. "[TRADOC] was reluctant to get engaged because we had not examined design sufficiently, to think it was ready for prime time [in doctrine] ... so we kept our distance and we reviewed it and offered comments".[29] Johnson was unaware if within the Marines' doctrinal community, sufficient designers had worked with Naveh to avoid reinforcing institutional norms and beliefs. Regardless, through collaboration with the Army, the Marines wrote the first design chapter for both land forces within doctrine. Despite the Marines authoring the chapter, the U.S. Army endorsed those concepts into formal doctrine and TRADOC would enforce integration therein of the new ideas. Design would become something the U.S. Army was now directed to understand and apply.

From 2004 through 2010, Naveh directly engaged with design students at SAMS and across the American military. He would continue to promote design via a postmodern mix of different disciplines, folded into iterative deconstruction and destruction of cherished military beliefs on war. Above all else, SOD challenged any single rationalization of complex security challenges in that even the idea of a problem is itself socially constructed and an illusion blocking one from systemically realizing other opportunities. Contrasting this with how *FM 3–24/MCWP 3–33* declared the design process, the tensions between these two design camps are crystalized. In this first doctrine, designers sought to "rationalize the problem – to construct a logical explanation of observed events and subsequently construct the guiding logic that unravels the problem".[30] While Naveh insisted on design transforming military thought and the very institution of warfare itself, the Army and Marines would instead indoctrinate design into something compatible and complimentary to what already existed and sustained the modern warfare perspective.

The fact that the Army's final design version also states it can "unravel the problem" again shifts the design practice away from the earlier drafts advocating a "problem-setting" appreciation that complex systems resist solution-mindsets[31] toward the "ends–ways–means rationalization." *FM 3–24/MCWP 3–33* in final publication, by declaring that designers can "unravel the problem", shifts military design away from the earlier draft version promoting a "problem-setting" orientation. Setting a problem within a systemic framing will include how various actors including the designers conceptualize what a problem might be, while "unraveling a problem" suggests a reinforcement of the modern military belief system covered in the second chapter. Systematic logic, reverse-engineering, analytical optimization, and formulaic rendering are contained within a Newtonian styling for modern warfare where even complex problems still might be tamed. Militaries speak of complexity in war, yet complex systems feature emergence that is "irreducible … unpredictable or unexplainable, to require novel concepts, and to be holistic".[32] *FM 3–24/MCWP 3–33* presents a complete failure to incorporate the fundamentals of complexity theory, let alone the other more challenging aspects of Naveh's SOD. In complex warfare, such mechanistic, causal logic merely illuminates the mismatch of natural science epistemology to the realities of complex, even chaoplexic warfare.[33]

In the earlier draft version of *FM 3–24/MCWP 3–33,* there was a paragraph that drew inspiration from SOD by introducing "the operational narrative" where designers use storytelling to express design goals to the organization.[34] Indeed, the term "end-state" never appeared in the entire original passage. The final publication would purge much of this and insert clear definitions of "end state" including joint doctrinal citations with a revised statement on how design must occur. For Army and Marine doctrine, "design begins with the identification of the end state".[35] SOD rejected this, as Naveh's expectation of a military engaging within any complex, dynamic system required them to move with a learning system of opportunity to destroy, create, reframe, innovate, and shock adversaries that remain shackled to their routinized beliefs and practices.

There are myriad substitutions of established, mechanistic, existing military thinking into the final publication of *FM 3–24/MCWP 3–33* that removed the bulk of SOD content. "Operational logic" would be removed, replaced with "commander's intent", an already existing military construct detached from SOD's emphasis on leaders crafting unique learning environments that required them to also un-learn and re-learn in novel ways. A leader using operational logic in SOD might move their organization in a range of experimental directions, probing and sense-making as they improvise. "Commander's intent" is entirely different, and is couched into a reductionist, hierarchical mode that reinforces an "ends-ways-means" mentality of reverse-engineered oversimplification in warfare.[36] With SOD, Naveh promoted "initial appreciation" that again ties back to experiential learning theories, yet this term also would be jettisoned from the final publication and replaced with "situational awareness", another standard planning term.[37] The final version also stipulated that the commander must only pursue design after firmly grounding the organization in familiar planning models such as "the factors of METT-TC [Mission, Enemy, Terrain, Troops Available, Time & Civilian Considerations]" with appropriate doctrinal references.[38] Planning would dominate design in this first doctrinal publication.

The elimination of most SOD-centric concepts and terminology continued to the very end of the draft design chapter. While the June 2006 draft version ends the chapter with "enabling a continuous cycle of design-learn-redesign to achieve the campaign's purpose",[39] doctrine writers changed the ending in the final publication to: "enabling a continuous cycle of design-learn-redesign to achieve the end state".[40] These edits should not be overlooked at why and how the American Marines and Army would generate two differing camps on military design, and within their doctrinal publication arms, the pragmatic advocates for design assimilation would win the opening battles in crafting formal military design doctrine, something Naveh largely frowned upon. Despite his misgivings on any standardization of military design, a doctrinal chapter signifies that the two land component forces recognized design as something important and began a process to attempt to describe how the organizations ought to go about designing in warfare. While SOD experimentation continued in isolated classrooms at Fort Leavenworth, the U.S. Army would chart ahead with new efforts to capture the military design narrative.

Army Design Methodology: Design Succumbs to Bureaucratic Resistance

Naveh's teams continued to work with small groups of advanced military students at SAMS and in other capacities while the much larger institutional bureaucracies began formulating ways to standardize some sort of military methodology across the entire force. General William Wallace commanding TRADOC through 2008 saw an expanded demand for design explanation and instruction following publication of the 2006 single chapter of design in a counterinsurgency manual. To move design as a concept into something formally associated with military planning, TRADOC needed to consider how to put design into planning doctrine.

Wallace would direct SAMS to take the conceptual set of design ideas and convert it into some clear model or methodology in the next version of *Field Manual 5–0, Operations Process.* Intense discussions followed at Fort Leavenworth on whether design required a chapter in the new planning manual, its own separate field manual, or perhaps explanation of design across various chapters of *FM 5–0,* which is considered one of the foundational doctrines in the entire Army publication library. Johnson, still actively part of the Army's design experimentation efforts, recalled that Wallace called SAMS and directed the new Director, Colonel Stefan Banach, to work with Army's doctrine writers to create a design doctrine.

This new design doctrinal initiative would expand design from a brief chapter in a counterinsurgency manual to either a full deliberate chapter treatment in *FM 5–0,* or potentially a stand-alone design publication with new formalized standing in TRADOC's doctrinal library. Banach would invigorate design development within the SAMS program with faculty and students while Naveh continued teaching SOD to SAMS seminars. This is where factions emerged within SAMS and across Fort Leavenworth. Different groups wanted control of the design narrative, and everyone understood that formal doctrine spoke loudest. The first contentious question would be one of brevity. Could Naveh's SOD, which required intellectual study equivalent of a multi-disciplinary master's degree, ever be consolidated into a single chapter using clear, institutionally compatible terminology and instructions? This would help define several design writing efforts at Leavenworth during this period and explain how things ended up in the final doctrinal version of *FM 5–0,* published in 2010.

FM 5–0 would arrive in the hands of all major military units and leadership worldwide in 2010 with a slim, fifteen-page chapter on design. The chapter by itself does not tell the story of the fierce battles waged at Fort Leavenworth between 2008 and 2010, nor does it explain much design. The chapter does represent the very best that large, centralized bureaucracies can do, to include forcing one position over all others through positions of authority over all other matters. Despite this ham-fisted approach to design, the many participants and several of the primary draft documents remain to tell the rest of the tale. The final design chapter found in *FM 5–0* would inspire most Army design education, training, and design practice for the following decade, yet the manner in how that design content became doctrine is a storied, and often discouraging account of bureaucratic rigidity, emotional reactions, and anti-intellectualism.

In the two-year period between Wallace directing a new design deliverable for Army doctrine and the publication of *FM 5–0,* multiple proposed design narratives, chapters, and monographs were developed and circulated in various attempts to influence the final design output. Indeed, many efforts reflect a high level of creativity, intellectual effort, and commitment to transforming the Army so that it could better address complexity in warfare. One example is an unpublished, draft design manual that was intended by the authors to become a stand-alone design publication for TRADOC. *Field Manual Interim 5–2: Design (Draft)* would in thirty-nine pages explore design theoretically in many of the earlier Naveh SOD directions.[41] Some SAMS faculty helped write this version, while Banach steered SAMS faculty and students to also produce a much larger *Art of Design: Student Text* manual that was several hundred pages long. Other outside parties would contribute their own design perspectives, yet TRADOC would ultimately reject all these endeavors.

Alex Ryan, SAMS faculty during this period and one of the lead authors of the draft *FMI 5–2* and the *Art of Design: Student Text* project, saw within TRADOC a desire to oversimplify design through some rationalization that anything useful for the entire army could not contain anything difficult to learn. This would immediately cast most of SOD and Naveh's multi-disciplinary approach as off-limits. It would even render the various SAMS design efforts in draft doctrine, monographs, and faculty student guides as useless, since they all continued to provide explanation of what would be exotic, unorthodox ideas entirely alien to the larger Army community. TRADOC wanted any new doctrine on design to recycle all previous planning terminology, share planning concepts, support all existing doctrine, and in turn prevent design from doing what it is intended to do. Chia explains how institutions trap themselves into reinforcing their paradigm in such ways that it "exerts a compulsive force upon their thinking. When a particular conception permeates throughout the thought collective and influences everyday life and idiom, any contradiction, therefore, appears unthinkable and unimaginable".[42] All planning concepts in doctrine must be correct, and therefore any new design content cannot possibly challenge these proven, reliable, valuable ideas, as the army reasoned.

TRADOC wanted incremental progress in advancing existing institutionalized ideas, not disruption or innovation that required thinking about thinking, un-learning, or challenging deeply held beliefs that design brought to the institution.[43] Naveh and his designers had provoked the dragon, and now the dragon would move to protect its lair and treasure. Chia and Holt again provide useful analysis on how this manner of institutionally directed change from the top usually backfires. "The more directly and deliberately a specific strategic change is single-mindedly sought the more likely it is that such calculated actions eventually work to undermine their own initial successes, often with devastating consequences."[44] TRADOC's overarching control of how design would be adapted without harming design would end up creating another planning method that retained merely the title of design.

Ryan recalled how "[leadership within TRADOC in the 2008–2010 period] wanted to simplify the [design] doctrine ... [they] wanted the doctrine to be maximally accessible and so pushed back on any new terms, use of footnotes, etc."[45] Ryan would characterize leadership anti-intellectualism at that period akin to a "we are a blue-collar army" mindset that would be para-doxical to Naveh's insistence on deep intellectual growth and disruption.[46] Ryan, along with multiple leaders and staff officers involved in the design developmental period between 2008 and 2010 point to General Martin Dempsey, then TRADOC commander as the primary force for insisting on any design content becoming oversimplified and assimilated into existing military planning processes. Dempsey would instruct doctrine writers and design advocates to transform the dense, obtuse prose of SOD into something every soldier in the entire army might readily digest and apply.

By 2009, Dempsey would advance the design indoctrination effort beyond earlier efforts by Fastabend and Wallace in the 2004–2008 period for the Army. Dempsey directed Lieutenant General Caldwell, then the U.S. Combined Arms Center (CAC) Commander at Fort Leavenworth, to take the overly elaborate draft design chapters already developed by SAMS faculty and students and simplified it down into a short chapter of fifteen pages. This consisted of different versions of the *FMI 5–2* draft design chapter authored by SAMS faculty, along with several other competing products. Steven Leonard, then a staff officer working special projects for Caldwell at Fort Leavenworth, recalls the tensions during this volatile period for TRADOC:

> I was working as the CIG/CPG director for LTG Bill Caldwell when he was the CAC [Commanding General], probably in the 2009 timeframe when the project landed on my desk via General Dempsey. There had been a fair amount of work done to reduce the document SAMS was using into something more palatable that could introduce design more broadly to the force. Those efforts had only managed to annoy General Dempsey … LTG Caldwell handed it to me and basically told me to translate it into something that would pass "the Dempsey test" … the 15-page document that ended up as a chapter in the 2010 edition of FM 5–0 was the result of my rewriting.[47]

The dense concepts found in Naveh's original security design praxis were now considered unnecessary and too esoteric for most army personnel aside from small groups of advanced students at Fort Leavenworth. Somewhere along the way, institutional preference for simplification of all new ideas would ultimately produce anti-intellectualist agendas, coupled with institu-tional bias in defending all existing concepts to avoid conflict within doc-trine. As the American Army runs on doctrine, the act of publishing new design concepts in doctrine would have long-term impacts. The updated *Field Manual 5–0, Operations Process* published in 2010 would provide the second formal design treatment with a fifteen-page chapter on design for the U.S. Army.

The army's design chapter did not appear to follow traditional TRADOC doctrinal publication processes. Four Star General Officers normally are presented on doctrinal content in formal briefings for decision, guidance, or endorsement at the end of the process. Instead, General Dempsey personally involved himself by checking the grammar, terminology, and logic of this document while restricting draft access to the rest of TRADOC's doctrinal review process, according to individuals directly involved in the design chapter construction.[48] Ultimately, the TRADOC Commander and a Colonel on his staff would craft, edit, and produce the entire design doctrinal content for the U.S. Army in 2010, largely without little engagement with the multiple designers, faculty, and educators at SAMS co-located at Fort Leavenworth, nor with Naveh and his SOD team.[49] This is not how TRADOC's doctrinal production operates normally, and illustrates the high tensions, politics, and emotions in the 2009–2010 period.

Leonard, who ultimately wrote the final design chapter for *FM 5–0,* stated that he struggled to reduce the SAMS *Art of Design: Student Text* project down from 337 pages to anything useful for everyday soldiers that may only have a high school diploma. Advocates of Naveh's SOD, as well as those such as Banach who promoted a hybridization of design that removed much of the postmodern baggage, insisted on trying to keep most of the deep theory within any design doctrinal draft. Leonard battled on these fronts while also remarking that General Dempsey was playing the English teacher role "battling me on semantics and semi-colons".[50] Most of the demanding theory and particularly any concepts outside of established military education and institutional values was purged, which would be most of the complexity theory, and anything even hinting of postmodernism.

Institutional Backlash: The Army Rejects Designers in Ivory Towers

If the 2004–2008 period in the American Army's intellectual areas of experimentation featured a high degree of curiosity, risk, and willingness to explore heretical SOD concepts, the backlash was also forming that would manifest in 2009–2012. As Naveh and SAMS faculty taught design in different ways, with different configurations of content and practice, some SAMS graduates would return to combat, eager to apply lofty design concepts to military commanders that wanted better operational planners. While some designers did make important contributions to military forces by disrupting the institution and fostering innovation, the ones that failed at design would create the loudest noise. Some of the SAMS graduates returned to the force with their new sophisticated design ideas were not understood, or they personally lacked the bridging skills to bring design together with planning in novel formats. Some were overconfident, even arrogant, and this put a black eye on the SAMS program.

Objections began emerging from Division and Corps level headquarters as well as influential senior military leaders on whether these difficult and abstract design concepts were helping or hurting SAMS graduates as they re-entered the force. By 2008–2009, SAMS leadership received numerous complaints from commanders in warzones that the graduates were unable to "accomplish basic tasks such as writing an order … graduates [of SAMS] often thought way too conceptually … some graduates refused to stop admiring the problem".[51] These criticisms were new and came from the highest levels of the Army and often right from the battlefields. SAMS suddenly had an identity crisis of sorts, with the same tension between two design camps within the school now spilling over into the entire army and beyond.

There were accusations that SAMS was producing navel-gazing philosophers and not operational planners, and that the esoteric design ideas could not translate from the whiteboard to the battlefield. Often, philosophical discourse is discouraged for practical thought, despite how practice often only can change with necessary abstraction and theoretical contemplation.[52] This is why designing in military contexts can be dangerous. Benson, as a SAMS director recalled that as early as 2005 at the Unified Quest exercise, some TRADOC senior leaders went up to him and poked him in the chest and shouted, "what the [expletive] are you teaching at SAMS? How can we have a [expletive] wargame when your people are not doing war?!!"[53] Even after leaving SAMS, Benson would continue to get random emails and phone calls from Division Commanders demanding to know "what the heck are you all teaching at SAMS?" Unfortunately for SAMS, the wagons would circle around TRADOC and the army as an institution, with design and anyone associated with a pure, Naveh-inspired version of it outside the perimeter. Timing proved disastrous for the SAMS director most associated with Naveh and design, Colonel Stefan Banach.

In the 2007–2009 period, Colonel Stefan Banach served as the eleventh director of SAMS, and spearheaded several design initiatives with Naveh as well as TRADOC and SAMS cadre. Banach shifted the SAMS curriculum to that of an intense design program atmosphere, with a great infusion of postmodern theory, complexity theory, systems theory, sociology, and extensive design exercises that would cut away at the traditional planning and military history focuses. Banach orchestrated significant design research projects including the draft *FMI 5–2* and the *Art of Design: Student Text* project where SAMS faculty including Alex Ryan and Naveh's SOD educators took lead roles. Banach would seek a new Americanized sort of design that selected aspects of SOD but attempted to be more palatable to TRADOC and orthodox planners insisting on protecting doctrine and planner terminology at all costs. This turned into a lose-lose situation, with SAMS graduates departing the course lacking the operational planning abilities demanded from the field. Instead, they were showing up speaking design in a manner that few could understand.

Design continued to suffer from a reputation of being elitist, in that only the gifted few could rise above ordinary planning and do as Naveh did. Naveh would contribute to this negativity by declaring his design concepts were not for "ordinary mortals".[54] In a TRADOC research report in 2012 on design integration problems, the authors highlighted the cultural perception that "only the 'elite' few can do [design]".[55] War college students outside of Leavenworth would pile on, declaring things such as: "design really makes no improvements on the current planning process … Design is at its basic core an academically arrogant idea".[56] One infamous published critic of Naveh's SOD is by Naval War College Professor Milan Vego, who justified a wholesale rejection of design with his interpretation of why the Israeli military eliminated it from their own force. Vego considered "the entire SOD concept elitist … other [Israeli] officers could not understand why the old system of simple orders and terminology was replaced by one that few could understand".[57] He prophesized that design would go the way of Effects Based Operations and fall away once military attention shifted to the next shiny object.

One matter infrequently addressed on Naveh's dense design concept is whether critics had invested sufficient study or time doing design. Naveh and others would cast blame at senior IDF leadership ignorant of SOD and those unwilling to invest in deep study to learn something outside of the familiar. Each case is unique, and likely many could counter-argue that SOD educators often failed to effectively convey these deep concepts in ways that registered with enough of the audience. In Vego's case, Johnson does recall that the professor attended one SOD seminar activity at SAMS in 2004–2005 when invitations went out to the war colleges. Johnson remembers that

> [Vego] appeared very uncomfortable, even agitated … [he] never went to any of the other SOD lectures that Naveh provided. We had invited him a number of times to sit in on briefings of what design was, from beginning to end … given by Shimon and he never participated in any of those. He departed after one day and never returned.[58]

Whether or not Vego grasped design concepts or had already closed his mind to the radical ideas is beside the point. If just one exposure to SOD could invoke such a reaction from deeply respected, experienced military educators, how could Director Banach and his faculty find a way to translate SOD into something acceptable yet still valuable to the entire U.S. Army?

In 2008–2009, things quickly spiraled into combative engagements across Fort Leavenworth on design, the future of SAMS, doctrinal publications, and bruised egos. Naveh was banished for a second time from a military schoolhouse in less than a decade, this time from SAMS after an argument with SAMS faculty. Ryan, then a newly assigned SAMS faculty member, remembers:

> there were several falling outs based on both personality and ideology. Jim Schneider, Tim Challans [SAMS faculty], and then Hubba Wass de Czege [first SAMS director], Shimon Naveh, and Rick Swain [contractors] severed ties with the design program they had created at SAMS due to disagreements with the SAMS Director [Stefan Banach].[59]

Other SAMS faculty there saw the tension between design purists and pragmatic planners as the proximate cause. Butler-Smith wrote that "[the] discontinuities in SAMS faculty and the incorporation of contractors … created two 'versions' of design, the varied perspectives of SAMS directors … and the lifecycle of senior officers interested in design" would all create tensions in the larger story of how design shaped and was shaped by the U.S. Army.[60]

By 2010, Naveh would only engage with select SAMS students studying design, and only meet covertly in coffee houses and restaurants at Fort Leavenworth away from prying eyes. Banach completed his assignment and was replaced with a new director, Colonel (later Brigadier General) Wayne Grigsby who quickly promoted a different philosophy on how design was to be educated at the school. It would be a "sea change" in style, beliefs, and guidance from higher leadership at TRADOC and encouraged by senior commanders frustrated with SAMS graduate performance over the last few years.[61] The SAMS program would rapidly shift under the Banach–Grigsby transition, with dramatic changes to course curriculum, lessons, source material, assigned readings, as well as the relationship with operational planning for Army officers.[62] Design would be firmly placed as an optional precursor to obligatory planning, and the manner in which army personnel would execute design thinking would be rigidly tied to doctrinal edicts. Naveh would move further out of orbit, and the army would take design into a different direction than the original creator intended.

The Army Plans Design in Detail: The Assimilation into Systematic Problem-Solving

After 2012, the U.S. Army shifted to an updated model for doctrinal publications and would redesignate their manuals to be named "Army technique publications" or ATPs. These would function as the more in-depth, extensive doctrinal products to decrease the size of the original field manuals. The shorter, executive-oriented documents would be called "Army Doctrinal Publications" or ADPs. The 2010 version of *Field Manual 5–0* would become two publications in the future, with *FM 5–0* core content turning into *ADP 5–0*, and the in-depth explanation of topics such as ADM becoming *ATP 5–0.1, Army Design Methodology,* which would be published in 2015. *ATP 5–0.1* advanced the depth of methodological detail, process for designing in the army, and expanded definitions within this publication so that army personnel could formally perform army design activities. The 2019 version of *ADP 5–0* would subsequently recombine *ADP 5–0* and *ADRP 5–0,* but *ATP 5–0.1* would remain since 2015 the current army design doctrinal publication of record.

Since its inception in the 2006–2010 period, the American Army pulled select design concepts from Naveh's original systemic operational design, while also attempting to distance itself from the dense concepts, emphasis on self-framing and eclectic philosophical components. Instead, the Army drew heavily on select terms and models from systems thinking while still retaining most all the legacy decision-making content that continued a Newtonian stylization of warfare into systematic formulas and analytical optimization. Systems thinking and complexity theory reject the Newtonian style and posit a complex, dynamic reality that cannot be simplified into simplistic, "problem-solution" optimized affairs except in isolated, localized contexts. The American army, reflective of the overarching western military paradigm explained in the second chapter, hosts a pragmatic orientation to problem-solving in planning and ADM. Rapoport explains this pragmatic orientation as follows:

> The pragmatic orientation is the problem-solving orientation. When the existing state of affairs does not correspond to a desired state of affairs, the problem-solving attitude turns attention to the concrete nature of the discrepancy. To formulate a problem means to spell out in concrete terms – that is, in terms that can be related to concrete observations – just what it is about the existing state of affairs that distinguishes it from the desired state. One then scans one's memory or experience or store of knowledge to see whether means are available to remove undesirables features or to add desirable ones to what is *given*. If these deletions or additions can be accompanied by manipulating matter, one resorts to technological solutions. If changes are required in relations among people, solutions can be sought in the realm of politics. One takes account of where one *is* and at each step decides where one wants to be next, given the constraints of the situation and the means at one's disposal. The past is examined only to see which means have worked and which have not in (presumably) similar situations. The horizon of the future is limited by how much can be expected to be accomplished starting from where one *is*.[63]

Rapoport essentially describes how the army has created ADM to operate subservient to a broader, institutionalized, pragmatic orientation to warfare that permeates all planning, strategy, and critical self-reflection. The "three ball chart" exists pragmatically where the difference between the existing and desired state of affairs becomes the environmental frame for ADM. Problem formulation follows the concrete observations held by the army's belief system on describing what all warfare consists of, and ADM doctrine stipulates that planning teams refer back into established historical record and shared institutional knowledge (doctrine) to use "ways" and "means" within institutionally approved manners that depict objectively oriented formulas for decisions and actions. Only historically validated, observed, and scientifically rendered concepts are approved for ADM planning teams to consider, and the operational approach in ADM positions the current environment, declared problem-solution dynamic, and the way forward to a single future ends in a linear-causal, mechanistic process to follow.

Below, the "Army Design Methodology" graphical depiction is how the Army seeks soldiers to conceptualize design for military activities. First depicted in an earlier but similar graphic in the draft design document *FMI 5–2* in 2009, the "three ball chart" would also be depicted in another close approximation in the design article written by Banach and Ryan in 2009 that was published in *Military Review*.[64] Both are placed side-by-side below. Aside from repositioning each of the balls differently from the Banach and Ryan article, *FM 5–0* in 2010 would replicate the graphic and virtually all the design contents from the Banach and Ryan-published article.[65] Interestingly, the "three ball chart" would be eliminated in the 2015 *ATP 5–0.1*, yet the underlying concepts and how ADM is expected to be executed would remain unchanged from this first brief chapter's framework established in 2010 (Figure 4.1).

Perhaps unintentionally, the geometric composition of square shapes and design relationships promoted in Naveh's first version of SOD praxis would continue into the eventual spherical and linear depiction used by the U.S. Army. This could relate to Ryan's and Banach's extensive exposure to Naveh's first SOD methodology and how he conceptualized it as covered in the previous chapter. Or one might chalk up the popularity of basic geometric shapes for conceptualizing warfare as an institutionalized preference dating back popular military works such as Sébastien Le Prestre de Vauban's geometric treatment of siege warfare, artillery, and engineering operations.[66] The Newtonian style of thinking first addressed in the second chapter is expressed by the army with these simple, static, and mathematically proportioned models. Army planners can extend all existing planning logic that remains Newtonian upon the new design construct without any disruption. Tsoukas provides additional detail on why such a logic is appealing in the following:

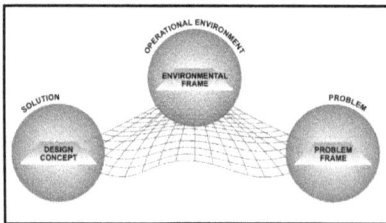

From the Banach, Ryan "The Art of Design: A Design Methodology", Military Review Article, 2009.

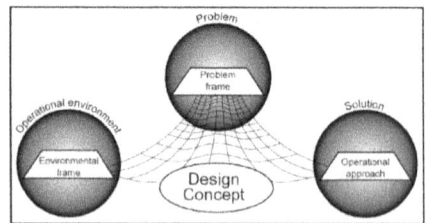

From the U.S. Army's *Field Manual 5-0*, Chapter 3 (2010) final design doctrinal chapter.

Figure 4.1 First and Second "Three Ball Chart" Configurations of Army Design. Source: US Military (Public Domain).

The Newtonian style of thinking operates by constructing an idealized world in the form of an abstract model, in order to approximate the complex behavior of real objects. For example, Newton's laws of motions describe the behavior of bodies in a frictionless vacuum – a mathematically handy approximation, good enough for several real-life occasions. Moreover, the core of the Newtonian style consists of two assumptions. First, the extremal principle; namely, that the objects of study behave in such a way as to optimize the values of certain variables. And, second, prediction is possible by abstracting causal relations from the path-dependence of history.[67]

Newtonian styled warfare is easier to understand, making it ideal for large, technologically oriented forces such as the American Army. Even the Israeli military would battle with Naveh oversimplification of SOD into something more appealing, such as a Newtonian treatment. Naveh's original SOD graphic suggests some acquiescing to military demands in this vein, as his other draft graphics and briefing slides become incomprehensible in their dense compositions. That the Army would build upon early SOD simplified geometric constructs is in keeping with the 2004–2009 period of intense design experimentation between Israeli and American designers. How the three-ball graphic went from Ryan and Banach's published article into the 2010 design chapter in *FM 5–0* is easier to answer. Leonard, tasked to write the design chapter and translate some of the earlier SAMS draft work, would simply take the graphic right from the recently published article and insert it along with much of the underlying content.[68] There is little question on where the Army's design methodology and this first influential graphic originate from, but there does remain the question of how Ryan, Banach, and SAMS contributing faculty moved from Naveh's SOD into what became ADM between 2004 and 2010.

Bob Johnson offers one tantalizing clue, although there continue to be many arguments on who deserves what credit in the arrival of Army design, with significant disagreement across myriad personal accounts. Johnson, while studying SOD at OTRI in Israel with Naveh and his original design team, produced several graphical conceptualizations of how the Army might adapt SOD in some readily digestible way. Johnson states he sought an "Americanized version of Naveh's ideas" and presented this to TRADOC staff.[69] Unpublished until now, the below graphic suggests that several of the American officers working with Naveh throughout the 2004–2008 period, in Israel and in America, might claim some parental status to what ultimately became the three-ball chart and how the U.S. Army conceptualizes military design activities (Figure 4.2).

Johnson's original design graphic shown in Figure 4.2 suggests that Naveh and his designers, in conjunction with American students and faculty, had certainly discussed various ways to translate the difficult SOD methodology into something more amenable to the American military institution. Johnson's

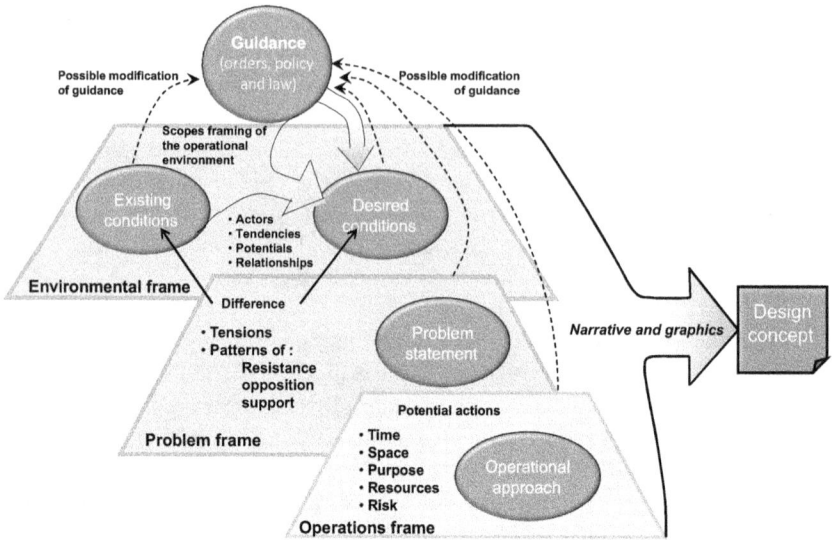

Figure 4.2 TRADOC Prototype "Americanized Design" Concept from 2004.
Source: Original graphic from Bob Johnson.

graphic contains the same three frames that would become the ADM meth-odology, complete with many of the design concepts that would later define *FM 5–0*'s design chapter, and the subsequent 2015 stand–alone design pub-lication. One common theme throughout the Army's two decades of design experimentation and indoctrination efforts is emphasized in contrasting this early draft graphic with the three-ball chart one in 2010. The U.S. Army insists on simplification, clear definitions, and absolute convergence of any new concepts to the established military decision-making methodology, ter-minology, models, and theory. When able, the Army reduces, standardizes, and eliminates dynamic, paradoxical, or otherwise challenging content. In other words, prioritization of Newtonian stylings subsequently blocks any incursion by new disciplines such as complexity theory, systems thinking, or postmodern ideas. Select terms and models may be assimilated, but only in complete adherence to a singular, mechanistic manner of understanding war.

Between the 2004 first stages of American design experimentation and the formal production of a stand–alone design doctrine by the U.S. Army in 2015, there appears to be a consistent pattern juxtaposed between the 2004 Johnson draft design graphic and the design procedural narratives in *ATP 5–0.1*. This "Americanized version of Naveh's ideas" as Johnson framed it can be stated as Army Design Methodology or ADM for the rest of this chapter, and unless specified applies to most of the developmental period of design experimentation. The Army would move in a direction away from SOD, but collectively move as a mostly cohesive group, despite individual differences and nuances. Returning to Ryan's analogy of two design camps, the Naveh purists would be excluded (or recuse themselves) from how the pragmatic planning camp would create this Army design. The only design camp able to shape ADM effectively would be those that maintained the shared conceptualization found in the three-ball chart and other close variations. This also would bleed over into how the U.S. Marine Corps would craft design doctrine, covered in the next chapter.

Naveh created SOD as a complete alternative way to frame how an organization understands war, freeing it to reconceptualize how to pursue warfare in dissimilar, emergent, and entirely innovative future expressions. Naveh's design would inspire novel planning, but it exists above and beyond the limits of modern military planning or institutional strategies. Otherwise, without this philosophical place of origin outside of military decision-making methods paired to select theories and models, Naveh's SOD practitioners *would be unable to consider beyond* the dominant institutionalized war frame. This is where ADM differs from SOD. The American adaptation of SOD would reverse this configuration, forcing all design into a subordinate role that still must support overarching planning methods, nested in a single, unquestioned war paradigm. While SOD had no real limits, ADM practitioners would not be allowed to examine beyond the declared problem that the commander or a higher authority directed. SOD could question the military methods, conceptual models, underlying theories supporting them, and the belief system of the military institution that produced all ontological and epistemological frameworks therein. SOD could fundamentally transform an entire military, whereas the American army suggested that in unusual contexts, the organization might delay planning activities briefly to consider designing so that established plans might work more effectively.

In the first chapter of *ATP 5–0.1,* Army designers are reminded of existing doctrinal concepts such as "Mission Command", the operational planning process, and how design can be used in complex, ill-defined situations *provided they did not violate any of these pre-existing doctrinal constructs.* ADM existed to extend the imagination of the army unit far enough to consider previously unrealized opportunities, but everything designed would be done so in clear planning activities complete with a shared lexicon, complete with most models found in other parts of doctrine re-applied within design activities. Whereas SOD existed to shatter institutional barriers and unproductive illusions about warfighting, ADM would exist to protect the army's sacred cows, reinforce a particular war paradigm above any others, and clip the wings of any critical thinking that might provoke institutional discomfort. Tsoukas and Vladimirou summarize this organizational knowledge challenge as follows:

> Our understanding of organizational knowledge will not advance if we resign ourselves to merely recycling commonsensical notions of

knowledge for, if we were to do so, we would risk being prisoners of our own unchallenged assumptions, incapable of advancing our learning.[70]

ADM exists to protect the institution, which functions as a centralized hierarchy in that modern militaries are products of the Industrial Revolution and the Age of Enlightenment, but also hark back to the premodern origins covered in the second chapter. ADM doctrine declares: "Conceptual planning corresponds to the art of command and is the focus of the commander with staff support".[71] Naveh too would place the General Officer as the primary designer, but he would also correlate the high rank and authority of a senior military designer with the ability to reform the organization, defy the institution, and chart entirely novel courses away from the legacy system of warfighting as they wished. SOD remained outside of military doctrine and any formalized decision-making or planning processes because Naveh saw the dangers of institutional overreach; the desire to control and preserve can easily drive an institution to regulate activities such as creative and critical thinking, which negates any true manifestation of except in institutionally conservative ways.

Army design doctrine would offer "there is no one-way or prescribed set of steps to employ ADM".[72] Yet this term appears isolated from the many pages of prescriptive design directions within *ADP 5–0.1* that force designers to use existing planning concepts, terminology, and orient design toward the same reverse-engineered, systematic logic that permeates all army processes. There can be a variety of institutionally compliant ways to perform design, if indeed there is even a requirement to do so without moving immediately into deliberate planning activities. Any design work would need to conform to preexisting planning requirements, further divorcing ADM from the original SOD intent. SOD could explore, clarify, challenge, and potentially change the epistemological stances of the military institution concerning warfare and even war itself. ADM would only be permitted to enhance existing planning efforts in compliance with unquestioned epistemological frameworks that formed the foundation to how the American Army understood itself (form) and how it existed to function.

For ADM, problems exist to be identified, analyzed, paired with an acceptable and institutionally validated solution, and then solved to advance toward a pre-configured future end-state. Yet this systematic logic of "problem-solution" is best associated with simple, not complex systems.[73] Ackoff would explain how complex systems demand designers to work not toward solution, but *dissolution* of a perceived "problem" where the design of an emergent, new system configuration means that what is thought of as problematic now is dissolved, yet simultaneously the emergent system brings forward previously unimagined, even more challenging problems through the design itself.[74] Naveh oriented SOD activities toward systemic dissolution, hence the "systemic" and "operational" meanings in SOD's name. ADM would associate all possible problems, including ill-structured or "wicked" problems with some future solution that would unlock progress toward strategic goals or objectives.

Easier problems required only sufficiently sound planning activities, and for the unusual or difficult matters, a commander could direct a planning team to design creatively, still oriented in a "problem-solution" configuration. Rittel would add to Ackoff's expansion of how humans conceptualize aspects of complex reality as "problems" by asserting that organizations terminate engagement with a problem for only three conditions. People run out of time, resources (money), or patience.[75] Stone confirms this from a policy perspective where politically (and by extension, war) problems are never "solved", and are ever-changing, never-ending affairs.[76] Problem solving is systematic thinking, whereas designing with complex, never-ending problems requires systemic logic. This abstract level is rarely accessible to designers seeking some conforming "problem-solution" systematic formulation of institutionalized behaviors. ADM defines a design problem so rigidly that, unlike in SOD, a design team is compelled into convergent processes that validate existing institutional norms in warfare. Planning assimilates design in the U.S. Army, so that all possible manifestations of design must occur within clear planning rules and beliefs.

Continuing with the epistemological decision that "design" in U.S. Army contexts is a subordinate step to an all planning and decision-making endeavors, *ATP 5–0.1* refers to the formation of designers to practice design not as "design teams" as used in most all other design communities, but as planning teams.[77] The planning team, directed to execute design praxis, is epistemologically oriented by the ATP doctrine to be problem-solution oriented, and additionally focus on an environment within which the projected design problem is understood to be existing.[78] Environments are held in an external framing where the army organization and individuals performing design or planning exist outside of the environment being examined. Complexity theory and systems thinking place the observer within any such environment, in that a complex, dynamic system learns and responds as one senses and probes within the shared system. At an epistemological level, the army declares any environment requiring military critical analysis as something to be isolated, such as how natural scientists conduct experiments in a laboratory in controlled conditions. Yet design is not accomplished through analytical means. Schön and Rein offer: "design, whether of social service systems or social policy, is not a process of problem solving governed by criteria of technical-rational analysis".[79] Designers cannot remove themselves systematically, in that as Naveh titled his military design with "systemic" explicitly because designers are themselves part of the environment.

ADM's environmental frame becomes an act of analytical optimization, systematic reduction into formulas of inputs and outputs, with an assumption that even complex systems can be flash frozen, isolated, reduced, and then re-assembled with new understanding to gain some sort of enhanced control.[80] Modern militaries as technical rationalists believe that adaptations of natural science concepts should carry with them the ability to render the subjective into something seemingly objective.[81]

Designers applying ADM are directed to frame an "operational environment" by anticipating what the desired end-state should look like, and model
"the natural tendency of the operational environment" by constructing
models of desired future states of other actors as points of comparison with
this desired end-state.[82] "Natural tendency" correlates to a nature of war,
reflecting a Clausewitzian inspired single war theory on war reflecting the
natural sciences, where Clausewitz would normatively define in a Newtonian
styling of "fiction" preventing what would otherwise be a system in sound,
predictable order. Rapoport summarizes how Clausewitz associates war not
with dynamic complexity but with humans capable of random, irrational, or
illogical behaviors messing up the otherwise smooth, universal tenets of war:

> There are two ways of viewing a model of anything: as a description
> of its "essence," i.e., of what it would be like if this "essence" were
> not subjected to disturbances and fluctuations; or as a description
> of what the thing modeled *ought* to be like, an ideal to live up to ...
> Clausewitz describes war as it would be if it were not for what he calls
> "friction," circumstances that prevent war from being what it "really is"
> in its essence. Clausewitz tries to stay on that level of idealized descrip-
> tion. Nevertheless, he is so fascinating by his supposed discovery of the
> "essence" of war that he slips into normative language, maintaining that
> war *ought* to be bloody.[83]

The Army, nested in such a war paradigm, insists that design must follow
a complimentary path where complexity in war is rationalized into a hard
science outlook, drawing again from Clausewitz and "friction" preventing
war from unfolding toward a foregone (analytical) conclusion according to
the inner logic of an assumed "nature" of war.

> Just as in the absence of friction physical bodies would move in strict
> accordance with the idealized mathematical models of theoretical
> [Newtonian] mechanics, so a war would develop in accordance with its
> "true nature", if it were not for human foibles such as failure to grasp
> opportunities, lack of will, considerations extraneous to the true war
> aims, or war weariness.[84]

ADM exists to continue this overarching army belief that war must be
rationalized into a mimicry of the natural sciences, with select consideration of social irrationality and uncertainty produced by humans within the
war environment. Within the war paradigm that the Army applies design
and planning, the ADM terminology remains physics-oriented, with a pro-
liferation of Newtonian metaphors, classical mechanics epistemology, as
well as engineering language and analytic constructs supporting objective
expectations about reality. Design directives such as "the planning team also
models the future *natural tendency* of the operational *environment*"[85] illustrate

this fusion of biological as well as physics-based concepts set within a Newtonian mindset.

ATP 5–0.1 moves through a sequence of design frames of the operational environment, isolating the problem (or problems), and then framing solutions with an optional iterative step for reframing. The operational environment itself differs from how Naveh and his designers articulated "operational" in Hebrew as explained in the previous chapter. The environment is treated in a rationalized reductionist approach to freezing complex systems to analytically optimize frozen portions of them at assumed foundational levels so that universal rules and patterns might be extracted. Available solutions known to the Army and usually tightly bonded to institutional self-relevance are shopped around to the identified problem sets. New problems, if discovered, are forced into clear definitions that previously existed so that institutionally preferred solutions (tied to organizational identity, purpose, belief system, and values) are used in a manner that validates an institution's overarching narrative on what it is, and how it must remain relevant to national security affairs.[86]

Reframing represents one aspect of iterative, divergent design praxis that extends from Naveh's original SOD, but ADM does not consider framing or reframing in the original Israeli mode of creative destruction.[87] In Army planning where design is positioned as a subordinate optional endeavor, staffs attempt to "help the commander determine progress towards attaining the desired end state, achieving objectives, and performing tasks."[88] Simple systems present a closed dynamic where one optimal or "best" solution exists, and once discovered, can be reliably repeated consistently to solve simple problems universally, anywhere, and forever. Complicated systems resist optimal solutions but do contain a range of sufficient solutions where "good enough" gains a contextual quality that might still be analyzed quantitatively, but also qualitatively. Complex and chaotic systems are ever-changing, perpetually shifting into new forms where previous solutions are obsolete or even counterproductive. Navigating a maze is difficult, but attempting this within a maze that reconfigured itself differently with every step would seem impossible, and also reflect complexity, chaos, and dynamic system behaviors.[89]

Further challenging to an army desiring systematically convenient dynamics in warfare, complexity expresses nonlinear and emergent developments that cannot be conceptualized or well understood in any sort of logic *prior* to the systemic drift toward that change. For militaries, warfare is composed of many interrelated systems where many indeed are simplistic or complicated, and this is where detailed planning, checklists, and best practices render superior results. Yet complex warfare systemically *cannot be broken down in advance* through any reverse-engineering, mechanistic, or systematic manner of preconfiguring ends to ways and means as Army decision-making doctrine directs. Any "end state" or objectives cast into a dynamic, complex future will be acts of pure fantasy, only appearing rational in how a military retains a single war paradigm that expects commanders and staffs to continue to operate as technical rationalists.

In real complexity, commanders, and their staffs (and their higher head-quarters) may have entirely wrong assumptions, the wrong assessment criteria, wildly irrelevant goals, and end-states, and are forced to conceptualize through these challenges using a single war paradigm that prevents any critical or creative reflection that threatens the institutional frame for war. Assessment in the U.S. Army's mode of design appears as a convergent cognitive practice, *not a divergent one*. Instead of expanding an ever-widening range of possible future opportunities, the assessment mode perpetually reinforces quantification and readjustments oriented exclusively on the solitary, pre-established "end" as established originally. Reframing in ADM becomes a perpetual act of institutional conformity, with each cycle of design activity fixed to a doctrinal process that retains all of the institutionalized components for planning built into any possible design framing effort. Reframing does not stimulate innovation, disruption, or creative experimentation. ADM reframes to converge design outputs that correlate with the desired planning conditions envisioned by the commander (or upward within the centralized hierarchy) before the design started.

Potentially, designers might have the wrong assumptions, wrong assessment criteria, wrong end-states, as well as an inability to consider alternative paradigms that employ entirely dissimilar logics on the same tangible, observable activities within the complex system. Naveh's original reframe praxis in systemic operational design was destructive to render the ability to self-assess and create.[90] SOD would challenge predetermined "ends" in military conceptualization so that designers could iteratively generate increasingly different designs of the entire system, including themselves and adversaries, so that novel opportunity unrealized by both is illuminated for new action. ADM is configured so that the desired military goal or "end state" articulated before any planning (or optionally, design) is started. Missions are tasked to units, and commanders begin with their initial intent. ADM does offer the possibility that a design approach may transform one end state or objective into something different, but only if this falls within the Army's mission command philosophy where the overarching command intent remains unchanged. New "ends" must correspond in the form and function of how the army realizes all possible "ends", rendering any design innovation as subservient to the institutional war paradigm upheld by the American army.[91] Schön and Rein offer:

> by focusing our attention on different facts and by interpreting the same facts in different ways, we have a remarkable ability ... to dismiss the evidence adduced by our antagonists. We display an astonishing virtuosity in "patching" our arguments so as to assimilate counterevidence and refute countervailing arguments.[92]

The army declares war as such, and through doctrinal convergence, insists on one manner to conceptualize all possible warfighting understanding.

Thus, if a new end state is designed as depicted in Figure 4.3, designers using ADM will repeat the planning cycle into deliberate action toward this new end state, if it is accepted by the command. The system is frozen, analysis is conducted upon sections of the larger, dynamic system, and once systematic logic renders a formulation of existing inputs and outputs of institutionally validated warfighting actions, ADM prepares the operational approach so that it nests with all aspects of the encompassing operational beliefs on how warfare is done properly. Army designers are forced to generate institutionally recognizable assessments and use all existing terminology, conceptual models, and rigidly obey things such as the elements of operational art and other unquestionable tenets of the modern military paradigm.

By 2022, the Army would change some of their ADM in the latest publication of FM 5–0 (also with a return to titling these publications as Field Manuals) once more. The army would distinguish a chapter on "army problem solving" and another chapter on "army design methodology" where both are processes pursued by planning teams.[93] In the design chapter, planners are still directed to "solve problems" that may require "new or original approaches", yet the framework for designers to pursue such novelty remains tightly wedded to a reductionist, Newtonian styled formulation of analytical, deductive reasoning.[94] Systematic logic once again demonstrated quite literally in the 2022 doctrine. Taking the "systems thinking" concepts and exact graphic from the 2015 version of *ATP 5–0.1*,[95] systems thinking for the army is depicted with different systems overlapping across system boundaries that are considered permeable, and "allows exchange of inputs and outputs".[96] Planning is directed to occur immediately, and only if the problems "are intuitively hard to identify [and assign known solutions against] or an operation's end state is unclear, commanders *may initiate* ADM before their headquarters engages in detailed planning [emphasis added]."[97] The doctrine also deliberately calls the design team a planning team, mentioning that planning teams may sometimes be referred to as a design team, and that planning teams and design teams might occur concurrently to inform one another. By 2022, the army would firmly entrench design as all but an indulgent affair for organizations that might consider innovative approaches but remain largely focused on convergent planning actions to systematically pair known solutions to what seem to be familiar problem sets.

Much of the 2022 design methodology remains entirely unchanged from 2010 to 2015, despite subtle insertions on narrative construction, team diversity, and different graphical depictions. The "three ball chart" is absent visually, but every aspect of it remains within the foundational terminology and direction on how ADM occurs procedurally. Further, the 2022 version of FM 5–0 places enemy and friendly center-of-gravity analysis within the design chapter as a required step in design, once again imposing a Newtonian

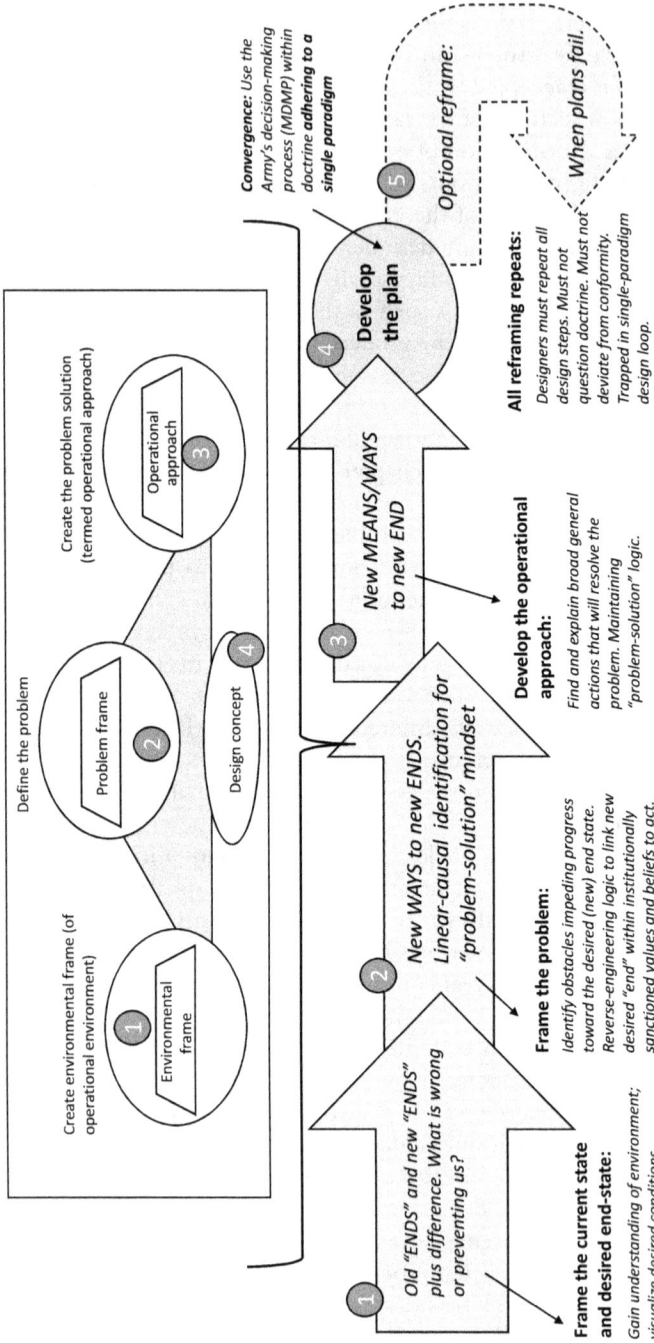

Figure 4.3 Epistemological Framing of Army Design (2008–2022).
Source: Author's Creation.

styled, reverse-engineering approach within an "ends-ways-means" technical rationalization of complex warfare. FM 5–0's design content in 2022 reflects a stagnation of design thought in army doctrine spanning nearly a decade, despite forces continuously engaging in complex security contexts such as Afghanistan, great power competition, and global counter-terrorism activities that do not share much with conventional land warfare of generations past.

Illustrated above, the "three ball chart" conceptualization of ADM that remains unchanged within the design instructional narrative through the most recent Field Manual 5–0 published in 2022, despite the elimination of the chart itself. The fourth chapter of that manual is an abbreviated summary for how the army will perform design activities and reproduces all the previous ADM content found in earlier doctrine and concept papers. Below the methodology, the author provides one possible way to epistemologically map what ADM normatively declares their design process is, how they go about doing it, and *why* the army insists innovation, critical and creative thinking must function in this particularly rigid form.

In the three epistemological arrows illustrated above, planning teams attempting ADM first seek to frame the original "desired end state" based upon the commander's initial vision or guidance. Design occurs only when a commander directs for such activities to be done, and only so that they lead to subsequent army planning (or in tandem). Thus, the commander already believes they have some sort of problem, and ADM pairs complex or unfamiliar problems with the purpose for attempting design. ADM begins with a preconfigured, albeit likely an abstract framing of "problem-solution" that is preventing the unit from moving onward toward their military goals. The obsolete "ends" is contrasted with a framing of the environment so that the core question of "what has changed, and what is preventing us from advancing to our goals" can formulate the challenges at hand. This in turn stimulates formation of some new "ends" where the problem–solution dynamic might be resolved using critical and/or creative options generated by the design team. The environmental framing presents both the present (legacy "ends" prior to design challenge realization by the commander) plus an emergent "new end" that matches institutionally accepted military concepts and patterns of behavior.

Desired conditions are factored into how ADM users approach framing the environment, normally through declared planning concepts used without design. ADM directs designers from the start to converge toward planning constructs, models, and theories such as "operational art" "decisive points", using Army "red teaming" concepts, and "centers of gravity analysis" that form both the planning activities for the army as well as design.[98] Army planners can then try to identify what difference exists between the current environmental frame and that of the future desired one so that a problem frame might be developed next. The ADM inquiry remains "ends" oriented, in that a single desired end state is the central requirement of this first step of designing the environmental frame. Yet the rational, pragmatic way the army conceptualizes any possible "ends" is perpetually reverse-engineered, where all activities must commence with some declared and future end.[99] Yet this runs against how complexity emerges in ways that reject engineering into the future. March and Olsen, cited by Schön and Rein offer:

> it is difficult to describe a decision, problem solution, or innovation with precision [as it comes into being], to say when it was adapted, and to treat the process as having an ending ... change develops meaning through the process by which it occurs.[100]

Israeli SOD explored disruptively without declared ends in mind, yet ADM would subscribe to the modern planning mode of technical rationalization through reverse-engineered "ends-ways-means" formulation.[101]

The second arrow advances the ADM process where the new "ways" are considered to achieve the desired future "end" by *solving the design problem*. ADM designers are expected to identify any conceptual or tangible obstacles impeding progress toward the organization's military goals so that they can reverse-engineer back from the future singular military goal and employ all institutionally sanctioned design methodologies and tools (such as elements of operational art) to systematically formulate a linear-causal path of action. ADM promotes what Schön and Rein criticize the rationalized logic within these arrangements.

> Practitioners tend to assume that the factors essential to the goals they pursue lie at least partly within their control. With their taken-for-granted assumptions, they tend to ignore the factors that lie beyond their control and the shifts of context that may distort the hoped-for outcomes of deliberate action.[102]

This problem–solution linkage underpins how ADM defines the "problem frame" and sequentially can move to the "operational approach" third step using technical rationalism on how modern warfare is supposed to unfold in time and space.

The third arrow links the "new means/ways" configuration to the new desired "end" where a "problem–solution" orientation maintains a sequential linkage of inputs to desired outputs (systematic logic instead of systemic thinking such as in Naveh's SOD). Planners using ADM conceptualize a movement the organization forwards from the current environmental frame into the desired future state by executing a proposed design activity, yet this also must be transferred first into deliberate planning activities that further reinforce institutionally directed actions and behaviors within the war paradigm. The army demonstrates a *positivist* epistemology on how complex warfare occurs with planners either designing new ways to accomplish objectives or using known and historically validated techniques that worked before in earlier conflicts through deliberate planning.

The positivist epistemology, explained by Hatch,

> assumes you can discover the truth about phenomena through application of the scientific method. Acceptable knowledge is generated by developing hypotheses and propositions on the basis of theory, and then testing these by gathering and analyzing data that allows you to compare the implications of your theory to external reality.[103]

The army asserts the modernist, rational perspective that war, even complex war requiring design, correlates a stable reality existing outside of human existence, where knowledge is verified through observation of those things. This independent observation quality means that no matter who observes an object, the knowledge they will generate will be objective, validated through natural science methods, and applicable universally. ADM channels "known and proven" planning beliefs toward a complex reality that still must obey an assumed static natural order of war, where characteristics of warfare might shift, but core laws and tenets remain objectively grounded so that future end states must relate to earlier "ends-ways-means" observations in past conflicts.

ADM nests all design praxis within an overarching requirement to move onto developing a plan; all design exists to reinforce existing planning activities, meaning that designers will be prevented from framing how and why the army forms its paradigm to direct such planning as this would challenge and disrupt ADM's explicit purpose in the subordinate (and optional) role to planning. This is the fourth step illustrated and involves design convergence toward all ADM outputs functioning to enable enhanced military decision and action. ADM as a design process is linear, mechanistic, nested in existing legacy constructs and doctrinal practices, and relies upon overarching single-paradigm beliefs and models that essentially strip design teams of any divergent, reflective, or disruptive qualities that challenge the institution. Doctrine is obeyed, rules are followed, and the army's belief system underpinning all theories, models, and methods are off limits to ADM activities without severe risk to the practitioners.

Thus, in the above epistemological framing of ADM, designers must toe the line to all Army doctrinal practice to even accomplish the goal of performing design for the army. One can critique the performance of a military unit doing the design or plan, but the logic underpinning the design/planning methodology itself is unassailable. *ATP 5–0.1* (reinforced once more in 2022 by the new *FM 5–0*) devotes an entire chapter on explaining to would-be designers that they must use existing military planning models and concepts. This is a continuation of the same pattern of forcing designers to exclusively use existing and institutionalized tools from planning seen in earlier army doctrinal publications. How these planning tools would be driven into all army design methodologies is revealed by some of the early design theorists inside the army's effort to institute some design content for training and education.

Johnson, playing an influential early role in how an "Americanized version" of Naveh's SOD might materialize, sought to resist TRADOC efforts to mandate planning techniques as part of design. "Centers of gravity" would be one of several planning models that Johnson would attempt to fight off without success. "I am more convinced than ever," Johnson stated in a recent interview,

> that the whole notion of "centers of gravity" is something that makes us feel good about it, but it has absolutely paid no dividend in terms of what we have been able to achieve within any of our operational efforts.[104]

Johnson would try to argue with the army, and later with the Joint Forces Command design writers also attempting to craft design doctrine for Joint Forces. The doctrine writers attempted to include COG models into design at the Joint Staff College for a project that would end up becoming Joint doctrine on design, despite these concepts existing independently of military design and created for operational planning activities. "I fought with them pretty significantly when I was leading that writing effort … that COGs had no place [in design]".[105] This pattern of militaries inserting already existent planning or strategic models, theories, and terminology (plus associated metaphoric devices therein) into design formation illustrates at a deeper level the institution's paradigm prioritizing their belief system and values over that of a demand for innovation, disruptive critical and creative thinking in complex warfare, Naveh and others would argue.

Despite his efforts and those of Naveh and proponents of SOD praxis, the Army, and Marines as well as Joint Forces Command, would continue to nest COGs and other classical operational art models into all design activities. Favorite institutional models would remain entrenched, along with classical mechanics language and metaphors used in detailed planning activities. Few of these concepts have a commercial or civilian equivalent, with "center of gravity" itself being a Newtonian metaphor adapted from the theoretical work of Clausewitz and others. Sidestepping the debate on whether COGs are useful or not for operational planning or strategic thinking, the primary concern here is whether these tools must be forced into military design activities also. Army designers by ADM methodology and doctrine are ordered *to use specific cognitive tools exclusively* in their endeavor to attempt to think differently, despite those very same cognitive tools already heavily used in traditional planning activities where little design is warranted (or desired).

ATP 5–0.1 applies an implicit teleological frame (a purpose-driven explanation of phenomena) on the conduct of conceptual planning and detailed planning over time in warfare. All war demonstrates, in the army's preferred paradigm, this broad relationship between conceptual and detailed planning that unfolds as below, based not on the unique contexts within complexity as design is intended,[106] but to normatively validate the rationalized, pragmatic way that war is expected to be using the dominant institutional frame.[107] This also establishes where design is supposed to function, and how detailed planning remains the dominant, causal factor in warfighting. In Figure 4.4, it depicts a gradual shift from conceptual planning (ADM) toward greater detailed planning where the commander and his staff employ the military decision-making process. This curve implies that initially in complex military situations, the commander and his staff need to invest in conceptual planning more, to gain a better understanding of the system behaviors. Subsequently, as time passes and the organization tests hypotheses and conducts conceptual planning iterations, they will "reframe the operational environment and the problem" to move progressively toward the desired end state. This illustrates the tension of denying alternative perspectives concerning reality and war so that the modern war paradigm automatically covers all conflict, everywhere, always (via a universal ontology). See Figure 4.4 taken from *ATP 5–0.1.*

There are a series of significant epistemological efforts depicted by the Army in this figure that help further frame why the U.S. Army considers design as it does, and how it directs planning teams and units to approach all complex, ill-structured, or confusing security challenges. The element of "time" is significant in the above figure because the implication that over time and the progressive accumulation of knowledge and experience, the

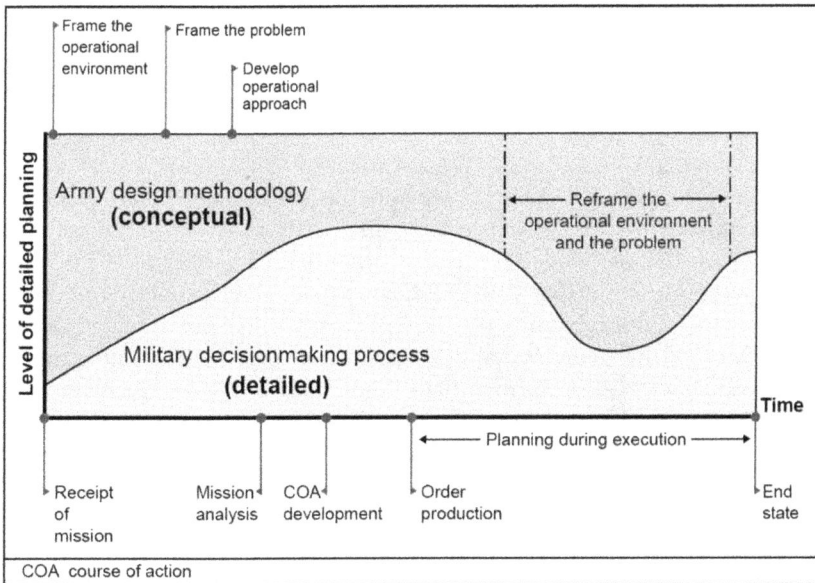

Figure 4.4 How the Army "Integrates" Design with Planning.
Source: US DoD (Public Domain).

organization will gain increased understanding, control, and prediction over a complex system. Yet systems theorists decry this assumption; no collection of knowledge permits greater prediction of a complex system due to how they behave, with dynamic complex systems interacting and learning so that even if the organization gained more knowledge, that knowledge may be entirely irrelevant in the changed system.[108] Naveh's SOD emphasized a per-petual cycle of "learning about learning" to encourage design teams to avoid this pitfall. Complex systems do not become static, nor will a military force become able to manage complexity and reduce it into something more sim-plified by gaining a design insight that is not already susceptible to change, drift, flux, or system adaptation.

Complex reality will not give any indication of linear causality, nor will most "end states" work or strategic goals be apparent except when within immediate reach.[109] Militaries confuse tactical, localized, or simplistic ends accomplishment with an assumption that tactical phenomena correlate at higher scales to what militaries consider operational and strategic activ-ities.[110] Furthermore, "end-state" is considered an illusion in many regards due to the important aspect of "emergence" in complex systems.[111] Original goals are abandoned by organizations, novelty occurs, and the subsequent goal that manifests later could never have been anticipated earlier.[112] Lastly, the above figure again places the Army's design as a subordinate element of the overarching planning methodology, where only at the beginning and potentially during necessary reframing phases would design be of military value. Suppose the graphic were inverted, so that as time passed on, the organization *might need to increase their conceptual planning through design.* This is never considered within ADM doctrine, as it would contradict how the army correlates abstract design thinking as an optional step in what is for-ever a technically rational process of engineering-oriented, Newtonian styled decision-making.

ATP 5–0.1 places the Army commander as central to the design effort, yet then enforces the same coordination and participation requirements and stipulations as that of detailed planning methodologies. Although *ATP 5–0.1* makes sweeping statements such as "Commanders are the most important participants in the operations process … commanders are integral to any ADM effort",[113] the Army design doctrine then prioritizes that commander's time as valuable enough to potentially limit their active involvement with design. Martin and others that have employed ADM in combat contexts report back that indeed, many senior commanders spare little or no time at all for design engagements, treating the process as an extension of traditional military planning with formal decision-briefings as the lone engagement period for designers and their sponsor.[114]

While this is likely true due to the institutional barriers to many commanders having the freedom and even the intellectual curiosity to pursue design efforts at the detriment of other pressing organizational demands, *ATP 5–0.1* continues in its second chapter to render design coordination requirements the same as planning ones. The Army commander may choose to actively lead a design team, or merely receive periodic updates and briefings, or send their surrogate instead to act on their behalf, or potentially incorporate the designers into their larger operational planning presentation instead. In practice, many Army commanders devote identical amounts of time to their role in detailed planning as they do in design. Army designers from the field have seen the typical commander participation as problematic, in that design is different from planning and requires far more dedicated and nuanced commander interaction.[115] Retired Major General Paul Lefebvre, a Marine officer, offered during an interview that: "Commanders today realize today that they do not have all of the answers … and are willing to [engage in design discourse] with planners … and you get a deepening of the commander's understanding of these problems".[116] This generational shift may represent the growth of design thinking within the U.S. Department of Defense writ large, and could influence how ADM might change in the next decade as well. Senior military leaders able to step outside of the dominant war paradigm may consider that conceptual thinking might need to become more than just an optional first step in the detailed military decision-making process.

A Lack of "Why-Centric" Inquiry: U.S. Army Single-Paradigm Design

ATP 5–0.1 covers a great deal of design material, yet unlike Israeli SOD, it lacks any self-reflection or "self-disruption" that Naveh placed as fundamental in his military design philosophy.[117] Ryan reflected on how the U.S. Army's integration of Israeli security design "altered design much more than design altered the institution … systemic operational design challenges [the Army's] entire worldview".[118] Experiential learning, as emphasized in Israeli SOD and drawing from earlier sociological theories, first addresses the "why" of human affairs, organizational frames, social paradigms, and why an organization approaches complexity in particular ways. More significantly, reflective practitioners explore beyond these institutional limits into what else is not even being considered that may present novel opportunity for disruptive change.

When there is a greater emphasis on "what" and "how" modes of inquiry, this indicates the presence of convergent planning methodologies that are well suited to describe and curate analytic optimization for the organization.[119] Efforts to maintain a rationalized, pragmatic mode of military design inquiry will wittingly or unwittingly remove themselves and their organization from the critical and creative examination. This omits the entire first focus of SOD where Naveh requires a deep design framing of the "self" that must address "why"-centric inquires. Although an empathy-based mode of design inquiry might also provide stakeholder tensions in core beliefs, paradoxical perspectives and multiple cultural or social differences within complex reality,[120] a designer can just as willfully project their own paradigm to interpret the "how" and "why" for empathetic appreciation toward vastly different belief systems out of epistemological and ontological incommensurability with different stakeholder frames. Commercial designers may fall into this trap, and as ADM is configured, army planning teams pursuing an environmental frame as their first design action will often subscribe to seeking objectivity and implicit validation of their institutional war paradigm. Problems are understood to be solved, and why one might approach the "problem-solution" formula with one technique or another. Questioning why the organization insists on this manner of framing reality, or why the decision-making methodologies, conceptual models, and selected military theories are normally off-limits, making for a convergent, rigid, unimaginative military.

For military designers to pursue "why-centric" modes of inquiry, the normal emphasis of deductive and inductive reasoning must be softened. Deductive reasoning, foundational in army planning doctrine and extended into ADM, uses existing, objectively oriented knowledge to draw conclusions, make predictions about the future, or construct explanations that justify the continued use of that logic. Deductive reasoning begins with the assertion of general rules, such as elements of operational art, which again stem from military mimicry of the natural sciences. Provided that there is a natural, unchanging order to war, the universal laws and patterns can be extracted by military theorists such as Jomini or Clausewitz, so that militaries can then move from the general war rule to the specific application. Inductive reasoning starts with the observations that are context based, and then through scientific analysis, one might reach a generalized conclusion. Actual scientific research occurs in this manner, while most military decision-making methodologies impose mostly deductive reasoning as part of how modern militaries apply pseudo-scientific actions addressed in the second chapter.

Abductive reasoning is best suited for complex, dynamic contexts where the system will not permit an organization to immediately carry over historical knowledge that is relevant. Complex systems perpetually change, adapt, and learn. What worked yesterday may today be entirely irrelevant in complexity, particularly in chaotic systems. War is expressed not in largely simplistic or regulated, complicated systems but in complex and chaotic ones, meaning that abductive reasoning often is important and largely absent from army design methods. ADM directs planning teams to apply deductive reasoning right from the start, with doctrinal stipulations on precisely what planning tools and processes that must be followed, and how the environment should be analyzed in objective, technically rationalized modes of inquiry. Abductive reasoning starts with admitting one does not know what is occurring, or whether the knowledge already in possession is of any utility in the emerging challenge that the observer is intricately woven into as well. Information at hand is incomplete, and as a designer iteratively engages with a complex, dynamic system, new relationships, and different, unrealized knowledge will come into being. Abductive reasoning is perpetually incomplete, intuitive, and best generated in creative activities that do not follow set patterns or historical norms. There is an improvisational, temporary, iterative, and flexible quality in abductive processes, which is how Naveh's SOD presented a different way to make sense of complex warfare beyond the rational limits of the modern war machine.

Abductive reasoning fixes upon systemic inquiry where asking why things are leads to larger systems becoming relevant instead of how deductive reasoning tends to move toward specifics and reduction of complexity. The more a design team considers "why", this moves toward additional "why-centric" discussions. Abductive design thinking leads not to more answers, but to more questions. This also is why American military commanders in the SOD experimental period of 2004–2010 tended to get frustrated with officers showing up to work on a staff with SOD experience. Instead of leaping into deductive or inductive processes, SOD designers would "admire the problem" and potentially get themselves into trouble with senior leaders that wanted operational planners, not designers. Yet without abductive thinking positioned prominently within a military design methodology, the designers end up doing as ADM is structured to perform. Design teams turn into planning teams, and design concepts are generated so that existing campaign designs and operational plans that were not working before the design may now somehow perform better in complex warfare. The institution itself is unchanged, and designers that deviate from the set methodology will face alienation or be instructed to repeat the process "by the books."

ADM attempts to standardize military design for the army so that the accusations that "designers are solving some other problem outside what they were directed to do" are less likely.[121] While SOD promoted a notion that designers could be heretics to their own institutions so that they could usher in necessary change, the army would craft design into doctrine so that designers must conform with the overarching institution. Planning is primary, the commander is central to all decision-making and intellectual affairs, and designers must use planning tools, models, and methods to generate some moderately unexpected or unorthodox options that still provide the commander with institutional familiarity. ADM is designed to explore unconventional ideas that still can be understood by conventional means. In *ATP 5-0.1,* planning teams are directed to use questions in their design endeavors that have no abductive reasoning, only deductive ones.[122] The suggested questions for designers to establish their environmental frame are exclusively "what-centric" with clear preference for generalized war models and methods to be applied toward. The questions below are almost preconditioned to stimulate ADM planners to use the prescribed planning tools to attempt to answer them, moving immediately into deductive reasoning. They remain conforming to the design methodology without moving the designer toward any epistemological consideration or reflexive practice. The six example suggestions from *ATP 5–0.1* are listed below. Similar examples are also in the 2020 *FM 5–0* design chapter:[123]

- "What are the other sources of power in Newland beyond General E?"
- "What are the limits to the drug money provided by the cartels?"
- "Is the society this homogenous?"
- "What are the different groups and divisions in the population?"
- "What are some of the other key international relationships or interests within Newland?"
- "Are there limits to military aid provided by Country X and Country Y?"

The series of example questions for Army designers to generate their operational frame demonstrate deductive reasoning, an institutionalized planning emphasis, and a desire to accelerate from brief abstract thinking into a solutions–oriented, rationalized execution phase as quickly as possible. As Paparone explains:

> Rationally derived meaning-for-action in Military Design also calls for an explicit objective, "end state." The reasoning goes this way: Break the desired end state (the prescribed meaning for action) into intermediate, contributing meanings-for-action (like tributaries would form rivers and lakes). The rationalist assumes the contributing meanings for action add up to the overall meaning for action (capped a campaign or strategy).[124]

The design questions above orient planning teams so that they apply deductive reasoning, use existing doctrinal planning tools, and follow the ADM methodology that leads directly to new planning options that conform to the entire army's dominant war paradigm.

Following ADM in sequence (referring once more to the epistemological illustration earlier), the design process culminates in an operational approach that is intended to take from the envisioned alternative future states of that operational environment under inquiry, and "envision the desired end state" through designing so the organization accomplishes that transformation.[125] Although *ATP 5–0.1* mentions multiple alternative futures, these remain limited to a single–paradigm interpretation of that design team, further directed toward doctrinal convergence with the required application of U.S. Army planning terminology, the causal "ends–ways–means" logic,[126] and the necessary utilization of existing non–design doctrinal constructs such as elements of operational art. This technical rationality

> views strategy as a search for the ways (courses of action) that best satisfy stated ends (objectives) within the assigned means (instruments of national power) … Strategy is then developed using a waterfall process that cascades from the top down.[127]

The single design "end state" selected from ADM's operational approach will not differ epistemologically from any of the other ones the design team develop that are not selected, or those the commander is preconditioned to recognize within established practices. This is a limitation of employing a design methodology that lacks self-reflection, paradigm awareness, or the ability to consider alternative perspectives on reality and appreciate tensions, interplay, and overlap between them.

ATP 5–0.1 illustrates its use of the functionalist paradigm for explaining Army design theory in one significant passage where the doctrine attempts to clarify "description, explanation and meaning".[128] Although this passage provides sufficient definitions for each term as well as brief summaries of deductive, inductive, and abductive reasoning, *ATP 5–0.1* prioritizes analysis as the mode for accomplishing the validation of meaning. Whereas Israeli systemic operational design would promote synthesis,[129] and earlier draft Army design chapters took from Naveh to frame "problem setting" and the elimination of strict "ends–ways–means" logic to open designers cognitively to systemic thinking, holism, and abductive reasoning, *ATP 5–0.1* promotes analytic optimization as the cognitive driving force for all of these concepts. Naveh sought to introduce an entirely new way to think and act in war, while the American Army (and most all others) preferred design to remain an extension of established linear planning logic.[130]

Epistemologically, the functionalist paradigm views complex reality as an analytic challenge. Or, as Ryan more colorfully puts it, "Army officers often characterize this more simply by saying 'we are a blue-collar Army' with 'an engineering mindset … give us a clear objective and adequate resources and we'll achieve the mission' "[131] War can be flash-frozen, isolated, and reduced down into fundamental elements where the proper analysis might be rendered. Then, the military thinker might re-assemble the entire complex challenge whole again, unfreeze it and have greater rationalized understanding, control, and risk reduction than before this proposed process.

Conclusions

Ultimately, ADM represents a victory of the dominant military paradigm to maintain strong institutional resistance toward disruptive ideas or alternative challenges to include different paradigms and paradoxical logics. If design is a postmodern mode of logic for the postmodern military form, then the modern military institution generates substantial opposition to it, often for self-relevance, preservation of core beliefs, and either a witting or unwitting devotion to a single social paradigm. The bureaucratic system for developing, reviewing, and maintaining doctrine remains a modernist manifestation of the functionalist approach to war that uses an objective, linear, mechanistic, and systematic manner of making sense of war via a Newtonian stylization. Army doctrine would be published and displace any creative, disruptive, and unorthodox design experimentation occurring within Army schools and in select units. While army designers may pursue other ways of doing design within the army, they do so at the risk of being labeled as heretics, and any failures will immediately be considered a violation of adhering to institutionally directed design methods found in published doctrine.

ATP 5–0.1 continued the tradition first set in the design chapter of *FM 3–24* in directing Army designers to use existing non-design concepts such as elements of operational art. The 2022 version of *FM 5–0* further illustrates a decade-long stagnation of army design thinking, now trapped within TRADOC's doctrinal grip. Whether there remains design innovation and experimentation at places like the SAMS program may not matter, if the larger bureaucratic institution maintains control of ADM as understood and written for doctrine that exists not to challenge or disrupt the institution, but to preserve it. Over the last two decades of army design experimentation and subsequent indoctrination into the force, design would go from a highly intellectual, improvisational, and unorthodox way to think disruptively about complex warfare into an extension of the army's central methodology for all decision-making. Army planning would assimilate design.

In the span of less than a decade, the U.S. Army would discover, experiment with, and eventually assimilate security design praxis to render a uniquely American Army variation unlike the parent design logic. To this day, there is fierce debate on whether the design "purists" or the design pragmatic integrators got things right. The American army is large, technologically advanced, and is perhaps the most powerful military instrument of power for land warfare in human history. It is also expensive to maintain, coordinate, and difficult to manage in terms of talent, education, recruitment, and retention. Militaries excel at increasing uniformity, reliability, repetition, and increasing efficiencies in approaching complex warfare and security challenges. This comes at a cost where the army, perhaps the best at standardization, doctrinal cohesion, and tactical excellence, is exceedingly poor at promoting innovation, creativity, critical thinking, and improvisation through designing in unrealized, unfamiliar, or emergent ways for future challenges. The army experienced failure and frustration during difficult counterinsurgencies in Iraq and Afghanistan, and those continue to plague the U.S. Department of Defense in 2022 after the spectacularly devastating fall of Kabul to Taliban forces in 2021.

The army in 2022 has its own branded design methodology that reflects robust doctrinal instructions, and the army regularly educates and trains forces with design across their professional schools, training centers, and in military deployments and activities. Yet the earlier less structured design period of 2004–2009 demonstrated an explosion of new, exotic, and also confusing and esoteric design experimentations. Naveh and his SOD designers were treated like mountain yogis or the oracles of ancient Greek temples by the army leadership desperate for new ideas in the first decade, but later would be viewed as elitist, tiresome intellectuals full of incoherent ramblings. Was the army right to streamline design and eliminate virtually everything potentially disruptive or paradoxical to the institutional norms and beliefs, or was the army guilty of anti-intellectualism once again, unable or unwilling to change in ways that questioned deeply cherished war frames? The debate is not yet over, and with continued frustrations still occurring over multidomain warfare, dynamic adversaries gaining peer-like or overmatch military abilities, and the expansion of warfare into cyberspace, space, and other areas of previously unavailable military responsibilities, the army may revisit design. In doing so, it may end up reframing and redesigning much more than an army doctrinal publication. Yet this story is still developing.

There remains a small population within the Army (and across other services) that carry forward the "purist camp" positions of Naveh's original vision, coupling deep sociological, philosophical, and multi-disciplinary approaches to military design. Mostly working in a grassroots, decentralized, yet well connected form, this eclectic design community continues to insert provocative design concepts into how and why the Army needs to reframe design, change the ADM indoctrination efforts, and also expand into how Army organizations might demonstrate design teams in action. This small movement appears intent on removing the authority concerning Army design from TRADOC and bureaucratic careerists, and instead restoring it to highly skilled military designers that use experimental, improvisational, plastic modes of design that escape control and formulaic assimilation of doctrine. Whether this design insurgency succeeds or if the Army continues to delegate design into an optional planning step will unfold in the next decade.

Notes

1 Sean McFate, *The New Rules of War*, First Edition (New York: William Morrow, 2019), 36.
2 Markus Mäder, *In Pursuit of Conceptual Excellence: The Evolution of British Military-Strategic Doctrine in the Post-Cold War Era. 1989–2002* (Bern, Germany: Peter Lang AG, 2004), 305–311.
3 Karl Weick, "Drop Your Tools: An Allegory for Organizational Studies," *Administrative Science Quarterly* 41 (1996): 301–313.
4 Deborah Stone, *Policy Paradox*, 3rd edition (New York: W.W. Norton & Company, Inc., 2012), 33.
5 Steven Leonard, Interview with Colonel (retired) Steven Leonard on Army Design Methodology on 06 Aug 2021, interview by Ben Zweibelson, mp3 Audio File, August 6, 2021, 23:03 to 24:19.
6 McFate, *The New Rules of War*, 30–36.
7 Kevin Benson, Interview of Colonel (retired) Kevin Benson, former SAMS Director, 04 AUG 2021 mp3 audio file, interview by Ben Zweibelson, mp3 Audio File, August 4, 2021, 3:15 to 3:45.
8 SAMS has a field grade officer cohort attending for one year and a smaller Senior Fellows cohort of senior officers that attend SAMS for a year and remain to lead field grade seminars for a second year.
9 Benson, Interview of Colonel (retired) Kevin Benson, former SAMS Director, 04 Aug 2021 mp3 audio file, 5:56.

10 Bob Johnson, Interview with Bob Johnson on U.S. Army Design 30 SEP 2021, interview by Ben Zweibelson, mp3 Audio File, September 30, 2021, 01:08:00.

11 Johnson, 09:45–10:10.

12 Johnson, 10:00–10:51.

13 Benson, Interview of Colonel (retired) Kevin Benson, former SAMS Director, 04 Aug 2021 mp3 audio file, 8:30 to 11:45.

14 Benson, 17:35.

15 Johnson, Interview with Bob Johnson on U.S. Army Design 30 SEP 2021, 16:30–16:40.

16 Dan Cox to Ben Zweibelson, "Fwd: Research Question on SAMS, 2009–2010 Period for Design," February 24, 2020, personal correspondence between Cox and the author.

17 Alex Ryan, "A Personal Reflection on Introducing Design to the U.S. Army". *The Medium* (blog), November 4, 2016, https://medium.com/the-overlap/a-personal-reflection-on-introducing-design-to-the-u-s-army-3f8bd76adcb2

18 Ryan.

19 Department of the Army and Department of the Navy, HQ United States Marine Corps, *Field Manual 3–24/MCWP 3–33.5, Counterinsurgency* (Washington, D.C.: Department of the Army, Department of the Navy, 2006), 4–1, www.hsdl.org/?view&did=468442

20 Headquarters, Department of Army, *Field Manual 3–24: Counterinsurgency (Final Draft – Not for Implementation)*, Final Draft version-not for implementation, TRADOC Field Manual Series (Washington, D.C.: Headquarters, Department of Army, 2006), 4–1, https://fas.org/irp/doddir/army/fm3-24fd.pdf

21 Headquarters, Department of Army, 4–2.

22 Department of the Army and Department of the Navy, HQ United States Marine Corps, *FM 3–24/MCWP 3–33.5*, 4–2.

23 Donald Schön and Martin Rein, *Frame Reflection: Towards the Resolution of Intractable Policy Controversies* (New York: Basic Books, 1994), 34–36.

24 Schön and Rein, 30.

25 Headquarters, Department of Army, *Field Manual 3–24: Counterinsurgency (Final Draft – Not for Implementation)*, 4–3.

26 Shimon Naveh, *Systemic Operational Design: Designing Campaigns and Operations to Disrupt Rival Systems (Draft Unpublished)*, Version 3.0, unpublished draft (Fort Monroe, Virginia: Concept Development & Experimentation Directorate, Future Warfare Studies Division, US Army Training and Doctrine Command, 2005), 7.

27 Jamshid Gharajedaghi, *Systems Thinking: Managing Chaos and Complexity, A Platform for Designing Business Architecture*, Third (New York: Elsevier, 2011), 8, http://pishvaee.com/wp-content/uploads/downloads/2013/07/Jamshid_Gharajedaghi_Systems_Thinking_Third_EdiBookFi.org_.pdf

28 James Mattis to Ben Zweibelson, "RE: Design for Defense Book PDF Manuscript," September 16, 2021.

29 Johnson, Interview with Bob Johnson on U.S. Army Design 30 Sep 2021, 22:20–22:38.

30 Department of the Army and Department of the Navy, HQ United States Marine Corps, *FM 3–24/MCWP 3–33.5*, 4–3.

31 Mary Jo Hatch, *Organization Theory: Modern, Symbolic, and Postmodern Perspectives*, Third Edition (Oxford, United Kingdom: Oxford University Press, 2013), 230.

32 Mark Bedau and Paul Humphreys, eds., "Philosophical Perspectives on Emergence," in *Emergence: Contemporary Readings in Philosophy and Science* (Cambridge: Massachusetts Institute of Technology Press, 2008), 9.

33 Antoine Bousquet, "Chaoplexic Warfare or the Future of Military Organization," *International Affairs (Royal Institute of International Affairs 1944-)* 84, no. 5 (September 2008): 915–929.

34 Headquarters, Department of Army, *Field Manual 3–24: Counterinsurgency (Final Draft – Not for Implementation)*, 4–4.

35 Department of the Army and Department of the Navy, HQ United States Marine Corps, *FM 3–24/MCWP 3–33.5*, 4–4.

36 Jeffrey Meiser, "Ends + Ways + Means = (Bad) Strategy," *Parameters* 46, no. 4 (Winter 2016): 81–91.

37 Department of the Army and Department of the Navy, HQ United States Marine Corps, *FM 3–24/MCWP 3–33.5*, 4–6; Headquarters, Department of Army, *Field Manual 3–24: Counterinsurgency (Final Draft – Not for Implementation)*, 4–6.

38 Department of the Army and Department of the Navy, HQ United States Marine Corps, *FM 3–24/MCWP 3–33.5*, 4–6.

39 Headquarters, Department of Army, *Field Manual 3–24: Counterinsurgency (Final Draft – Not for Implementation)*, 4–9.

40 Department of the Army and Department of the Navy, HQ United States Marine Corps, *FM 3–24/MCWP 3–33.5*, 4–9.

41 Headquarters, Department of the Army, *Field Manual Interim (FMI) 5–2: Design (Draft)* (Washington, D.C.: Headquarters, Department of Army, 2009).

42 Robert Chia, "From Modern to Postmodern Organizational Analysis," *Organization Studies* 16, no. 4 (1995): 583.

43 David Pick, "Rethinking Organization Theory: The Fold, the Rhizome and the Seam between Organization and the Literary," *Organization* 24, no. 6 (2017): 302–304.

44 Robert Chia and Robin Holt, *Strategy without Design: The Silent Efficacy of Indirect Action* (New York: Cambridge University Press, 2009), x.

45 Alex Ryan to Ben Zweibelson, "Re: Research Question on SAMS, 2009–2010 Period for Design (UNCLASSIFIED)," February 26, 2020.

46 Ryan, "A Personal Reflection on Introducing Design to the U.S. Army," November 4, 2016.

47 Steven Leonard to Ben Zweibelson, "RE: [Non-DoD Source] Design," August 4, 2021.

48 Steven Donnell, Interview of Mr. Steve Donnell by Dr. Ben Zweibelson on 28 Jul 2021 concerning Marine Corps Design Methodology mp3 audio file, interview by Ben Zweibelson, mp3 Audio File, July 28, 2021, 32:00 to 32:15; Leonard, Interview with Colonel (retired) Steven Leonard on ADM on 06 AUG 2021, 06:27 to 08:00.

49 Ryan to Zweibelson, "Re: Research Question on SAMS, 2009–2010 Period for Design (UNCLASSIFIED)," February 26, 2020. Ryan recalls that Lieutenant Colonel Steven Leonard was tasked to re-write *FMI 5–2* into a 13-page compressed design chapter for *FM 5–0*. Ryan states neither he nor any of the original *FMI 5–2* authors were consulted, and that Leonard's alleged instructions were to "crunch the FMI down to 13 pages."

50 Leonard, Interview with Colonel (retired) Steven Leonard on ADM on 06 Aug 2021, 07:49 to 08:02.

51 Daniel Cox, "RE: One Additional Question on SAMS, Design and 2006–2010 Period (UNCLASSIFIED)," March 5, 2020, personal correspondence.

52 Schön and Rein, *Frame Reflection*, xvii.

53 Benson, Interview of Colonel (retired) Kevin Benson, former SAMS Director, 04 Aug 2021 mp3 audio file, 15:30 to 15:50.

54 Yotam Feldman, "Dr. Naveh, or, How I Learned to Stop Worrying and Walk through Walls," online social media and news blog, *HAARETZ.Com* (blog), October 25, 2007, www.haaretz.com/misc/article-print-page/1.4990742

55 Anna Grome, Beth Crandall, and Louise Rasmussen, "Incorporating Army Design Methodology into Army Operations: Barriers and Recommendations for Facilitating Integration," *U.S. Army Research Institute for the Behavioral and Social Sciences*, no. Research Report 1954 (March 2012).

56 Wilburn McLamb, "The U.S. Army's Design Doctrine: A Solution to the Ills of the Operations Planning Processes?" (paper for partial satisfaction of master's degree, Naval War College, Newport, Rhode Island, May 2009), 14.

57 Milan Vego, "A Case Against Systemic Operational Design," *Joint Forces Quarterly* 53 (quarter 2009): 73.

58 Johnson, Interview with Bob Johnson on U.S. Army Design 30 Sep 2021, 32:15–32:56.

59 Ryan, "A Personal Reflection on Introducing Design to the U.S. Army," November 4, 2016.

60 Alice Butler-Smith, "Operational Art to Systemic Thought: Unity of Military Thought," in *Cluster 3* (Hybrid Warfare: New Ontologies and Epistemologies in Armed Forces, Canadian Forces College, Toronto, Canada: unpublished – Canadian Forces College internal document, 2016), 70. Unpublished draft provided to the author.

61 Wayne Grigsby et al., "Integrated Planning: The Operations Process, Design, and the Military Decision Making Process," *Military Review* XCI, no. 1 (February 2011): 28–35; Donnell, Interview of Mr. Steve Donnell by Dr. Ben Zweibelson on 28 Jul 2021 concerning Marine Corps Design Methodology mp3 audio file, 39:58 to 40:50.

62 Cox to Zweibelson, "Fwd: Research Question on SAMS, 2009–2010 Period for Design," February 24, 2020.

63 Anatol Rapoport, *The Origins of Violence: Approaches to the Study of Conflict* (New Brunswick, New Jersey: Transactions Publishers, 1995), 237.

64 Headquarters, Department of the Army, *Field Manual Interim (FMI) 5–2: Design (Draft)*, 18; Stefan Banach and Alex Ryan, "The Art of Design: A Design Methodology," *Military Review* 89, no. 2 (April 2009): 109.

65 United States Army, *Field Manual 5–0, Operations Process* (Washington, DC: Headquarters, Department of the Army, 2010), 3–7.

66 Sebastien le Prestre de Vauban, *The New Method of Fortification, as Practised by Monsieur de Vauban, Engineer-General of France. Together with a New Treatise of Geometry. The Fifth Edition, Carefully Revised and Corrected by the Original*, Fifth (London: S. and E. Ballard (reprint by Creative Media Partners), 1722); Henry Guerlac, "Vauban: The Impact of Science on War," in *Makers of Modern Strategy: From Machiavelli to the Nuclear Age*, ed. Peter Paret (Princeton, New Jersey: Princeton University Press, 1986).

67 Haridimos Tsoukas, *Complex Knowledge: Studies in Organizational Epistemology* (New York: Oxford University Press, 2005), 213–214.

68 Leonard, Interview with Colonel (retired) Steven Leonard on ADM on 06 AUG 2021, 8:23 to 8:43; Banach and Ryan, "The Art of Design: A Design Methodology," 109.

69 Johnson, Interview with Bob Johnson on U.S. Army Design 30 Sep 2021, 1:09:00–1:12:00.

70 Haridimos Tsoukas and Efi Vladimirou, "What Is Organizational Knowledge?," *Journal of Management Studies* 38, no. 7 (November 2001): 975.

71 US Army Headquarters, Department of the Army, *Army Design Methodology (ATP 5–0.1)* (Washington, DC: US Department of the Army, 2015), 1–3.

72 Headquarters, Department of the Army, 1–3.

73 Gharajedaghi, *Systems Thinking: Managing Chaos and Complexity, A Platform for Designing Business Architecture*, 11–13.

74 Russell Ackoff, "On the Use of Models in Corporate Planning," *Strategic Management Journal* 2, no. 4 (December 1981): 353–359.

75 Jean-Pierre Protzen and David Harris, *The Universe of Design: Horst Rittel's Theories of Design and Planning* (New York: Routledge, 2010), 154.

76 Stone, *Policy Paradox*, 36.

77 Headquarters, Department of the Army, *Army Design Methodology (ATP 5–0.1)*, 1–4.

78 Ofra Graicer, *Two Steps Ahead: From Deep Operations to Special Operations – Wingate the General*, Special Edition (Dayan Base, Tel Aviv, Israel: Israeli Defense Forces, 2015), 35–37.

79 Schön and Rein, *Frame Reflection*, vii.

80 Jamshid Gharajedaghi and Russell Ackoff, "Mechanisms, Organisms, and Social Systems," in *New Thinking in Organizational Behaviour*, by Haridimos Tsoukas (Oxford, United Kingdom: Butterworth-Heinemann Ltd, 1994), 290; Stone, *Policy Paradox*, 63–67.

81 Christopher Paparone, "Designing Meaning in the Reflective Practice of National Security: Frame Awareness and Frame Innovation," in *Design Thinking: Applications for the Australian Defence Force*, ed. Aaron Jackson and Fiona Mackrell, editor's manuscript pre-publication version, Joint Studies Paper Series 3 (Canberra, Australia: Defence Publishing Service, 2019), 90–91; Haridimos Tsoukas, "What Is Organizational Foresight and How Can It Be Developed?," in *Organization as Chaosmos: Materiality, Agency, and Discourse*, 1st Edition (New York: Routledge, 2013), 265; Gharajedaghi, *Systems Thinking: Managing Chaos and Complexity, A Platform for Designing Business Architecture*, 25–26.

82 Headquarters, Department of the Army, *Army Design Methodology (ATP 5–0.1)*, 1–4.

83 Rapoport, *The Origins of Violence: Approaches to the Study of Conflict*, 237.

84 Rapoport, 179.

85 Headquarters, Department of the Army, *Army Design Methodology (ATP 5–0.1)*, 1–4.

86 Carl Builder, *The Masks of War: American Military Styles in Strategy and Analysis* (Baltimore: John Hopkins University Press, 1989); Tsoukas and Vladimirou, "What Is Organizational Knowledge?"; Schön and Rein, *Frame Reflection*; Weick, "Drop Your Tools: An Allegory for Organizational Studies."

87 Naveh, *Systemic Operational Design: Designing Campaigns and Operations to Disrupt Rival Systems (Draft Unpublished)*, 12–24.

88   Headquarters, Department of the Army, *Army Design Methodology (ATP 5–0.1)*, 1–4.

89   Gharajedaghi, *Systems Thinking: Managing Chaos and Complexity, A Platform for Designing Business Architecture*, 51.

90   Naveh, *Systemic Operational Design: Designing Campaigns and Operations to Disrupt Rival Systems (Draft Unpublished)*, 12–24.

91   Christopher Paparone, *The Sociology of Military Science: Prospects for Postinstitutional Military Design* (New York: Bloomsbury Academic Publishing, 2013), 90–97.

92   Schön and Rein, *Frame Reflection*, 5.

93   United States Army, *Field Manual 5–0: Planning and Orders Production* (Washington, D.C.: Headquarters, Department of Army, 2022), 3–3, 4–1, https://armypubs. army.mil/epubs/DR_pubs/DR_a/ARN35403-FM_5-0-000-WEB-1.pdf).

94   United States Army, 4–2.

95   Headquarters, Department of the Army, *Army Design Methodology (ATP 5–0.1)*, 1–8.

96   United States Army, *Field Manual 5–0: Planning and Orders Production*, 4–3.

97   United States Army, 4–5.

98   United States Army, 4–1 to 4–22.

99   Paparone, *The Sociology of Military Science: Prospects for Postinstitutional Military Design*, 20, 90–94.

100 Schön and Rein, *Frame Reflection*, 55. The authors reference James March and Johan Olsen, *Rediscovering Institutions* (Free Press, Glencoe, Illinois, 1989), 62–63.

101 Paparone, *The Sociology of Military Science: Prospects for Postinstitutional Military Design*, 18–19.

102 Schön and Rein, *Frame Reflection*, xiv.

103 Hatch, *Organization Theory: Modern, Symbolic, and Postmodern Perspectives*, 12.

104 Johnson, Interview with Bob Johnson on U.S. Army Design 30 SEP 2021, 55:14–55:30.

105 Johnson, 55:14–55:30.

106 Schön and Rein, *Frame Reflection*, 41–43.

107 Schön and Rein, 27.

108 Protzen and Harris, *The Universe of Design: Horst Rittel's Theories of Design and Planning*, 53–56.

109 Chia and Holt, *Strategy without Design: The Silent Efficacy of Indirect Action*; Kenneth Stanley and Joel Lehman, *Why Greatness Cannot Be Planned: The Myth of the Objective* (Switzerland: Springer International Publishing, 2015); Richard Buchanan, "Wicked Problems in Design Thinking," *Design Issues* 8, no. 2 (Spring 1992): 5–21; Robert Chia, "Reflections: In Praise of Silent Transformation – Allowing Change Through 'Letting Happen,'" *Journal of Change Management* 14, no. 1 (2013): 8–27.

110 Paparone, *The Sociology of Military Science: Prospects for Postinstitutional Military Design*, 18–19.

111 Chia, "From Modern to Postmodern Organizational Analysis," 600.

112 Stanley and Lehman, *Why Greatness Cannot Be Planned: The Myth of the Objective*.

113 Headquarters, Department of the Army, *Army Design Methodology (ATP 5–0.1)*, 2–3.

114 Grant Martin, "A Tale of Two Design Efforts [And Why They Both Failed In Afghanistan]," *Small Wars Journal*, July 7, 2011, 16; Grome, Crandall,

and Rasmussen, "Incorporating Army Design Methodology into Army Operations: Barriers and Recommendations for Facilitating Integration," 21–23.

115 Martin, "A Tale of Two Design Efforts [And Why They Both Failed in Afghanistan]," 13–14.

116 Paul Lefebrve, Interview with Major General (retired) Paul Lefebvre MARSOC CDR on Marine Design Theory 17 Aug 2021, mp3 Audio File, August 17, 2021, 00:05:10 to 00:05:38.

117 Naveh, *Systemic Operational Design: Designing Campaigns and Operations to Disrupt Rival Systems (Draft Unpublished)*, 7–14.

118 Alex Ryan, "A Personal Reflection on Introducing Design to the U.S. Army," in *Cluster 2* (Hybrid Warfare: New Ontologies and Epistemologies in Armed Forces, Canadian Forces College, Toronto, Canada: University of Ottawa and the Canadian Forces College, 2016).

119 Protzen and Harris, *The Universe of Design: Horst Rittel's Theories of Design and Planning*, 151–162.

120 Schön and Rein, *Frame Reflection*, 47; Dennis Gioia and Evelyn Pitre, "Multiparadigm Perspectives on Theory Building," *Academy of Management Review* 15, no. 4 (1990): 584–602; Marianne Lewis and Andrew Grimes, "Metatriangulation: Building Theory From Multiple Paradigms," *Academy of Management Review* 24, no. 4 (1999): 672–690.

121 Cox, "RE: One Additional Question on SAMS, Design and 2006–2010 Period (UNCLASSIFIED)," March 5, 2020; Ryan, "A Personal Reflection on Introducing Design to the U.S. Army," November 4, 2016; Butler-Smith, "Operational Art to Systemic Thought: Unity of Military Thought."

122 Given the questions provided, one could argue that select examples might lead to inductive reasoning. Yet any inductive reasoning proposed would need to sidestep the declared ADM requirements to apply the general, universal tenets such as operational art, COGs, or other deductive planning constructs. Virtually all ADM opportunities to pursue questions as suggested in doctrine incur a deductive orientation by default.

123 Headquarters, Department of the Army, *Army Design Methodology (ATP 5–0.1)*, 3–4; United States Army, *Field Manual 5–0: Planning and Orders Production*, 4–11.

124 Paparone, *The Sociology of Military Science: Prospects for Postinstitutional Military Design*, 107.

125 Headquarters, Department of the Army, *Army Design Methodology (ATP 5–0.1)*, 3–5 to 3–8.

126 Meiser, "Ends + Ways + Means = (Bad) Strategy," 81–84.

127 Ryan, "A Personal Reflection on Introducing Design to the U.S. Army," November 4, 2016.

128 Headquarters, Department of the Army, *Army Design Methodology (ATP 5–0.1)*, 3–9.

129 Graicer, *Two Steps Ahead: From Deep Operations to Special Operations – Wingate the General*, 33–36.

130 Mäder, *In Pursuit of Conceptual Excellence: The Evolution of British Military-Strategic Doctrine in the Post-Cold War Era. 1989–2002*, 299–302.

131 Ryan, "A Personal Reflection on Introducing Design to the U.S. Army," November 4, 2016.

5 Marine Design Methodology

From Innovation to Indoctrination in Two Decades

The U.S. Army and the U.S. Marine Corps are land forces within the Department of Defense, yet culturally, organizationally, and in terms of military role and purpose, both institutions pride themselves on being quite unlike the other. Yet both played significant roles in how Afghanistan and Iraq would unfold in counterinsurgency operations after 2001, often taking near-identical roles, missions, and responsibilities in these conflicts. Both military organizations would also become frustrated with how outdated concepts, insufficient doctrine, and a lack of creative and critical thinking skills were preventing any military success aside from clear, tactical, and localized achievements. It is in this overlap that the Marine design story begins.

The U.S. Marine Corps first engaged with design concepts when the U.S. Army and the Marines combined intellectual forces in the drafting and publication of *Field Manual 3–24/MCWP 3–33.5*. As explained in the previous chapter, both the Army and Marine Corps found early security design concepts interesting and potentially the new way of thinking in complex warfare that was desperately needed in the spiraling chaos in Iraq and Afghanistan. While other services focused on air, sea, space, or even cyberspace, the two services with most of the physical responsibility for security in Iraq and Afghanistan would be the only forces specializing in land warfare.

The U.S. Marines have an interesting and storied history with discovering Naveh's unique way of conducting design for defense applications, and it overlaps and interplays with the U.S. Army in many areas. Naveh would engage with both organizations at nearly the same time in the 2003–2004 period, often at the same events. Naveh would engage with audiences consisting of multiple services, yet SOD had a particular pull for those forces engaging in the most complex, dynamic security contexts which often involved populations, ideologies, and complex defense demands. Lieutenant General (retired) Paul Van Riper provides insight into how senior Marine leadership discovered Naveh's ideas, and quickly moved to experiment with them for potential inclusion into Marine transformation endeavors:

> I began consulting with U.S. Army Training and Doctrine Command (TRADOC) in the late 1990s and was involved in the "Army After

DOI: 10.4324/9781003387763-6

Next War" or AAN Games and the succeeding "Army Transformation War Games" (ATWG). Naveh was invited to observe one of those games in 2004. I had met Naveh previously when he took part in a seminar sponsored by Andy Marshall at the Office of Net Assessment and I was familiar with his 1997 book, *In Pursuit of Military Excellence: The Evolution of Operational Art*.[1]

Van Riper would receive an early SOD presentation from Naveh and immediately fly to Israel to attend an in-depth, week-long workshop at OTRI where Naveh provided a deeper explanation of his theories and methods. Van Riper, a retired Marine Lieutenant General by 1997 would in the early 2000s be one of the first Marines to discover Naveh's ideas and cross-pollinate them with both the Marines and the U.S. Army. Brigadier General David Fastabend, the U.S. Army Deputy Director for Futures, would work with Van Riper in a consulting capacity and delve deeper into SOD. Van Riper recalls this was pursued:

> In a deliberate and structured manner; [Fastabend] wanted to avoid the negative reactions the Joint Forces Command had provoked with its unsupported assertions about the efficiency of Effects Based Operations (EBO). As I recall, Brigadier General Fastabend laid out a five-year study and evaluation plan, which was to begin with the use of SOD in the 2005 ATWG. This was to be parallel or a satellite effort to the game's conventional planning process.[2]

Van Riper would also work periodically with U.S. Army TRADOC theorists between 2005 and 2009 evaluating SOD in Army exercises. He took to advising the Marines on how to integrate design into Marine deliberate planning and decision-making as part of his role of being a senior mentor and retired general officer.

The Army would shift the study of SOD into the U.S. Army School of Advanced Military Studies (SAMS) and employ Naveh there as a visiting faculty member with SOD-specific exercises and evaluations. To continue design collaboration, the Marines sent two Marine Corps School of Advanced Warfighting (SAW) students to join the U.S. Army SAMS design exercises in 2005–2006. Then Lieutenant General James Mattis would do this based on Van Riper's recommendation and personal intuition that Naveh's ideas had merit. The U.S. Army would in conjunction with the Marines collaborate on design research into Naveh's theories and methods, spanning ten months in 2005–2006. General Mattis would task John Schmitt, then a Marine serving under his command, to write a white paper on some of the findings and conclusions from this deep dive. This became one of several core Marine design initial documents that would influence future paths to take.

Schmitt's paper, entitled "A Systemic Concept for Operational Design", would draw heavily from systems theorists such as Senge, Popper, Checkland,

and Bertalanffy, with additional sources spanning complexity theory, sociology, organizational theory, and military doctrine. Schmitt's work would also be devoid of any of Naveh's postmodern sources or concepts, indicating that like the U.S. Army, the Marines had wittingly or unwittingly discounted Naveh's emphasis on postmodern ideas.[3] Schmitt's research would however feature significant recommendations introducing systems theory, complexity theory, and alternative decision-making constructs to challenge the Marine Corps. A similar pattern would reoccur in the Australian Defense Forces, as covered in the seventh chapter. Systems thinking and complexity theory, given their scientific qualities, seem to be more palatable to military audiences than that of postmodern philosophy.

In the same fertile period of 2005–2006, the Army and Marines embarked on publishing a new version of the counterinsurgency manual (*Field Manual 3–24/MCWP 3–33.5*) as both ground forces were organizationally frustrated in Iraq and Afghanistan. As the last chapter described, the Army and Marines divided writing responsibilities between them, and this joint publication venture would be the first between the two organizations. As with the story of the Army and design, the Marines would also go through a strange, controversial, and at times irreconcilable series of decisions on what design should mean to the Marines, how it would function, and where it would be positioned for institutional education, training, and practice.

Mattis recalled that he and General Petraeus divided up the writing assignments of chapters for *Field Manual 3–24/MCWP 3–33.5* between the U.S. Army and U.S. Marine Corps. Due in part to Mattis' exposure to Naveh's design concepts, Mattis prioritized Marine assimilation of design thinking to challenge and disrupt the fixation on "Effects Based Operations" where the Marines had, in Mattis' opinion, become dependent on linear-causal, reductionist modes of warfare conceptualization. "The only chapter we Marines specifically requested was [chapter 4 of FM 3-24/MCWP 3-33.5], Design. So, the initial draft was done at Quantico," according to Mattis. "On Army chapters [of the field manual project], Leavenworth kept us informed and took our input, and we did the same, taking Army input for the chapters written by the USMC".[4] It was collaborative, but Mattis would emphasize that the first draft of [chapter 4 of FM 3-24/MCWP 3-33.5], "Design" in *Field Manual 3–24/MCWP 3–33.5* was primarily authored by the Marines. Van Riper contributes that Schmitt's design white paper also served as an important initial study for the first explanation of Naveh's design in the new counterinsurgency manual.[5]

While the Marines are significantly smaller than the U.S. Army, the ground responsibilities remain as difficult and ever-changing in warfare. As the amphibious and mobile ground forces for the U.S. Navy, the Marines over the last few decades of constant utilization in multiple warzones and hot spots around the globe have felt the same sort of pressure as the Army on whether existing doctrine, decision-making methodologies, theories, and mental models are still sufficient in contemporary warfare challenges. In the last chapter, we saw the powerful influence that the Army SAMS program had across the Department of Defense and across the Joint and international community. SAMS had a larger impact than the smaller SAW program, yet both would generate deep design experimentation and produce the military intellectuals for both land forces. Both programs also would influence other military education across their forces with considerable, lasting impacts on design thinking.

In April 2006, Major William Vivian published a draft white paper entitled "Operational Art and Complexity: Learning and Deciding When Confronting Wicked Problems" for the Marine Corps University.[6] This sixteen-page unpublished document was the result of Vivian being invited to observe the U.S. Army's Title 10 Wargame held in the Norfolk/Tidewater area. According to Steven Donnell, a key Marine design doctrine writer and educator,

> given the location, JFCOM [U.S. Joint Forces Command] attended/ participated too. One of the key presenters was Naveh, for an encore performance after his SOD brief at the same event a year earlier in 2005. It was during this time frame that things in [Operation Iraqi Freedom) were not going well … both the Army and the USMC, as healthy, self-correcting institutions, began to look elsewhere for insights into how did things get this way and how can we do better.[7]

Vivian's white paper summarized Naveh's SOD and made recommendations on how the Marines might incorporate it. It reinforced earlier recommendations by Schmitt's white paper for the Marines as well, as well as the intuition of senior leaders and mentors such as Mattis and Van Riper. Naveh's design soon would become popular, albeit in niche circles and with mostly the intellectual crowd of military professionals.

Vivian's white paper soon made the rounds and quickly caught the eye of senior Marine Corps leadership. Lieutenant General James Mattis, then commanding the Marine Corps Combat Development Command (MCCDC) in the summer of 2006, liked the paper and wanted to use it to advance the SOD ideas into the Marines. Mattis wanted to change how Marines thought in complex warfare, and he felt Naveh's SOD has the potential to provide this needed transformation. MCCDC is the Marine version of the U.S. Army's Training and Doctrine Command, although smaller and combining other educational and research capabilities. According to Donnell, "Mattis person-ally emailed Vivian extolling the virtues of his paper. [He] told Bill [Vivian] how much he was looking forward to all the USMC Gazette articles he would write in his career".[8] Mattis then emailed every key Colonel and leader at Quantico and directed a new bi-weekly design session to commence, facilitated by Van Riper in his senior mentor, retired role. Donnell was also directed by the Marine Air-Ground Task Force Staff Training Program

(MSTP)[9] Director to be his representative at those sessions and report back to him on where this design idea would move to next.

Mattis likely already knew about Naveh's SOD, as he attended the 2005 Unified Quest military exercise where multiple schools and military programs competed in war games. He also had Van Riper engaging with him about SOD in the 2005 period as well; these multiple efforts likely overlapped. Mattis wanted to disrupt and dismantle the convergent, linear thinking that EBO was pulling the Marines toward at the time. Army SAMS Director at that time, Colonel Kevin Benson recalled this as follows:

> Mattis and [Lieutenant General William] Wallace were both bitterly oppose to Effects Based Operations. I had conversations with both of them … they were looking for something that had some theory, some substance, because both of those men viewed EBO as linear; bomb until we get what we want … they did not have a concept that they could put up to oppose EBO until OTRI came up with SOD, and [SAMS founding director] Wass de Czege was the one to introduce them to it. Naveh gave them an amazing presentation. The only guy I felt that half-way followed what Naveh was talking about was Mattis.[10]

Wallace, then the U.S. Army Combined Arms Center (CAC) Commander, was with Mattis at the "Unified Quest" exercise as it was a Joint affair. Naveh would brief Mattis and other senior leaders on SOD, creating in the Marines and the Army an initial burst of interest in 2005 that would carry over into subsequent years of increased experimentation. By 2006, Mattis made up his mind and steered the entire U.S. Marine Corps toward a collision course with the strange ideas coming from Israel on reconceptualizing complex warfare. In committing to this transformation, Mattis would encounter institutional resistance from those opposing such radical actions, and also deep skepticism on what SOD really suggested. This put Mattis' SOD experiment into direct opposition with the EBO proponents which had strong numbers inside the Marine Corps. Major General (retired) Lefebvre recalled:

> Mattis got mad with the difficulty in moving the Marines away from [Relative Combat Power Analysis and EBO logic] … he said he wanted thinking leaders. That was when they started playing around with the first step in the Marine Corps Planning Process. Mattis concluded that the design process was the way to go.[11]

During these bi-weekly design sessions in the 2006–2008 period, Donnell would back-brief the Director and the MSTP team about design and related topics such as systems theory. These strange ideas were disruptive, and for MSTP, like TRADOC for the Army, these radical concepts not only required a steep learning curve, but the learning caused one to need to "un-learn" much of what was held in high value. MSTP essentially owned the Marine

Corps Planning Process (MCPP) and the cornerstone for how the Marines conceptualized all decision-making in warfare. Like the Army's TRADOC, the Marines did not take kindly to criticism on their foundational manner of making sense of war. The institutional antibodies for design grew, and in a common quality of centralized hierarchical bureaucracies, many decided to "wait the commander out".

Donnell, there inside of MSTP, recalled that once Mattis departed his position to take command of the First Marine Expeditionary Force (I MEF), "all of the excitement for design left with him and his personal senior mentor, Van Riper".[12] The institution would strike back, and despite the high rank and powerful influence of Mattis, he alone could not progress the design effort without more Marines seeing the same benefits. Lefebrve reiterated this as follows:

> If [Mattis] liked it, everyone loved it. And that lasted as long as he was there … after he left, and the Marines went to Iraq, we got into a "follow the yellow brick road" [linear-mechanistic planning] mentality and design sort of fell off.[13]

Yet despite the internal politics, the match was lit, and a small group of Marines were becoming more interested in design.

In this same period of military inquiry and examination of various competing theories on warfare, three major camps emerged inside the Marines. Traditional military planners were the first population, followed by the advocates of EBO as a replacement to traditional planning problems. The military design group would constitute the third, and arguably smallest group in this new dynamic. In this strange mix of stakeholders in how the Marines ought to transform, EBO would gain traction inside the U.S. Joint Forces Command (JFCOM) and also in the U.S. Air Force, particularly in the fields of military intelligence, targeting, and aviation in the mid-1980s through the late 1990s. By 2006, as covered in the third chapter, the Second Lebanon War featured an internal struggle between Naveh's SOD proponents and the dominant population of EBO defenders. Mattis, Van Riper, and other senior leaders witnessed these institutional struggles, and Mattis would become a major critic of EBO except for in clear, kinetic (tactical-localized) applications. He would move to shift the Marines away from EBO toward SOD, but in changing command from MSTP, he would lose that battle but gain a much larger, louder platform. Mattis would take command of JFCOM, at that time the top advocate for pushing EBO across the services.

In this interesting twist of fate, Mattis' first action upon taking command of JFCOM in 2009 was to terminate EBO in a formal memorandum banishing the theory to the wastebin.[14] JFCOM could no longer use EBO except in specific targeting applications, and even the phrase "EBO" was now forbidden. He saw greater promise in some version of Naveh's ideas as the new way forward for military forces and sought to clear the path of obstructions.

Donnell recalled that as the U.S. Army in 2009 moved to codify their design methodology into formal doctrine,

> that was what Mattis really leaned on hard … in order to pressure the Marine Corps when he was at [U.S. Joint Forces Command] to get back and reengage on the design as what was happening [when MSTP retrograded after Mattis' departure].[15]

The Marines would move in the same directions as the Army, also encountering similar institutional tensions on what design should be in an "Americanized form", and how their service ought to frame this construct. Donnell recalled, "those that opposed the ideas of design knew that when Mattis left MSTP, they knew that once he left Quantico, that MSTP was anti-design".[16] Institutional defenders for a period shielded their established practices and beliefs from any design, halting any design inclusions. Yet in the summer of 2009 after renewed pressure from Mattis at the flag officer senior level, the Marines restarted design integration to coincide with a major doctrinal revision of their core planning methodology. Soon, different factions began digging in to advocate or oppose various suggestions on design inclusion.

Marine design experimentation moved gradually until the 2010 publication of the Marine Corps Warfighting Publication (MCWP) 5–1, *Marine Corps Planning Process*. In that publication's foreword signed by Lieutenant General George Flynn, the Marines acknowledge that "practical application [of the Marine Corps planning process] has also revealed that portions of the planning process and MCWP 5–1 require clarification or elaboration … among these, *design* has emerged as a term requiring further emphasis".[17] Taking a more explicit step than the U.S. Army, the Marines would also rename the first step in their planning (previously called *mission analysis*) as *problem framing*. They did this to emphasize the design concepts being added to how the Marines were expected to make sense of warfare, and unlike the Army, the Marines *required* design thinking in every single planning endeavor. In the first chapter of the revised planning publication, the Marines would position design as central to their own doctrinal philosophy of maneuver warfare (Figure 5.1).

The 2010 planning doctrine would go on to introduce how the Marines should understand design in several pages in the first two chapters, generally matching the brevity of the U.S. Army's design chapter in their Field Manual 5–0, *Operations Process* from the same year.[18] The Marines would hold to the U.S. Army orientation of linking design to solving complex problems in warfare. Marines defined design as follows:

> Design is the conception and articulation of a framework for solving a problem … Design provides a means to learn and adapt and requires intellectually versatile leaders with high-order thinking skills who

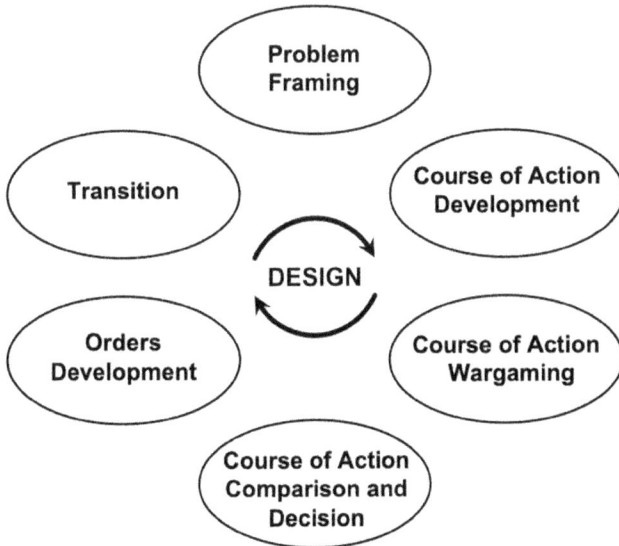

Figure 5.1 How the Marines Conceptualize Design and Planning (2010–2022).
Source: MCWP 5–1, 2010 edition.

actively engage in continuous dialogue and collaboration to enhance decisionmaking [sic] at all levels.[19]

Like the Army, the Marines would shed any postmodern theory as well as most of the complexity theory out of security design, focusing instead on assimilating many concepts and terminology from complexity theory and systems thinking. Van Riper, in 2021, lamented this failure to intellectually expand their frame as Naveh envisioned:

> At its heart, SOD relies on discourse learning to an understanding of the logic of a problem from which a potential solution emerges naturally. The Corps, regrettably, was and remains attached to an unsound rule-based planning process to which it has melded a flawed understanding of design.[20]

Mattis also reflected on how the Marines would struggle in shifting from a mechanistic, linear planning conceptualization toward something different. In a 2021 interview, he offered that design might even be a "precursor (state-setting) effort enabling proper planning ... coupled to a focus on the enemy and Boyd's OODA loop [model], in an application that meant it was not linear".[21] Design would, if incorporated properly, loosen the grip that this mechanistic, Newtonian styled planning had upon the Corps. Mattis felt design for the Marines would become "near constant even to the point of being a catalyst to variations on a theme, so to speak, of any plan, to include a watchful eye on assumptions that call for a change in the plan (starting with design)".[22] Mattis would seek divergent thinking, disruption, and critical self-examination so that Marines did not remain trapped in a convergent, linear, non-reflective mode of warfighting. Eventually, the Marines would make significant changes in how they approached design, but these would not occur until well after the original transformation efforts of Mattis and others.

Unlike the Army's design chapter in *FM 5–0,* the Marines in 2010 selected brief military examples to try to illustrate how design can inform mechanistic, systematic planning activities, and that design is not just the first step, but something continuous and iterative. One example draws from the Korean War involving General MacArthur's Eight Army, with the other from the American Civil War against the Confederate Armies. The 2010 Marine design doctrine advocates that design is "a continuous activity and must never be viewed as an isolate event occurring only during problem framing. It occurs throughout the planning-execution-assessment continuum".[23] This is more in keeping with Naveh's original SOD and may link back to some earlier design recommendations first made by Vivian in his white paper. Vivian stressed the iterative aspects of design and complex systems where

> the decision maker continues to learn from the implementation of the solution; understanding of the problem continues to evolve … the decision maker could then re-enter the beginning of the cycle and construct an updated solution or mental model, based on the information gained from implementing the original choice.[24]

This emphasis on weaving SOD design concepts throughout the traditional military decision-making process would appear to retain in this first Marine design doctrine in 2010. Yet the SOD name would be ignored, and the Marines unwittingly would call their new concept "operational design" instead, which already was the well-understood title of campaign planning for Joint Forces and Services dating back decades. Donnell recalled: "SOD was intimidating to [the early Marine doctrine writers tasked with writing the content]. It became important to them that this new design would not be called SOD, but just be called 'operational design' instead".[25] Institutionally, the U.S. Department of Defense now formally had multiple doctrines across different services calling different things by the same name.

Aside from variations on that first graphic positioning design as central to how the Marines move from problem framing to orders development and transition, there were no other design graphics or illustrations on design itself. The doctrine declares: "The goal of design is to achieve understanding largely through critical thinking and dialogue – the basic mechanism of design. The ability to address complex problems lies in the power of organizational learning through design".[26] Readers may notice that like the Army, the Marines prioritize critical thinking within design but avoid any reflective practice. There is no appreciation of the Marine paradigm for warfare, or how designers might gain self-awareness of such a thing. Ryan's critique of the Army's inability to bring this critical part of Naveh's SOD into design is applicable here. How does an organization introduce the design of a new war paradigm if one cannot even start a discussion that addresses various paradigms?[27] The Marines would lack the specific illustrations and conceptual models such as the "Three Ball Chart" found in the Army design efforts, leaving the manner of how their design unfolded arguably vaguer and more ill-defined.

The Marines placed far more emphasis on problem framing, critical thinking, and discourse than previous doctrinal publications. How the commander and the staff went about thinking and discussing a complex, ill-structured problem would become paramount in design, unlike traditional planning where hierarchical, bureaucratic structures made such things clearer. The 2010 Marine design content was, like the Army's 2010 efforts, confined to a few pages and nested within a much larger planning document. The Marines did place a clear division between design activities and those of planning, stating: "having engaged in a design dialogue with his planners and staff in order to gain insight into the problem, the commander provides his initial intent and guidance in order to direct continued actions in the planning process".[28] Despite stating earlier in the publication the importance of design occurring well beyond the initial problem framing phase, the rest of the publication would contradict that notion, and render design a first step in what was otherwise the same Marine deliberate planning process that Mattis, Van Riper, and other leaders sought to transform.

The 2017 Revolution: Marines Brand Their Own Design Methodology

A seven-year gap in any overt design thinking would occur in the Marine Corps, spanning from 2010 to 2017. In an interesting overlap, a similar period of design stagnation befell the American Army, and also the Australian Defense Force (covered in the seventh chapter). It is difficult to determine whether design merely continued internally, at education programs or within units with military designers, or if the organization writ large took a break from deep institutional reflection and transformation. During that period, the Coalition drew down in Iraq, and the Afghanistan surge of forces quickly transitioned into a new effort by the Obama Administration to accelerate hand-over to the Afghan Security Forces. Military deployments and combat rotations diminished, and the emerging near-peer competition between China, Russia, and over previously uncontested geographical regions such as the Arctic, low-earth orbit, and the South China Sea became some of the emerging new challenges.

In March 2017, the United States Marine Corps published the first version of their own military design methodology. This was not a formal doctrine, rather it represented draft or conceptual material that the Marines might consider. The new design document would be similar to the Army Design Methodology found in the 2010–2015 doctrinal publications, but also share some inspiration from Joint operational design concepts using earlier campaign planning constructs. Donnell, a doctrine writer during this period, recalled: "this was a strategy where the Marines could get a design method out to the forces … in a pre-doctrine format," so that Marines had a cleared methodology to apply instead of the short design chapter from 2010.[29] The informal authority of a Marine pamphlet would complicate efforts to impose this particular design approach across the Marine Corps, particularly with those Marines already using a looser, more SOD-esque design approach. Colonel Travis Homiak, a SAW graduate and familiar with Naveh's design, was familiar with the publication in 2017 but discounted whether it had institution-wide impact. In his own words,

> I believe that publication is legitimate [actual doctrinal material published and current in the Marines], but I don't think that there is necessarily any institutional horsepower behind it … [These] pamphlets do not carry the institutional force of a *Marine Corps Doctrinal* or *Marine Corps Warfighting Publication* (*MCDP* or *MCWP*).[30]

Despite questions on the doctrinal authority to shift design toward the ADM style, this pamphlet would be a critical step in the Marine Corps moving to formal design doctrine, as it would in 2020. Several SAW graduates would write this 2017 document, taking significant cues from the SAMS design program and existing Army Design Methodology.[31]

The Marine Corps would couch their design methodology firmly within the epistemological framework of how it understood decision-making through the MCPP. Marine Design Methodology existed to compliment the established detailed planning process, and not the other way around. The *MSTP 5–0.1* document entitled *Marine Corps Design Methodology* is a doctrinal pamphlet, meaning that it is taken not as the rule, but as a suggestion. *MSTP 5–0.1* would mimic many of the Army Design Methodology content and processes, demonstrating a shared institutional war paradigm that reinforces the same critiques offered in the second and fourth chapters. American militaries seek to freeze complex systems, isolate sections, reduce them, determine rules systematically, and then reassemble the system to continue a reverse-engineered, Newtonian styled mode of warfighting.

The 2017 Marine design doctrine, however, does differ from Army in that it unapologetically references *where it came up with some design content*. Their introduction cites the Army's existing design doctrine and Joint operational planning as the core sources for their design method. The Army design doctrine makes no such mentions, which is indicative of most all military

doctrinal publications generally. That the Marines would make this exception suggests they realize that design is a new, yet also controversial idea.
Perhaps, Marine doctrine writers felt it necessary to emphasize the larger
history of design thinking so that any initial skepticism within the Marine
Corps is blunted. To accompany the roll-out of this new concept, doctrine
team members went on the road with elaborate slide presentations to explain
the new design content.[32] That said, the publication does not detail Naveh or
the earlier Marine design experimentation in the 2004–2010 period.

The Marines replicated most aspects of the Army's sequential problem-
solving design construct, almost to a fault. Why this occurred is unknown,
but likely the result of Marines studying design within the Army schools such
as SAMS and returning to the organization to share the ideas. Additionally,
many Marines would read about design in articles or in conferences and at
war games, getting more of the Army Design Methodology content than
the less understood, infrequently published Israeli SOD. This Marine design
would not just emulate what the Army did to extract certain concepts from
the dense, intellectually intimidating SOD content, but the Marines would
also brand their design a naming device that was single-service centric.
The U.S. Army had "Army Design Methodology", and five years later the
Marines claimed theirs with "Marine Corps Design Methodology" as well.
Coincidently, the U.S. Navy in 2013 would publish their own design in
Navy Planning (Navy Warfare Publication) *NWP 5–01*, appendix D. Calling
their concept "Design", the Navy stipulates that unlike operational design,
military design is "an optional methodology that may be used in concert
with operational art prior to and in conjunction with mission analysis".[33]
The Naval design methodology directly mimics ADM, to include recycling
the "Three Ball Chart" with minor wordsmithing. The Naval design effort
appears disconnected from this Marine effort, and that the Marines make
no mention of this Naval design doctrine suggests the Marines ignored it as
much as Naval officers did.[34]

The Marines conceptually in 2017 made a significant pivot from their earlier
2010 framework on design being iterative and implicit throughout all Marine
planning and executing. Instead, this new suggested doctrine directed that
design activities be sequenced as *the first step* for a marine organization to execute within a predetermined linear and sequential planning methodology.[35]
It also declared that design "[does] not occur anywhere else in planning or
execution".[36] For Marine designers, they would deal with complexity, solve
it, and apparently move right into traditional Marine planning once that obstacle is overcome. Readers should recall that the U.S. Army originally in
the 2005–2009 period positioned design distinct from planning processes,
drawing from Naveh's belief that design and planning are independent of
one another yet complementary for different purposes. For Naveh, design
never ended, nor did planners turn from design to planning; a fusion of novel
design led to novel planning with iterative, improvisational, ever-changing
adaptations occurring through both efforts. Yet in 2010, the Army would

first assimilate design into a dominant planning framework, suggesting that Marines exposed to Army Design Methodology in 2010–2017 likely took this rearrangement back to the Marine Corps. Naveh's earlier emphasis on SOD design fundamentals would largely be lost to both organizations in their design indoctrination phases.

Israeli design doctrine published in 2015 also shifted design into a precursory step to operational planning, but that document was not widely circulated outside the IDF and is unlikely to have impacted Marine design development.[37] Nonetheless, this pattern of subordinating design to military planning demonstrates deviation from earlier Naveh SOD, and suggests that the American Army's prominent design efforts in 2010–2015 may have pulled Marine design thinking away from Mattis' original intent toward some fusion of design terminology and entrenched Marine beliefs in war remaining in a systematic, reductionist, Newtonian style. The loss of original design and this continued institutional fixation on legacy modes of warfare would in 2021 continue to frustrate senior Marine leaders and retired mentors. Lefebvre reflects on this tension with:

> [In 2021] we [still] do not have the professional military education right. There is no doubt about it. You can develop all the warfighting concepts in the world you want, but until you fix the way we think about the adversary … you are not going to fix this.[38]

Lefebvre would take issue with a continued singular sort of adversarial framing exclusively done through the traditional MCPP decision-making methodology and legacy campaign design techniques.

In another interesting variation, MSTP Pamphlet 5–0.1 presents the need for *tactical* design thinking, in keeping with marine culture that focuses on small unit, decentralized and remote operations. The Marine's tactical emphasis is unique, in that the Army and also the Israeli Defense Forces focus on higher operational levels for action, with the IDF centering design entirely at the general officer level specifically. Neither the Army or the IDF discourage design at tactical unit levels, but Army doctrine suggests that smaller, subordinate units may lack the specialized staff or numbers to support dedicated design activities as an operational level organization might. Marines differ in this, as their design and their operational planning address tactical activities prominently, to include being the only military force that declares a "tactical center of gravity" for operations and detailed planning.

The Marines would not incorporate the original Army "three ball chart" graphic, yet they would assume most of the methodological processes and narrative found in Army Design Methodology doctrine. Thus, the chart is eliminated, but the organizing logic of Army design extends directly into Marine design. The 2017 Marine design pamphlet would conceptualize Marine design visually for the first time, depicted below where design and planning are woven together as a process flow chart. As a conceptual document, *MSTP 5–0.1* would propose the Marines describe the current and desired states of the operational environment, define a problem set, and produce an operational approach for planners to then conduct a detailed analysis and coordination with. While the figure below relies upon a linear, sequential, systematic arrangement of blocks and arrows reading from left to right, it also retains virtually all aspects of Army Design Methodology down to each of the "three balls" reformed as linear activities placed in order of execution (Figure 5.2).

In Figure 5.2, Marines integrate a linear planning methodology (the MCPP) where the design activities are front-loaded into the mission analysis section of their detailed planning process flow. There is an adaptation beyond the Army's design where Marine planners work to contribute to design efforts through a fusion of planning teams doing traditional activities to inform the design team. The bottom element of "Intelligence Preparation of the Battlefield" or IPB is depicted to occur *simultaneously* with Marine design. This differs from Army Design Methodology that declared design a separate, precursory, and optional activity from planning. The Marines would assimilate design into MCPP so that designers and planners collaborated each time a Marine organization needed to commence any decision-making activity.

There is another subtle distinction between Marine and Army design in that the Marines appear to apply systems theory logic with greater appreciation of the inability to reverse-engineer precise "end states" in complex systems. *MSTP Pamphlet 5–0.1* softens the original Army design language by discussing *desired states* that are defined as "a product of Design that represents

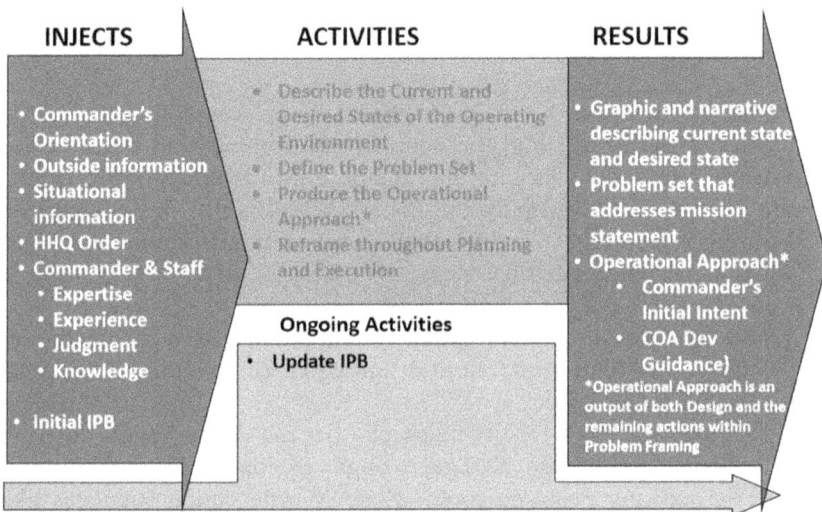

INJECTS

* Commander's Orientation
* Outside information
* Situational information
* HHQ Order
* Commander & Staff
  * Expertise
  * Experience
  * Judgment
  * Knowledge
* Initial IPB

ACTIVITIES

* Describe the Current and Desired States of the Operating Environment
* Define the Problem Set
* Produce the Operational Approach*
* Reframe throughout Planning and Execution

Ongoing Activities

* Update IPB

RESULTS

* Graphic and narrative describing current state and desired state
* Problem set that addresses mission statement
* Operational Approach*
  * Commander's Initial Intent
  * COA Dev Guidance)

*Operational Approach is an output of both Design and the remaining actions within Problem Framing

Figure 5.2 How the Marines See Design in Their "Process Flow" to Action.
Source: Marine Doctrine.

a feasible set of conditions at a future time, within a zone of tolerance, that are more favorable than the current state".[39] Although the remainder of their design methodology returns to a Newtonian styled, mechanistic manner of detailed planning, this difference demonstrates a Marine appreciation of complex systems, emergence, nonlinearity, and potentially what Chia and Holt recognize as "emergent spontaneous order" in complex socially constructed reality.[40]

MSTP Pamphlet 5–0.1 deviates from the earlier Army inspiration by requiring Marine designers to "describe their design using a narrative and a graphic".[41] This is not found in Army design doctrine and suggests that Marine designers may have considered some of the Army studies on problems with ADM implementation that included concerns that design ideas were not effectively conveyed from design teams to planners.[42] The emphasis on Marine design deliverables incorporating both a graphic and a narrative are required to "provide a clear, concise, and familiar way of portraying this information to a decision maker ... [while also] enhancing the understanding of the environment".[43] Ryan's observations on Army design remain applicable to the Marine version, in that if the Marines insist on design narratives and graphics, but the same design doctrine stipulates institutionally accepted military terminology, models, and constructs as the only tools for completing that task, Marine designers are channeled into producing institutionally safe (thus hardly critical or disruptive) deliverables.

Thus, Marine design flexes divergently from traditional military plans by requiring narratives and illustrations, but not so far that any possible design deliverables are not immediately understood by Marine planners using legacy terminology, concepts, and planning models. Marines design as technical rationalists within the overarching world of Marine planning beliefs. Ryan explains:

> Technical rationalism combines a naïve realist epistemology with instrumental reasoning ... [but] the dominant institutional culture does not have the time or patience for philosophical distinctions. Doctrine limits the professional language of the Army [or Marines in this context] ... Naming is framing. It is difficult to escape the institutional paradigm when you can't change the language.[44]

Marines forced to design with specific constructs vetted by the institution simply will be unable to break out of designing anything but an institutionally acceptable design result.

MSTP Pamphlet 5–0.1 repeats much of what the Army design doctrine insists military designers do. Designers ought to conduct design by using most of the existing planning tools, where Marine design stipulates like the Army for designers to consider mission variables in a reductionist, categorial "Mission, Enemy, Terrain, Weather, Troops, Time" modeling. The design doctrine follows the Army lead and correlates complex military challenges with a "problem-solution" formularization, showing a preference for a Newtonian stylization of war over that of complexity theory.[45] Rittel expands on this misunderstanding of designing in complexity where one cannot attempt to "understand the problem" nor should "solution" be expected as some ultimate completion of the design task. Complex problems have no stopping rule and reject these efforts to preconfigure action prior to enacting with the perceived problem in dynamic, ever-changing, and un-testable ways.[46]

The doctrinal pamphlet features numerous graphics that illustrate linear-causal diagrams where "causal loops" are mapped in an objective, analytically optimized fashion.[47] This presents an "ends-ways-means" and potentially a return of "Effects Based Operations" thinking into this representation of Marine design, demonstrating a serious rift with Mattis' and Van Riper's original vision. Donnell, part of the 2010 and 2017 Marine design doctrinal efforts, declared that these systematic, reductionist graphics were omitted from the 2010 Marine design doctrine due to sufficient opposition to the EBO proponents:

> In the 2010 version, we killed those graphics. Those diagrams were pushed to the back to an appendix because we felt they were too linear. In the 2017 version, everyone fought and wanted them back. But we lost the battle and they got put back in.[48]

The technical rationalist population within the Marines would take back conceptual territory lost in the 2010 first effort to indoctrinate Marine design.

The 2020 Publication of MCWP 5–10, Marine Corps Planning Process

By 2020, the Marines would formally insert Marine Design Methodology into their doctrine with a new publication of the Marine Corps Warfighting Publication (MCWP) 5–10, *Marine Corps Planning Process*. Shifting back to the original 2010 Marine doctrine, the 2020 design chapter declares that "while design occurs throughout problem framing, design is an enduring activity not confined to the problem framing step".[49] Just as the Marines prominently placed design in the foreword of earlier 2010 planning doctrine, this 2020 version would immediately state that "the use of design over the last decade suggests that design is more than conceptual planning which establishes aims, objectives and intentions".[50] However, this formal doctrine for Marine design quickly moved to a declaration that design is used to solve complex problems, thus illustrating a continued institutional rejection of complexity theory, original SOD, and an entrenched war paradigm rendered in a Newtonian stylization of systematic, reductionist decision-making:

A more critical role of design is to promote understanding of the current situation as a basis for broad solutions. While design establishes the nature of the problem, the inclusion of a design methodology in this revision aids commanders, staffs, and planners in determining the problem set and a framework *for solving them.* The publication's design methodology reflects a belief that sufficient complexity can exist at all levels of warfare and across the conflict continuum to include tactical situations that will require an understanding of the set of problems that hinder movement from the current state to the desired state of an operational environment [emphasis added].[51]

Thus, despite a decade of Marine examination and consideration of security design praxis, they remain in 2020 wedded to a systematic warfare logic *that seeks to link inputs to desired outputs while maintaining all legacy warfare models and theories.* Despite the Marines stating in their 2020 planning doctrine that "planning seldom occurs in a straightforward, linear manner", they continue to offer up models, methods, and theory that permit precisely that sort of linear, systematic reasoning for warfare that is described and intended to be executed (and self-evaluated) in a sequential, doctrinally adherent manner.[52] Problem framing is positioned as the most important step in how the Marines construct warfare solutions to perceived warfare problems in order to reach desired future states (through identified military objectives and goals). Design is positioned as a commander-driven process where staff will conduct design activities following the Marine Design Methodology. This 2020 doctrine continues to emphasize "time and risk" as enduring, critically important factors that influence all aspects of design, planning, and execution.[53] The Marines repeat nearly the same Marine Design Methodology graphic as the 2017 design pamphlet, but in this updated version they remove the parallel "intelligence preparation of the environment" planning activity, making the design praxis the sole activity for this "problem framing" step in the overarching MCPP methodology (Figure 5.3).

The 2020 version of Marine Design Methodology thus remains largely unchanged from their earlier 2017 version except for minor adjustments and an updated, even simpler graphical depiction for how Marines should design. *MCWP 5–10* states the goal of design is to "achieve understanding gained largely through critical thinking and dialogue – the basic mechanisms of design … Group dialogue, when conducted within the proper command climate, can foster a collective level of understanding not attainable by any individual within the group".[54] Whether Marines can effectively do this beyond the rigid, institutionally directed rules and directives in *MCWP 5–10* is undetermined, but arguably difficult to do critical or creativity thinking that disrupts, challenges, or transforms the organization outside of clear boundaries for Marine design.

The rest of the design section in the new 2020 Marine planning doctrine revisits much of the draft 2017 pamphlet. Marine design continues to be a

INJECTS	ACTIVITIES	RESULTS

• Commander's orientation
• Outside information
• Situational information
• HHQ order
• Commander and staff
  ⊙ Expertise
  ⊙ Experience
  ⊙ Judgment
  ⊙ Knowledge
• Initial intelligence preparation
  of the battlespace
	Design Methodologies
	Supported by Staff Actions
	
	• Describe the current and
	  desired states of the operating
	  environment
	• Describe the problem set
	• Produce the operational
	  approach
	• Reframe throughout planning
	  and execution
		• Graphic and narrative
		  describing current state and
		  desired state
		• Problem set
		• Graphic and narrative
		  conveying the operational
		  approach
		  ⊙ Mission statement
		  ⊙ Commander's intent
		  ⊙ Course of action
		    development guidance

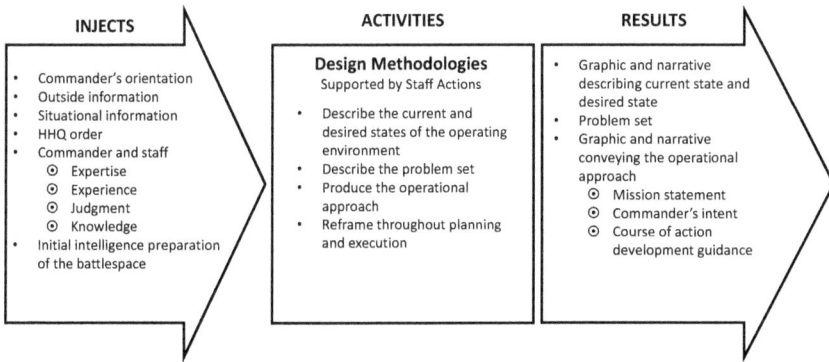

Figure 5.3 The 2020 Update to How Marines Frame "Problems" with Design.
Source: Marine Doctrine.

largely a process of planning effects, articulated using a blend of design terms and planning models and theories. Marine designers are tasked to determine what the correct set of problems is, and then design solutions that carry directly into established planning practices. Despite *MCWP 5–10* declaring that "planning seldom occurs in a straightforward, linear manner", the Marine doctrine continues to offer up models, methods, and theory that permit precisely that sort of linear, systematic reasoning for warfare that is described and intended to be executed (and self-evaluated) in a sequential, doctrinally adherent manner.[55] Problem framing is positioned as the most important step in how the Marines construct warfare solutions to perceived warfare problems in order to reach desired future states (through identified military objectives and goals). Design is positioned as a commander-driven process where staff will conduct design activities following the Marine Design Methodology. This 2020 doctrine continues to emphasize "time and risk" as enduring, critically important factors that influence all aspects of design, planning, and execution.[56] Design becomes another regimented, rationalized activity set within the overarching Marine institutional paradigm advocating a singular way to interpret complex reality.

In one interesting departure from the Army design inspiration, Marines discuss a "problem set" that is not intended to be some convergent process of isolating a single primary design problem as the Army directs. Marine design doctrine states: "it is unlikely that the design methodology will expose a single problem to solve. In reality, when engaging complex systems, many problems will emerge".[57] Yet the same marine design doctrine demands the integration of COG analysis into the design problem set, illustrating the Army pattern of forcing institutionalized beliefs on war and traditional planning concepts as mandatory design tools for designers to think differently. How Marines might think differently while shackled to rigid planning constructs is not addressed in the doctrine. In another confusing statement, *MCWP 5–10* states: "design establishes paradigms whereas task analysis is paradigm accepting".[58] This violates most every definition of social paradigm theory in multiple ways and implies either a fundamental misunderstanding of social paradigm theory in general, or perhaps an accidental Marine doctrinal word insertion.[59] As "paradigm" is not included in the Marine glossary, it remains unclear what the doctrine even means. Tasks and analysis activities occur within some paradigm, and design is expressed through considering one or more social paradigm. The doctrine writers appear out of their element, unintentionally emphasizing Naveh's particular zeal that designers must intellectually develop themselves across many disciplines so that they can use ideas from sociology without breaking them in the act of using them.

Marines maintain the centralized hierarchical form of modern military bureaucracies and continue to hold an epistemological stance shared by the Army on how commanders have both the authority and the deep knowledge of warfare to perpetually be the sole arbitrators of what the design team should and should not do. Further, the Marine commander is not involved in the design activities either, as the doctrine directs the design team to brief the commander once they finish the design. "[This] will also allow for an early opportunity to revisit design if the commander does not agree with the problem set".[60] Placing the commander central to decisions on complexity is once more a rejection of complexity theory, Naveh's SOD, and a continued extension of the Newtonian styled framing of modern warfare. Doctrinally, *MCWP 5–10* confuses readers in a declared emphasis on "group dialogue, when conducted within the proper command climate", the commander's "personal involvement and leadership in the planning process",[61] but then follows with directing that design teams: "should brief the commander on the design results … It will also allow for an early opportunity to revisit design if the commander does not agree with the problem set".[62] If commanders are personally involved, it is strange that they would still require briefings and potentially disagree on a problem set that they directly shaped.

Thus, *MCWP 5–10* lurches back and forth between clearly planning methods and concepts to that of design ones, seemingly written by two entirely different populations of doctrine writers not paying close attention to what the other group is adding to the document. The brief paragraphs addressing design reframing are especially problematic. The earlier *MSTP 5–0.1* pamphlet stipulated that design "is reexamined routinely during planning and throughout mission execution *when* significant changes to the operating environment occur [emphasis added]".[63] This particular use of "when" suggests that while complex systems are dynamic, nonlinear, and feature emergence that cannot be predicted with any prior planning or analysis, there is some sort of logical evaluation done by the commander and/or staff on when systemic changes require either additional planning, or deliberate reframing through design. Neither *MSTP 5–0.1* nor *MCWP 5–10* clarifies this, implying that the organization leadership somehow tacitly know when design or planning requires revisit.

MCWP 5–10 does provide a linear–causal illustration of when reframing might occur, yet the explanation remains so vague as to mean just about anything could create reframing conditions.[64] Reframing is tied to change, poor understanding, and organizational surprise through some sort of failure (in planning or action). The 2020 doctrine takes much from the *MSTP 5–0* pamphlet, including the design reframing rationale where if the Marines need to reframe any design or planning portion of their decision-making process, they will analytically determine which step needs to be revisited, and that revisit is governed each time by the overarching doctrine. Whether designing or planning, the teams must repeat the sequence of activities, use the same conceptual tools, produce the same expected deliverables, yet somehow produce something different and superior to the last lap with the same exact requirements and stipulations.[65]

Just as the Army and Israeli SOD design methodologies were mapped epistemologically to examine one way to frame the "why" of these approaches to complex warfare, the Marine efforts to craft their own design methodology is treated below. Either the 2017 *MSTP 5–0* graphic or the simplified *MCWP 5–10* illustration suffices, as both employ near identical theories, conceptual models, methods, and terminology for design. As the Marines drew heavily from Army Design Methodology and demonstrates little if any direct influence by Naveh's SOD approach, the below epistemological framing shares far more with the Army than with how Israelis continue to pursue design thinking. Since the U.S. Navy also mimicked ADM in 2013 as mentioned earlier in this chapter, there appears to be significant overlap and commonality between Army, Marine, and Navy design methodologies. This seems to be a far cry from the original intent of Mattis, and indicates the pattern of design assimilation into traditional, Newtonian styled planning frames spans across multiple military services (Figure 5.4).

The above illustration provides similar epistemological framing as earlier design methods in this book and represents one of many interpretations possible. Marine design doctrine exists to explain how fellow Marines are expected to conduct design activities along with integrating design into traditional Marine planning. It is also nearly identical to the previous chapter's U.S. Army epistemological illustration except in two regards. A new horizontal arrow is added on top where a non-design activity of Marine linear planning occurs simultaneously. This reinforces the particular Marine expectation that Marines should design and plan in parallel, with tacit interaction and discussion between the two groups frequently as they both move through their flow of information, process, and operations.[66] A vertical arrow is also added between the two parallel planning and design sequences, illustrating the interplay, overlap, and implied tensions between the Marine planners and designers. Marine design constructs new "ends-ways-means" constructs that provide a proposed operational approach for planners to further test and

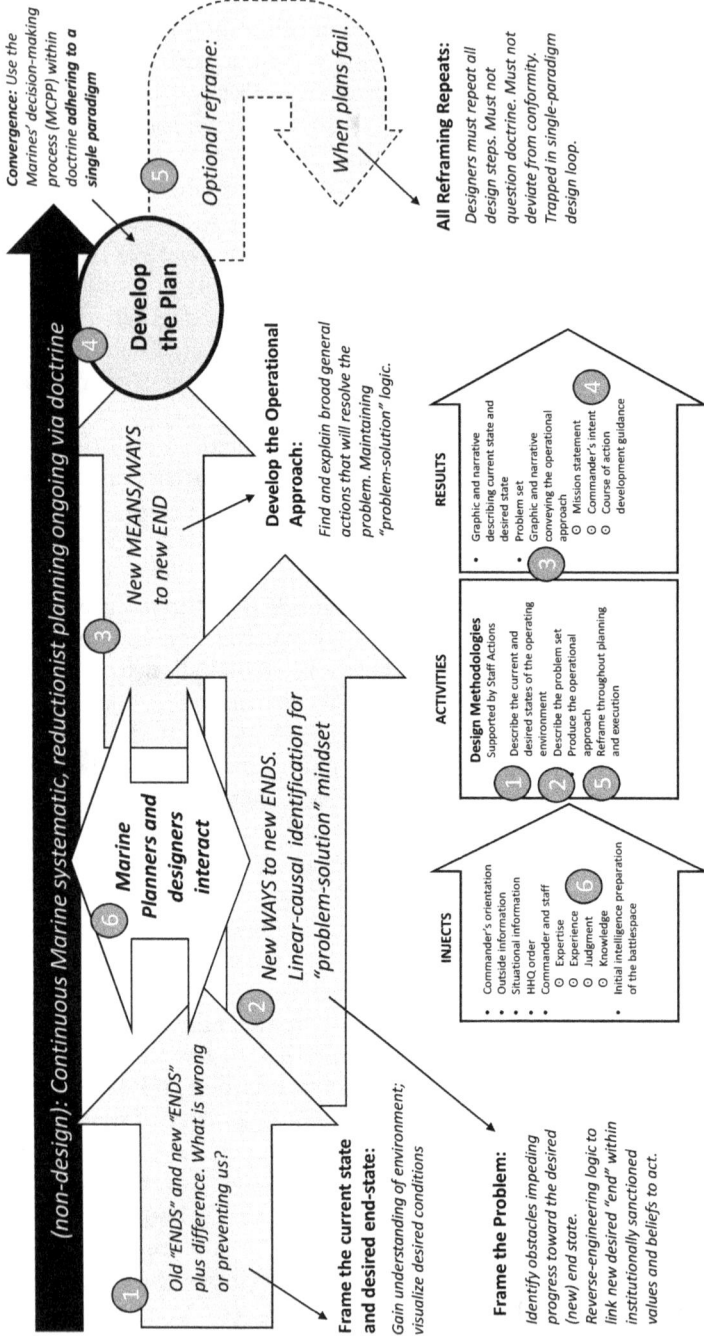

Figure 5.4 Epistemological Frame of Marine Corps Design Methodology.
Source: Author's Creation.

refine in "course of action" development and analysis. This mimics Army Design Methodology, but both the Army and Marines break from Naveh's SOD in the assimilation of design into unchanged planning methods.

Marine design thinking is believed by the institution to pair analysis with systems thinking so that planned courses of action can be made more effective against an adversary. The 2017 design pamphlet framed the design of more complete battlefield understanding with design: "through a careful study of the effects of a course of action on the enemy system".[67] Naveh, in a 2019 interview, rejects the pairing of military design activities to the validation or improvement of COAs. "The whole idea of COAs is wrong, totally wrong. Design cannot be based on this method – these analytic tools. Design is a synthetic tool, not an analytic tool. This is a huge mistake".[68] Courses of action operate convergently, with all COAs moving toward the very same desired end state or goal in military planning. Design works *divergently*, meaning that the more design prototypes and iterations pursued, the further a design team might move into uncharted areas and unrealized conceptual terrain for the commander. Epistemologically, Marines offer a design methodology that reinforces the very aspects of linear, systematic decision-making that original SOD sought to disrupt.

Marines, sharing the same convergent efforts for design as the Army, tightly bond design activities with COA development, with *MCWP 5–10* using "COA" 394 times in the 205-page doctrine. Curiously, the Marine problem framing states that "the operational approach requires the input and synthesis of both design and the staff actions".[69] Marine decision-making mirrors all Department of Defense planning methodologies where multiple COAs are generated, analyzed, and eliminated so that a singular COA is decided as the one way forward for military action. This is planning, but not how design functions. Naveh offers: "Decision making is not selecting one way, one course, one opinion – it is about synthesizing; getting to a higher level of understanding which means that you [the general] is supposed to do the thinking yourself".[70] Naveh uses "synthesis" in the appropriate context of systemic logic, whereas the aforementioned Marine doctrinal passage misapplies it in a systematic (known inputs lead to desired outputs sequentially) mode. Marine design in 2020 functions not as a disruptive agent to pulling the institution out of group think, but as an assimilated agent of enabling further planning just as the institution desires it. The disruptors are neutralized, and design in 2020 is unlike the original Naveh inspired vision of General Mattis and others in the earlier 2004–2009 period of Marine experimentation.

Flying Too Close to the Sun: Marine Innovation Efforts Crash to the Earth

Over two decades have passed since the first Marines encountered military design. The early days of listening to an eclectic, retired Israeli General spout on about postmodern theory, architecture, and systemic shock in complex warfare would first inspire Marines and stimulate their curiosity. Institutionally, Marines have always had a reputation for improvisation, small unit flexibility, and a penchant for experimenting quickly to adapt to changing contexts. The Marines also are associated with rigid, mechanistic, and systematic ordering of military affairs, to the point that Marines are extremely protective on what enters their institution, whether it is new technology, resources, organizational relationships with outsiders, and also new ideas on war itself. Thus, the Marine Corps is a complex fusion of extremes, with a "can-do" motivation to overcome obstacles in unique, clever ways, along with strict regimented codes of behavior that do extend into not just what words are used, but which ideas are permitted. Marine design experimentation features large helpings of both sides of the Marine institutional dynamic.

Mattis would reflect in 2021 on the last nearly two decades of design experimentation within the U.S. Marine Corps and observe that:

> The discussions and arguments about adapting Design and defining it, from Quantico happy hours at the club to [Command Post Exercises] and in Camp Pendleton and at sea seemed right to me. They were vigorous discussions, which are generally signs of healthy, unregimented thinking.[71]

Mattis' unique senior position during many of the earlier Marine experiments offers an important perspective in this retrospect. He saw the integration of military design as something that did not require vigorous convincing; rather, it would be quickly incorporated because it was the right thing for the Marines to transform toward in warfighting. Mattis went on to add:

> Departing Quantico for command of I Marine Expeditionary Force (42,000 sailors and Marines in Southern California/Arizona) and MARCENT (forces in Iraq, AFG, Somalia, etc.) I found Design already in use in the Fleet. I did not need to sell or push it. I even witnessed both a company and battalion commander having a spirited argument with their regimental commander in the field in which the total argument was on the Design of the Helmand campaign post-Marjah's takedown. The detailed argument (it went beyond a discussion) revealed officers from captain through colonel who were applying it in stride. I witnessed a less argumentative but still detailed discussion between a Battalion Landing Team commander and his Marine Expeditionary Unit commander dealing with the Yemen situation when I was the CENTCOM commander. I did not witness the pushback, for example, that had led me to question the JFCOM embrace of EBO and their attempts to embed that in USMC doctrine and that we successfully rebuffed while I was at Quantico in close concert with General Wallace.[72]

Mattis echoes other senior Marine leaders that saw across the last two decades of design improvisation, experimentation, and eventual inclusion into Marine doctrinal publications the requirement for Marines to think differently both formally and informally. As two camps developed in the Israeli Defense Forces and subsequently the American Army, the same would occur in the Marine Corps. Some saw design in a pure, tacit, and highly intellectual way that had no limitations except the curiosity of the practitioner. This is visible particularly in the mid-2000s when officers such as Vivian were learning about design theory, and other Marine innovators sought to apply SOD methodologies in war games and exercises. Senior leaders like Mattis would champion the design ideas as the larger institution would shift into how the Marines could incorporate them into established practices, behaviors, and policies. Many of these design advocates enjoyed the radical, disruptive nature of Naveh's SOD in that it required militaries to un-learn while learning, and also re-learn as they iteratively explored well beyond institutional limits. This also had a powerful, disruptive, and seemingly elitist atmosphere that would not necessarily play into the entire institution.

The Marines would experience within their MCCDC a near-identical pattern of design resistance as the Army's TRADOC experienced in generally the same period between 2008 and 2015. Here, the second camp entrenched and fought to assimilate select less-offensive components of design into something that would enable traditional planning to remain unmolested by the design heretics associated with Naveh's SOD movement. As covered earlier in this chapter, MCCDC would even resist General Mattis' efforts, and subsequently the 2010, 2017, and 2020 versions of Marine design doctrinal content would shift away from the earlier Mattis-led, loosely bounded design practice to that of a regimented, systematic, institutionally dominated mode of creative planning. This tension would extend across the entire Marine design experience, and likely remains a powerful obstacle today. Donnell in 2021 reflected on the decade-long endeavor to bring security design into Marine warfare decision-making:

> So, during the ten years as we went from the original pub [(MCWP) 5–1, *Marine Corps Planning Process*] that was more of an abstract discussion of design to a more detailed form … we left the task analysis part in there, which is to gather all of your specified and implied tasks … determine those that are essential, that will form your mission statement; that is very formulaic … and I am the one that recommended we have to change the first step. It cannot be mission analysis, we do not get missions, we get tasks. And it is from the tasks that you determine your missions. And the power of the [Marine] planning process is that you get to determine your mission based on your understanding of the environment … there is a constant tension [on design in doctrine] where people say they need more, they want the checklist … and then there is the crowd that says, "I don't need that." Mattis, Smith, Van Riper and others that said, "tell

me what you want me to do and why, and don't waste my time telling me how.[73].

Van Riper would count himself with Mattis and other proponents of the Naveh SOD movement in that the freedom, disruption, and radical offerings of design in that mode seemed to best be able to shock the Marine institution into reform. Van Riper did not insist on the dense, postmodern content and heavy intellectual study as essential to mass inculcation to design, but new ideas from unorthodox areas outside the traditional Marine institutional boundaries seemed essential. He would also lament the institutional rigidness of the Marines (institutionally, not individually) and their inability to integrate many of the necessary design ideas into essential reforms for the Marines in a vital period of confusing, frustrating, ever-changing warfare. In 2010, Van Riper recounts how he was asked by several Colonels at the Marine Corps Combat Development Command to informally review the latest *MCWP 5–1*, *Marine Corps Planning Process* that would first feature any design for the Marines. Later he was asked by Lieutenant General George Flynn to make a formal review of the doctrine. Van Riper recalls in 2021 what transpired:

> None liked my critique because I found the document failed to do more than tag operational design to the front of the existing tedious and inwardly directed Marine Corps Planning Process. Moreover, the manual was poorly written. In short, Marine Corps doctrine writers were unable to separate conceptual planning—the realm of design— with the need for systemic thinking from rule-based functional and detailed planning. They fell victim to the default Marine Corps view on planning, that is, an analytical and systematic view. Though there is a low-level general understanding of nonlinear systems in the Marine Corps, few commanders and staffs can use this understanding to aid in the solving of complex problems because the Corps is wedded to a form of systems analysis in its decision-making process.[74]

Today, the Marines may be at a saturation point where enough younger officers and career professionals have risen in rank toward the top of the organization and carried with them some interest or familiarity with design. Some have Naveh and SOD exposure, others a version of ADM, and still others, a general interest and curiosity in how design, in some broad or interdisciplinary sense, might stimulate change in military thought. Most seem to agree that an intellectual investment is required, if only to realize the limits of the Marine institutionalized war paradigm, and how one might lead Marines out of that quagmire. Donnell observed that by 2021:

> Some of those SAMS and SAW grads have made it in the Marine Corps, to the point where they do make a difference and have an influence, and

they do understand [design] and they do not roll their eyes and consider it just some sort of philosophical discussion.[75]

This will be revisited in the Conclusion, where a grassroots, long-term movement of military design innovators appears to finally be making a significant tidal change across multiple services. Often, this spans a generation of military professionals or more, in that those willing to adapt new ways of thinking must rise to sufficiently high levels of authority, and/or enough time must pass for the rigid institutional defenders to fade away. Kuhn described this as the paradigm shift in scientific applications, but it equally applies to sociological paradigm and organizational theory.[76]

A small, but nonetheless present and active group of Marines continue to pursue design activities outside of the doctrinally sanctioned rules and simplified procedures. This population might be better associated with the current custodians of earlier Marine design visions, the practitioners that intuitively listened to senior commanders such as Mattis while rejecting the ham-fisted efforts of doctrine writers. There may also be causation in these same Marines attending advanced military schooling where they experienced Naveh's SOD firsthand, or through indirect exposure. One example is with Colonel Travis Homiak, a Marine within the special operations community and a School of Advanced Warfighting (SAW) graduate. During his senior planning assignment (G-5 plans, G-3 operations) for the Marines Special Operations Command (MARSOC), Homiak led multiple design efforts and integrated design into their organization's operational activities, generating design white papers for the MARSOC commander. He was their lead author for the first MARSOC doctrinal publication and their first futures concept.[77] He recalls on the free-spirited, improvisational context of Marine design as executed by some designers as follows:

> What I do know is that the Marine Corps mantra concerning design as something that happens before planning dates back to around 2010 with the first re-publication of MCWP 5–0 which has morphed into the 2016 MCWP 5–10 (The Marine Corps Planning Process). The publications very much contained the idea that design is something that all of us do innately when we come up with a plan. The MCWP describes design as "the conception and articulate of a framework for solving a problem," (MCWP 5–10, 1–3). The purpose being "to achieve a greater understanding of the environment and the nature of the problem in order to identify an appropriate conceptual solution" (MCWP 5–10, 1–3) … From my experience and applying design to the problems of MARSOC, as I think you already know, we pretty much went our own way in how and where we applied design to our specific challenges.[78]

The Marines had a surge of initial security design interest early in the 2000s when it was unable to achieve lasting success on new battlefields where opponents deviated from past warfare norms. That initial excitement triggered a reaction of institutional skepticism, fear of change, and resistance to any legitimate challenging of established theories, mental models, and methods already indoctrinated into prescribed Marine behaviors and decision-making processes. Just as in the Army, Marines attached to institutionalized practices and beliefs feared the disruptions that design created could destroy things that mattered to what they felt warfare required. Lefebvre recalled:

> What you risk is the baby being thrown out with the bath water, which is what could happen with design … [by 2021] there have been some recent changes at Quantico [center for Marine education] where Generals have been switched around … and you are bringing up a younger group of combat experienced officers that, over the course of the next two years you could see a regeneration.[79]

With innovation, new ideas have a historic age of zero, thus as innovation emerges it is impossible to differentiate between game-changing approaches and catastrophic failures. This risk is understandably terrifying to an organization tasked with risking lives to defend national security and execute dangerous foreign policy missions. However, the organization unwilling to innovate will ultimately be dragged there in defeat by an adversary able to design new advantages ahead of those unable to imagine beyond institutionalized limits. Select Marines like Homiak demonstrate some of the earlier design vision and experimental, improvisational qualities urged by Naveh for disrupting a rigid institution unable to let go of outdated ideas and practices. Perhaps over time, these mavericks may influence the next generation of Marines. In Homiak's case, he now commands the Naval Reserve Officer Training Corps (ROTC) unit at the Virginia Military Institute at the time of writing this book, in direct contact with the next generation learning how to conceptualize warfare.

Throughout the Marine experimentation with Naveh's design, the organization followed a strict, somewhat anti-intellectual regimen beyond even that of the Army. "Anti-intellectual" is used specifically in how the Marines demonstrated a clear pattern of including some disciplines and theories, but categorically rejecting others without any investment in studying why Naveh's SOD fused them all together. Marine design experimenters all eagerly adapted complexity theory and systems thinking without question. While select groups within SAMS studied postmodern ideas of SOD heavily for the Army (yet did not get any content into doctrine),[80] the Marines abandoned postmodern concepts at the onset. Schmitt, as one of the attributed authors of the early design chapter for *Field Manual 3–24/MCWP 3–33.5* which directly inspired future Marine design praxis, stated in correspondence with the author the following:

I didn't cite any postmodern sources [in the Marine design monograph "A Systemic Concept for Operational Design"] because I did not read any of them. As I said, I don't believe you have to go through them to arrive at the design destination. I was already very familiar with systems theory, Herbert Simon, Russell Ackoff, and Checkland's [Soft Systems Methodology]. I felt that was a very firm foundation.[81]

Schmitt reinforces this anti-intellectualist selectivity that military professionals appear to demonstrate upon exposure to Naveh's SOD. Few objected to deep study and intellectual development focusing on a scientific discipline, military history, or even architecture or psychology. Yet the postmodern theories foundational to Naveh's original praxis would be ignored, discouraged, or avoided. Schmitt is but one example of numerous military professionals dedicated to studying parts of Naveh's SOD, particularly those that reinforced other academic research they had already conducted, again in scientific disciplines.[82] Schmitt would dive into the parts of SOD he recognized and valued, and in turn, present those selections to the Marines as "military design" in an adaptation of Naveh's original SOD. In correspondence with the author, Schmitt reflected on the negative reputation of Naveh's postmodern infused concepts: "If postmodernism is why [many] dismiss design, then that reinforces my belief that it was a serious and avoidable mistake to base design on postmodernism".[83] This narrow form of anti-intellectualism reinforces Ryan's and Butler-Smith's observations made earlier of similar Army avoidance of the difficult postmodern content, as well as Israeli resistance as mentioned by Graicer.

Schmitt's actions reflect a microcosm of what would occur across the Marine enterprise as numerous other Marines made the same judgment on SOD. The postmodern content would be ignored and eliminated, yet the systems thinking and complexity theory, along with basic design processes would be advocated. Those responsible for Marine planning doctrine would also do the same in yet another round of selective anti-intellectualism, this time dismantling much of the systems thinking and complexity theory so that the original Newtonian stylings of traditional, systematic Marine planning would remain undisturbed. The first wave of SOD omissions occurred in the 2004–2010 period, with postmodern content ripped out. A second wave purged most all complexity theory and actual systems thinking, replacing most of it with assimilated buzz words, misleading graphics, and the ever-dominating Newtonian style of mechanistic, linear–causal military decision-making. Along the way, the purist design camp would decry the marginalization of Naveh's foundational concepts, while pragmatic planners argued the futility of introducing elitist, obtuse, and incomprehensible content to a force that needed everything clear and simple to understand, including complexity.

Ultimately, Marine design in 2004–2009 was unscripted, flexible, creative, and also executable by a small population of intellectuals that were at war with the doctrine writers. While some would continue to pursue an unorthodox Marine form of design, they would do this outside of institutional graces, and in contradiction of published design doctrine. Others would ignore even the basic design, preferring to continue traditional Marine planning methods unchanged, also bucking the institution in a conservative manner. As Van Riper reflects in 2021: "The Corps, regrettably, was and remains attached to an unsound rule-based planning process to which it has melded a flawed understanding of design!"[84] Naveh's ideas would be so challenging as well as controversial (and perhaps paradoxical in many considerations) that even champions of SOD in the Marines and Army would select certain aspects and deny or decline others. This would create a disruptive, uneven, and potentially convoluted entry of design into both organizations, and possibly delay the acceptance and assimilation of design as well.

Gradually, the Marines formalized their design into a set methodology that would draw more extensively from indoctrinated terms, concepts, methods, and models already utilized by Marine planners. Draft doctrine led to further institutional battles, while Marine design education occurred in various forms, spread across their schools and training centers where design would pulse between rejection and cautious examination. Marines that traveled into other services or nations would gain some other design exposures and bring these ideas back as well. A grassroots movement blurring many versions of military design would grow. By 2020, the formalized Marine Design Methodology held tightly to earlier U.S. Army models and language, indicating that both the American land forces would only accept so many exotic and radical ideas from the Israelis. Yet even this design doctrine inspires curiosity, introspection, and the desire for some to expand beyond the institutionalized barriers to consider what other designs might be beyond the pale.

Today, despite the 2020 formalized Marine design doctrine remaining rigid and wedded to linear planning logic, the U.S. Marines have a foothold in design thinking held by a younger, more combat experienced and inquisitive cohort entering senior levels of command and influence. Where the Marines go in the next decade after the catastrophic fall of Kabul in August 2021 remains to be seen. Their journey from the start of the 9–11 inspired "Global War on Terror" would take one path that this chapter explored. Meanwhile, the story of Naveh, SOD, and the Israeli Defense Force would continue beyond these experiments with the American Army and Marine Corps. Naveh would be cast out of the IDF and blamed as a contributing failure in the Second Lebanon War, but times changed. Eventually, the IDF would ask him to return, and resume teaching SOD within their defense forces.

Notes

1 Paul Van Riper to Ben Zweibelson, "Re: Design for Defense Book PDF Manuscript," September 20, 2021.
2 Van Riper to Zweibelson.
3 During this research, some Marines involved stated they did not understand post-modernism, and agreed that they also did not invest much time into studying the concepts. In opting to eliminate the concepts they failed to understand readily, the Marines demonstrate a similar pattern seen in the IDF and the U.S. Army where due to unfamiliarity with concepts well outside the military frame, they unwittingly began to misinterpret SOD itself.
4 James Mattis to Ben Zweibelson, "RE: Design for Defense Book PDF Manuscript," September 16, 2021.
5 Van Riper to Zweibelson, "Re: Design for Defense Book PDF Manuscript," September 20, 2021.
6 William Vivian, "Operational Art and Complexity: Learning and Deciding When Confronting Wicked Problems" (Marine Corps University, Command Staff and College, Quantico, Virginia: Marine Corps University, April 2006).
7 Steven Donnell to Ben Zweibelson, "Re: Replying from My Gmail," July 27, 2021. Personal correspondence with the author.
8 Donnell to Zweibelson.
9 Commandant of the Marine Corps, "Marine Corps Order 1500.53A: Marine Air-Ground Task Force Staff Training Program (MSTP)" (Headquarters, U.S. Marine Corps, Department of the Navy, August 30, 2002), www.marines.mil/portals/1/publications/mco%201500.53a.pdf
10 Kevin Benson, Interview of Colonel (retired) Kevin Benson, former SAMS Director, 04 AUG 2021 mp3 audio file, interview by Ben Zweibelson, mp3 Audio File, August 4, 2021.
11 Paul Lefebrve, Interview with Major General (retired) Paul Lefebvre on Marine Design Theory 17 AUG 2021, mp3 Audio File, August 17, 2021, 00:34:00 to 00:34:30.
12 Donnell to Zweibelson, "Re: Replying from My Gmail," July 27, 2021.
13 Lefebrve, Interview with Major General (retired) Paul Lefebvre on Marine Design Theory 17 Aug 2021, 00:34:45 to 00:35:01.
14 James Mattis, "USJFCOM Commander's Guidance for Effects-Based Operations," *Joint Forces Quarterly* 4th Quarter, 2008, no. 51 (2008): 106–108.
15 Steven Donnell, Interview of Mr. Steve Donnell by Dr. Ben Zweibelson on 28 Jul 2021 concerning Marine Corps Design Methodology mp3 audio file, interview by Ben Zweibelson, mp3 Audio File, July 28, 2021, 1:07 to 1:20.
16 Donnell, 10:10 to 10:26.
17 U.S. Marine Corps, *MCWP 5–1: The Marine Corps Planning Process* (Washington, DC: Department of the Navy, 2010), ii, www.marines.mil/Portals/1/MCWP%205-1.pdf
18 United States Army, *Field Manual 5–0, Operations Process* (Washington, DC: Headquarters, Department of the Army, 2010).
19 U.S. Marine Corps, *MCWP 5–1: The Marine Corps Planning Process*, 1–3.
20 Van Riper to Zweibelson, "Re: Design for Defense Book PDF Manuscript," September 20, 2021.

21 Mattis to Zweibelson, "RE: Design for Defense Book PDF Manuscript," September 16, 2021.
22 Mattis to Zweibelson.
23 U.S. Marine Corps, *MCWP 5–1: The Marine Corps Planning Process*, 1–4.
24 Vivian, "Operational Art and Complexity: Learning and Deciding When Confronting Wicked Problems," 6.
25 Donnell, Interview of Mr. Steve Donnell by Dr. Ben Zweibelson on 28 JUL 2021 concerning Marine Corps Design Methodology mp3 audio file, 1:03:00 to 1:04:00.
26 U.S. Marine Corps, *MCWP 5–1: The Marine Corps Planning Process*, 2–1.
27 Alex Ryan, "A Personal Reflection on Introducing Design to the U.S. Army," *The Medium* (blog), November 4, 2016, https://medium.com/the-overlap/a-personal-reflection-on-introducing-design-to-the-u-s-army-3f8bd76adcb2
28 U.S. Marine Corps, *MCWP 5–1: The Marine Corps Planning Process*, 2–3.
29 Donnell, Interview of Mr. Steve Donnell by Dr. Ben Zweibelson on 28 Jul 2021 concerning Marine Corps Design Methodology mp3 audio file, 13:19 to 13:32.
30 Travis Homiak to Ben Zweibelson, "Re: USMC Doctrine Question!," January 12, 2020.
31 Homiak, a SAW graduate and author of several Marine doctrinal publications, was not involved in the writing of *MSTP 5–0.1*, and was only vaguely aware of it when interviewed in 2020 by the author.
32 Donnell, Interview of Mr. Steve Donnell by Dr. Ben Zweibelson on 28 Jul 2021 concerning Marine Corps Design Methodology mp3 audio file, 15:00 to 15:30.
33 Office of the Chief of Naval Operations, *Navy Planning NWP 5–01*, Edition December 2013 (Norfolk, VA: Department of the Navy, 2013), D-1, http://dnnlgw ick.blob.core.windows.net/portals/10/MAWS/5-01_(Dec_2013)_(NWP)-(Promulgated).pdf?sr=b&si=DNNFileManagerPolicy&sig=un5q%2FWUW 21Qzq52MmQ7KMfD%2FhHMdj%2Frp1xJSur5TF58%3D
34 Naval culture has a long history of dismissing doctrine, particularly naval content. The author, teaching design to multiple seminars of Naval students since 2019, has yet to find a student or faculty member at the Naval War College or the Naval Post-Graduate School that knows of the 2013 design annex, or has used it. Ironically, NWC uses Army Design Methodology at least in 2021–2022 to teach design. Prominent Naval theorist Mahan even recounts senior leader discouragement when attempting to write naval doctrine in the late 1890s. For historical context of Naval skepticism in doctrinal content, see: David French, *The British War in Warfare 1688–2000* (Cambridge, Massachusetts: Unwin Hyman Ltd, 1990), 34–36; Markus Mäder, *In Pursuit of Conceptual Excellence: The Evolution of British Military-Strategic Doctrine in the Post-Cold War Era. 1989–2002* (Bern, Germany: Peter Lang AG, 2004), 22–23; Alfred Mahan, *Mahan on Naval Warfare: Selections from the Writings of Rear Admiral Alfred T. Mahan*, ed. Allan Westcott (Mineola, New York: Dover Publications, Inc., 1999), x.
35 U.S. Marine Corps, MAGTF Staff Training Program Division, *Marine Corps Design Methodology* (Quantico, Virginia: United States Marine Corps, 2017), 1.
36 U.S. Marine Corps, MAGTF Staff Training Program Division, 7–8.
37 Israeli Defense Force Doctrine and Training Division, "Design: Learning and Knowledge Development Processes for the Development of Concepts at the General Staff Headquarters and the Major HQ Levels" (Israeli Defense Force Training and Doctrine Division, November 2015), 4.

38 Lefebrve, Interview with Major General (retired) Paul Lefebvre on Marine Design Theory 17 Aug 2021, 00:17:20 to 00:17:45.

39 U.S. Marine Corps, MAGTF Staff Training Program Division, *Marine Corps Design Methodology*, 2.

40 Robert Chia and Robin Holt, *Strategy without Design: The Silent Efficacy of Indirect Action* (New York: Cambridge University Press, 2009), 27.

41 U.S. Marine Corps, MAGTF Staff Training Program Division, *Marine Corps Design Methodology*, 2–4.

42 Anna Grome, Beth Crandall, and Louise Rasmussen, "Incorporating Army Design Methodology into Army Operations: Barriers and Recommendations for Facilitating Integration," *U.S. Army Research Institute for the Behavioral and Social Sciences*, no. Research Report 1954 (March 2012).

43 U.S. Marine Corps, MAGTF Staff Training Program Division, *Marine Corps Design Methodology*, 2–3.

44 Ryan, "A Personal Reflection on Introducing Design to the U.S. Army."

45 U.S. Marine Corps, MAGTF Staff Training Program Division, *Marine Corps Design Methodology*, A-4.

46 Jean-Pierre Protzen and David Harris, *The Universe of Design: Horst Rittel's Theories of Design and Planning* (New York: Routledge, 2010), 154–157.

47 U.S. Marine Corps, MAGTF Staff Training Program Division, *Marine Corps Design Methodology*, 20.

48 Donnell, Interview of Mr. Steve Donnell by Dr. Ben Zweibelson on 28 Jul 2021 concerning Marine Corps Design Methodology mp3 audio file, 1:22:00 to 1:23:00.

49 U.S. Marine Corps, *MCWP 5–10: Marine Corps Planning Process* (Washington, DC: Department of the Navy, 2020), 10–11, www.usmcu.edu/Portals/218/CDET/content/other/MCWP%205–10.pdf

50 U.S. Marine Corps, iii.

51 U.S. Marine Corps, iii.

52 U.S. Marine Corps, 2.

53 U.S. Marine Corps, 9–10.

54 U.S. Marine Corps, 10.

55 U.S. Marine Corps, 2.

56 U.S. Marine Corps, 9–10.

57 U.S. Marine Corps, 14.

58 U.S. Marine Corps, 20.

59 The word "paradigm" is unfortunately a popular military "buzz word" and frequently is misapplied in a range of Department of Defense documents, policy papers, and doctrine.

60 U.S. Marine Corps, *MCWP 5–10: Marine Corps Planning Process*, 14.

61 U.S. Marine Corps, 10.

62 U.S. Marine Corps, 16.

63 U.S. Marine Corps, MAGTF Staff Training Program Division, *Marine Corps Design Methodology*, 8.

64 U.S. Marine Corps, *MCWP 5–10: Marine Corps Planning Process*, 19.

65 U.S. Marine Corps, MAGTF Staff Training Program Division, *Marine Corps Design Methodology*, 8.

66 *MSTP 5–0* specifically uses "design methodology process flow" while *MCWP 5–10* alternates between "planning process", "information flow", "flow of operation", and "framing process" throughout the doctrine.

67 U.S. Marine Corps, MAGTF Staff Training Program Division, *Marine Corps Design Methodology*, 13.

68 Shimon Naveh and Ofra Graicer, Naveh audio 15OCT2019 34min22sec Day 2 morning.MP3, interview by Ben Zweibelson and Nathan Schwagler, personal interview, October 15, 2019, 28:40.

69 U.S. Marine Corps, *MCWP 5–10: Marine Corps Planning Process*, 16.

70 Naveh and Graicer, Naveh audio 15OCT2019 34min22sec Day 2 morning. MP3, 15:42.

71 Mattis to Zweibelson, "RE: Design for Defense Book PDF Manuscript," September 16, 2021.

72 Mattis to Zweibelson.

73 Donnell, Interview of Mr. Steve Donnell by Dr. Ben Zweibelson on 28 Jul 2021 concerning Marine Corps Design Methodology mp3 audio file, 49:43 to 54:25.

74 Van Riper to Zweibelson, "Re: Design for Defense Book PDF Manuscript," September 20, 2021.

75 Donnell, Interview of Mr. Steve Donnell by Dr. Ben Zweibelson on 28 Jul 2021 concerning Marine Corps Design Methodology mp3 audio file, 19:07 to 19:18.

76 Thomas Kuhn, *The Structure of Scientific Revolutions*, 3rd ed. (Chicago: University of Chicago Press, 1996); Robert Chia, "Teaching Paradigm Shifting in Management Education: University Business Schools and the Entrepreneurial Imagination," *Journal of Management Studies* 33, no. 4 (July 1996): 409–428.

77 Homiak was lead author for *MARSOC Pub 1 MARSOF* in 2011, and *MARSOF 2030* published in 2018.

78 Homiak to Zweibelson, "Re: USMC Doctrine Question!," January 12, 2020.

79 Lefebrve, Interview with Major General (retired) Paul Lefebvre on Marine Design Theory 17 AUG 2021, 00:41:49 to 00:42:05.

80 The author, a SAMS student in 2010, studied Naveh's postmodern content along with select others. Naveh was banished from the SAMS campus, so these few students had to meet with him outside the classroom.

81 John Schmitt to Ben Zweibelson, "Re: Design for Defense Book PDF Manuscript," September 21, 2021.

82 Schmitt conveyed to the author that he had completed advanced education in systems thinking prior to learning about Naveh and SOD. He found the content already familiar to him as a validation of SOD, but discounted the unfamiliar postmodern theory as unnecessary and potentially confusing.

83 Schmitt to Zweibelson, "Re: Design for Defense Book PDF Manuscript," September 21, 2021.

84 Van Riper to Zweibelson, "Re: Design for Defense Book PDF Manuscript," September 20, 2021.

6 The Design Phoenix Rises from the Ashes

Israeli SOD Reborn

Brigadier General (retired) Shimon Naveh between 1995 and 2013 led the intellectual movement within the international military community to generate the first military design theory, process, and experimentation in combat. Previous chapters explained the start of this journey within the Israeli Defense Forces, and the sudden exile that Naveh and fellow Systemic Operational Designers experienced with political fallout over the Second Lebanon War. SOD would be exported into both the American Army and the Marine Corps along the same two-decade timeline, although each organization would differ in what they did with the design ideas. By 2009, Naveh and his remaining SOD educators found themselves again cast out of the institutions that originally welcomed them with open arms. Naveh would no longer be welcome at the U.S. Army School of Advanced Military Studies (SAMS) at Fort Leavenworth, and by mid-2014 his design contracted work with the U.S. Special Operations Command (USSOCOM) would be terminated at the Joint Special Operations University (JSOU) due to similar political and ego-related bruising.[1] Naveh's blunt, often harsh manner of teaching design suggested either some cultural, or possibly some translation concerns between Israeli and American military pallets. The other unusual pattern demonstrated throughout this period is how each time Naveh's SOD group were removed from one organization, they already had open invitations to resettle in a new one.

By 2013, something changed in the Israeli Defence Forces' perspective on design thinking and General Officer education. Naveh was invited back to teach once more in Herzliya[2] to Generals that were students in their senior education program, and he was asked to bring his latest evolution of systemic operational design thinking. Naveh refused to formalize his SOD so that it could enter doctrine, and throughout the first decade of the 2000s, he would reframe his SOD construct periodically, frustrating organizations and schools that wanted stability and uniformity in the courseware and instruction. By 2013, Naveh already moved from what Graicer termed "SOD 1.0" into "SOD 2.0", and was tinkering with a third evolution. Later still in 2022, Naveh would apparently be developing a fourth version in 2021–2022.[3] Naveh sought to make SOD a manifestation of what SOD intended to

DOI: 10.4324/9781003387763-7

achieve; designing with design while designing that design itself, mid-design. For artists and improvisational spirits, this was bliss. For most modern military institutions, this presented nightmares to doctrine writers, educators, training centers, and students expecting familiar "checklist-style" information. Yet despite these andragogic obstacles, the Israeli Defense Force that infamously purged itself of SOD in 2005–2006 opened the door once more, inviting the original design heretic back in.

According to Beaulieu-Brossard, the Israeli Defense Force senior leadership observed "a lack of generalship and did not see an equivalent academic alternative [available to teach] than bringing Naveh back to nurture generalship".[4] Apparently, just as the Army and Marines experienced a burst of design innovation in the 2003–2010 period followed by stagnation for a decade, the IDF experienced some version of this, offset to a shorter and earlier period due to Israeli geo-political and security contexts. By 2013, the IDF recognized this intellectual gap, and in keeping with Naveh's original premise that SOD must be taught to senior officers and the top leadership so that the generals can create and orchestrate designs in warfare, he and his team were invited back to the General Officer School near Tel Aviv, Israel. Since 2013, Naveh and Graicer taught small groups of Israeli senior leaders from across their defense forces and also in various intergovernmental agencies in Israel, providing a form of SOD that according to both SOD educators is far more applicable and potent than the earlier forms. Like all innovators and educators, Naveh and his primary SOD collaborator, Dr. Ofra Graicer, found through many repetitions the better ways to teach and demonstrate these difficult warfare concepts.

Throughout the last twenty-five years of experimenting, teaching, and reframing military design, Naveh advocated heavily against shackling design into doctrinal form. Instead, design should be expressed within a community of practice more like art where theory and practice are fluid, discourse is encouraged, and alternative perspectives are welcomed. The tension between improvisational design artists and the institutional doctrine writers presents a challenge for how militaries might balance the two extremes. Military design exists to disrupt, challenge, experiment, reframe, and transform those aspects of a military's organization, form, and function so that innovation occurs in complex warfighting. Such a design cannot be simplified into set processes, nor does it seem feasible for design, as a multi-paradigmatic, flexible meta-approach to war to be subjugated into an operational planning step, optional or not. Of the military organizations Naveh, Graicer, and the SOD team worked with, it seems the more flexible thinking special operations forces (USSOCOM) had, according to Graicer, the greatest impact upon SOD evolution over the years.[5] Military artistry and the ability to improvise would at institutional levels be in tension with efforts to standardize and indoctrinate, with special operations culture tilting toward the iterative variety of SOD more than large, conventional forces. This tension would play out across the American militaries and also within the Israeli Defense Forces, and later in places such as Canada and Australia. Naveh, forever the design heretic, would explain his perspective on this tension in an interview in 2019:

> In order to be able to disrupt yourself, you need an external mirror. Or provide yourself a cognitive external scaffold which will enable you to look at yourself from the outside … to identify your biases, to identify the exact boundaries of your identity, to appreciate more critically your interests, etc. Once you identify them, it is easier to deconstruct or reinterpret or change them.[6]

SOD's framework for self-disruption and change moves much more like artistic expression as well as how improvisation for jazz musicians, stand-up comedians, and other creative artists functions best. Graicer, Naveh's primary design collaborator over the years and an Israeli army officer herself, would develop this idea of "self-disruption" folding in postmodern concepts to the complexity theory applications that SOD prioritized over any Newtonian stylizations in traditional military strategy and planning. They draw inspiration in the postmodern idea of "counterthoughts", which Deleuze and Guattari describe as: "violent in their acts and discontinuous in their appearances, and whose existence is mobile in history … these are the acts of a 'private thinker', wherever they dwell … they destroy images".[7] Potentially, the military form of design differs from commercial design in one additional way beyond those outlined in the first chapter. Military design is anti-doctrinal, in that militaries (not industry) seek to formalize into pseudo-scientific styled doctrine all decision-making processes, models, theories, and terminology for clear goals of uniformity, standardization, and prediction. Military design rejects what can only be understood as assimilation into the institutional frame that design seeks to disrupt.

Why Did Systemic Operational Design Remain Attractive Yet So Controversial?

Even by 2013 when Naveh would be invited back into the IDF, military design could no longer be accused of being some temporary fad, or that design was merely "old wine in new bottles" and a recycling of earlier military ideas with new terms and packaging. Indeed, by 2021 sufficient military design research, theory, literature, and case studies accumulated for a first formal military literature review to be published in an international design journal.[8] From the mid-1990s and dense, intellectual origins, SOD would weather storms of institutional outrage, confusion, rejection, and later assimilation. By 2013 the U.S. Army, Navy, and the U.S. Marine Corps published some version of military design in doctrine, with most of the advanced military educational programs in the U.S. Department of Defense instructing some version of design to students. In an interesting development within the IDF and their doctrine center, in 2015 the Israelis would also publish a design doctrine for their force. The brief twenty-five page document is titled *"Learning and Knowledge Development Processes for the Development of Concepts at the General Staff Headquarters and the Major HQ Levels"* and was published by their J3 Head of Doctrine and Training Division, Brigadier General Moti Baruch. Baruch, himself a graduate of the second General Officer's design course taught by Naveh and Graicer beginning in 2013 at the General Officer program, did not involve Naveh or Graicer in the development of this design doctrine. At it is unrelated to SOD directly, and there is limited evidence of any IDF-wide impact of the document, it is only mentioned here and not a focus of this research. Naveh would continue to see splinter efforts, particularly of former students such as Baruch, make the design doctrinal transition at the disappointment of the original SOD creator. Despite a pattern of design doctrinal attempts across the globe, Naveh also would witness a growing network of SOD enthusiasts, particularly those that shared his rejection of design in doctrinal form. International communities of military designers continue to examine and experiment with the intellectually more rigorous design, frequently in anti-doctrinal ways, particularly in the last decade.[9]

Yet despite this apparent success of military design writ large, Naveh continued to promote an intellectually demanding methodology where postmodern ideas remained essential to the student journey. Little of this would carry over into the Americanized versions of military design in doctrine, with minority populations of American designers able to approach design in the way that Naveh envisioned. Military institutions in general resisted any assimilation of the dense, difficult, and extremely paradoxical SOD as it challenged the foundations of their modern military paradigm. Postmodernism does to Modernism what it is designed to do, and SOD used postmodern concepts to break into deeply cherished modern military beliefs on war.

This opens military design up to accusations of it being too abstract, undefinable, non-scientific, or less valuable to military practitioners when a suitable analytically optimized checklist is readily at hand with precise terminology and a proposed history of hypothesis validation supporting it. Some critics of military design point to the lackluster accomplishments of the IDF in the 2006 Hezbollah War as proof of SOD being a flawed concept, and that design's overemphasis on postmodern theory and language is apparently incompatible with mainstream military practice and formal literature such as doctrine. "Postmodernism places a special emphasis on language, claiming that its ability to communicate the nature of reality is an illusion … postmodern thinkers jeopardize the integrity of the US constitutional system because they are naturally anti-institutionalists".[10] The two camps as Ryan described this tension would battle over what form and function military design would take formally, and often the pragmatic planning team won. The design purists would continue, but often in exile, branded heretics, or performing in the shadows.

Being an anti-institutionalist within what are likely some of the most institutionalized of all possible human enterprises is itself paradoxical. Naveh's SOD is appropriately accused of being "heretical" for good reason. Militaries correlate more to formal religious institutions than anything else, and as covered in the second chapter, military doctrine harks back not to some pure natural scientific origin but far earlier ascientific manifestations. Yet the postmodern ideas championed in SOD are provocative, disruptive, and suggest some of the perpetual trouble Naveh and his designers experienced over decades with multiple militaries. Just as Giddens recommends postmodernism remaining within the styles or movements for "literature, painting, the plastic arts and architecture",[11] proponents of the modern war frame feel such concepts have no business in the serious and entirely scientific pursuit of war.[12] This contributes to some conclusions on why the IDF purged SOD in 2005–2006 leading to the eve of the Second Lebanon War, why the U.S. Army and Marine Corps would remove all postmodern content when they moved to formalized design in their doctrines, but it does not explain why in 2013 the IDF would welcome Naveh and SOD back in. Could an Israeli grassroots design movement have survived, or perhaps SOD ideas permeated through into a new generation of leaders? Or possibly the original smears that "SOD caused the IDF strategic and operational failings in the 2006 conflict" had finally rung hollow?

The 2006 Lebanon conflict criticisms leveled at SOD and Naveh should not be strongly correlated with these other tensions concerning design, postmodern intellectualism, and self-disruption of the modern military professional. Causation and correlation are distinct for this reason. The significant political and educational issues occurring within the IDF at senior levels prior to 2005 are separate from the institutional preference to value explicit forms of military knowledge over tacit ones. Checklists and rigid drills make for ideal solutions to simple warfighting challenges, but they never translate into appropriate responses to complex or chaotic ones. Militaries need a blend of explicit and tacit knowledge, along with experienced professionals able to develop and manage how to combine both into novel, creative, and feasible approaches. This never occurred in the IDF in the 1995–2005 window of original SOD experimentation. Given that few IDF generals attended the SOD education, with even fewer grasping the concepts deeply, the IDF lacked any military design momentum except in isolated, individual efforts. This alone makes the argument that SOD caused the Second Lebanon War problems a weak, likely unprovable stance.

Naveh's original downfall in the IDF would feature political aspects, and also intellectual arguments on warfare that likely threatened power, influence, and even identity of prominent parts of the organization. Naveh, perpetually a brash and confrontational personality, did himself no favors by refusing to inculcate his SOD concepts into even some theoretical papers or documents widely available to his organization, or to expand the availability of these radical concepts to multiple organizational levels within the IDF for faster distribution. As other military design educators would discover, some-times the senior military officers are at times the wrong target for radical and transformative concepts where more junior leaders are overlooked.[13] Often, Naveh seemed to focus SOD on the wrong population in the IDF, yet this is also foundational to his vision that design is done centrally by the senior leader.

On the commercial side of design education and training, many human-centered design courses and programs cater to college-age students and uni-versity programs, as well as basic design modules for entry-level members of the work force. Large companies such as IBM invested significant resources pursuing an "enterprise-wide" basic design education model not just for senior level executives or mid-management, but across their entire organ-ization.[14] Many militaries opt to avoid this with military design and focus on senior or mid-level officers. Naveh would teach exclusively to IDF generals through 2005, but within the U.S. Army and Marine Corps, he switched to teaching mid-grade officers. This likely was due to where those militaries desired the SOD education rather than any change in Naveh's frame on design education. Later with U.S. Special Operations Command, Naveh and his SOD contracted team would even teach to junior officers and non-commissioned officers, again largely a forcing function of USSOCOM edu-cation and training standards, and not any new perspectives from Naveh. Once invited back into the IDF, Naveh would return to his original premise that SOD, at least in original Israeli security applications, must be taught to general officers in small-batch, customized, ever-changing ways. No doc-trine could contain SOD, nor could facilitation be done in a standardized, factory-style mode of mass education.

Flash Back: How SOD 2.0 Came into Existence Far from Home

If the 1995–2005 period represents the first phase of SOD commonly represented in Naveh's earliest geometric configuration presented in the third chapter, the second phase of SOD development would develop over-seas, within the new intellectual home that the U.S. Army offered to Naveh and some of his SOD team. The U.S. Army SAMS would become the experimental design studio for Naveh to reframe SOD, explore new ways to articulate it, and also wrestle with his American hosts on how one might "Americanize design" without breaking it. However, Naveh and his designers struggled with American goals to simplify SOD and place it into doctrine. This created a "dual track" mentality where SOD purists worked on new ways to develop SOD into deeper, refined formats such as the Prolegomena document, and other groups attempting to insert select SOD concepts into less intellectually rigorous "Americanized" variants.[15] SOD would develop, as would Naveh's experience in teaching and implementing it. SOD would also undergo concerted efforts to simplify, reduce, and capture it within

American military doctrine. Ultimately, Naveh and the design purists would move in the opposite direction that TRADOC and the Army's educational programs wanted for military design.

The SAMS program in Fort Leavenworth, Kansas is essentially the U.S. Army's intellectual flagship where select field grade officers get some of the most intense military advanced education; SAMS has historically been the Army's incubation lab for new concepts on planning, strategy, and innovation in warfare. The SAMS leadership and faculty invited Naveh in 2005 to visit and provide design education, first in the original form of "Systemic Operational Design" and later as a generic, more edited version termed "design" for the American Army. Prior to Naveh's formal integration into SAMS, his design concepts were already being initially circulated to isolated seminar discussions by Dr. Jim Schneider, a SAMS faculty member and friend of Naveh who had embraced design through earlier encounters with Naveh.[16] The timing was perfect for radical ideas in that the American military writ large were by 2005 increasingly frustrated with counterinsurgencies and failing foreign policy efforts in Iraq and Afghanistan and becoming desperate for new ideas concerning warfare, security, and defense. New ideas were wanted, and SOD showed potential. The fourth chapter of this book explained the growth of design into the American Army from the American perspective. This chapter completes that story by explaining the Israeli side of that period, including how and why Naveh and his team would develop SOD into a new version despite their own hosts rejecting both options.

The shift from Israel to America in this new period of design education featured an important expansion into designing for strategic reasons. As Graicer explains, "once we moved to the States ... we had to make our journey into the realm of strategy because before that, in Israel, we didn't bother talking about strategy ... we didn't need it; we are not an empire ... [Israel] is not running the world".[17] Once they began considering strategic aspects of designing for security challenges, they began using some metaphoric devices associated with navigation to address the ever-changing epistemological shifts in culture, societal values, language, and how the world is perceived. "We chose the notion of "drift" to keep with this metaphor of navigation for [the designer] for "re-compassing" yourself in the world."[18] Graicer explains that in order to understand this SOD notion of drift, the designer must cognitively map a wide range of social, cultural, as well as doctrinal, institutional, and political directives at the strategic to the operational levels and how it is all shifting and transforming while also pulling you along within a current.[19] Getting design students to contemplate *how reality is not fixed and uniform but transforming and fluid* would move SOD into its next form. Here, the contemplation of a complex reality where security challenges occur would be expanded into strategic levels beyond the original first form of SOD created for an operationally oriented yet strategically uninterested Israeli security program.

Between 2005 and 2009, Naveh was deeply involved in the SAMS program for the design applications, mentoring and advising students, educating cohorts in SOD, and supervising select students (as well as interested faculty) on their yearlong research monographs if they were studying security design.[20] During this period, SAMS students produced a surge of design-themed monographs, papers, exercises, and research. Students explored deep philosophical concepts, including reading difficult readings such as Deleuze and Guattari's "A Thousand Plateaus" during their year of study on design, strategy, and operational planning.[21] Naveh continued to draw heavily upon postmodern concepts of nomadic ways of thinking that included many sophisticated ideas within Deleuze and Guattari as well as other noted postmodern works as he reframed SOD into a new version.[22]

As Naveh's work was both foreign (conceptually) and filled with esoteric language and challenging, self-disruptive concepts,[23] American military academia at Fort Leavenworth struggled to match these ideas against the Newtonian styled concepts, theories, and methods that defined the modern war paradigm taught in their education and training endeavors. While this intellectual struggle unfolded, Naveh would move to refine, prototype, and transform the SOD as he was explaining it, causing further confusion and concern for the Americans. Select intellectuals could break through and grasp the ideas, while the majority could not. As a smaller population of design purists accompanied Naveh and his core SOD team on reframing the original SOD into the next version, a larger population of mostly American educators and leaders at SAMS and across TRADOC would move to extract select useful elements of the first SOD, to generate a separate Americanized version far removed from the original SOD.

Naveh and his small group of security design advocates would from 2005 to 2009 shift their original SOD toward this second version, experimenting and modifying with mostly American military students, faculty, and some international students. Many of the changes were topical, in that Naveh would conceptualize the same basic SOD ideas but present them in a dizzying array of new slides, white board drawings, and metaphoric devices to explain the SOD emergent process of self-disruption and experimentation. With several new inclusions of concepts such as "drift" and the refinement of some postmodern ideas so that they were better understood by a military audience, Naveh would represent SOD 2.0 in a new geometric configuration shown below. Foundationally, it still retained much of the original SOD methodology, but would be facilitated and articulated in this new presentation. The epistemological mapping provided below on the right side will remain unchanged, as Naveh did not transform any SOD core philosophical beliefs, nor did his team advocate for any intellectual simplification or substitution of the original concepts. This would gradually exacerbate the divide between SOD enthusiasts and a larger population of pragmatic planners seeking a design compatible with institutionalized decision-making methods (Figure 6.1).

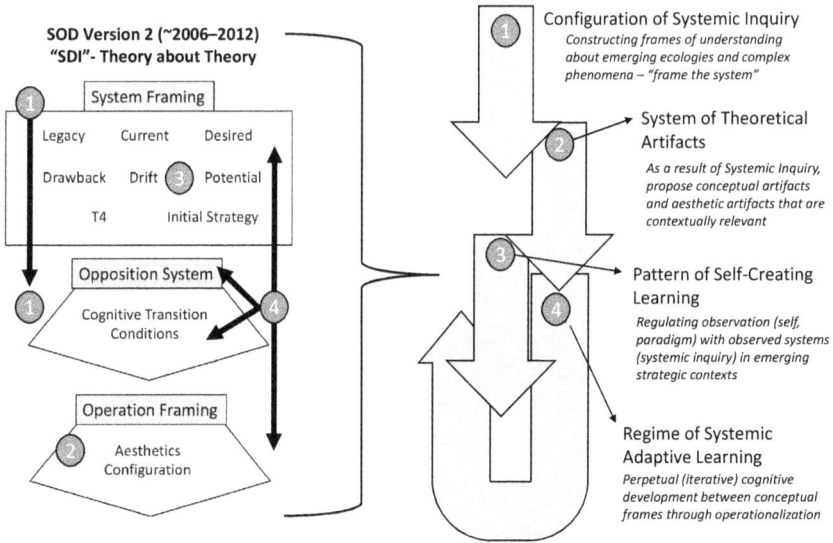

Figure 6.1 Epistemological Framing of SOD 2.0 (2006–2012).
Source: Author's Creation.

The SOD "version 2" methodology and geometric graphic depicted above on the left side comes from Graicer's published research and is one of the only publicly available graphics discussing the second phase of SOD development.[24] Again, Naveh and Graicer would spend the majority of 2006–2012 in a self-exile from the IDF for design education. Both would continue living in the country and shifting to the intellectual export of systemic operational design abroad, traveling back and forth while building an American network of collaborative designers. This change of how SOD was presented occurred likely for two non-correlative reasons. Naveh and his team were perpetually experimenting, and Naveh as an artist would leap from one conceptualization to a new one, often expecting his audience to move at the same speed. Hypothetically, Naveh potentially created dozens of SOD prototypes, with only this one gaining popularity at that point to gain the formal title of "SOD 2.0". The second reason is that the SOD movement likely needed a re-branding effort, to distinguish it from the original SOD and distance the design efforts in America with earlier criticisms of SOD 1.0 in the Second Lebanon War. This would coincide with the American pressures to "make design more easily understood, and help us put it into doctrine" that would become a stronger forcing function as time went on. SOD would get a fresh paint job and new wheels, but maintain most all of the original parts underneath.

The idea that SOD needed some public relations help is feasible, but difficult to verify. Officially, the Israeli military never actually banished SOD

advocates, although there were forced retirements, reassignments, and specific targeting of personalities during the IDF purge of SOD on the eve of
the Second Lebanon War. There were formal declarations by leadership to
cancel SOD education and practice within the IDF. Israeli media would, in
the aftermath of the conflict, print highly critical accounts of SOD, based
on the interview of the same SOD opponents that commenced the purges.
Whether this influenced Naveh to change SOD or if he already sought to
reconceptualize it through new experiences is unknown.[25] Naveh, ever the
creative experimenter, would change the SOD methodological conceptualization from the earlier geometric metaphoric device to this new depiction
on the left side of the above figure. He would also begin to teach SOD differently, with SAMS students getting a mix of earlier SOD concepts as well as
this "version 2" depending on where and when they interacted with Naveh's
educational outreach between 2005 and 2010.

SOD 2.0 would be exercised, experimented with, and further developed
over the 2006–2012 period where Naveh, Graicer, and select others facilitated
and educated design mostly in the United States, with at least one contracted
SOD exercise with Australian commandos in 2009.[26] After Naveh and his
team fell from grace at Fort Leavenworth in 2009, they would continue
developing SOD 2.0 with new clients such as the U.S. Special Operations
Command, and within various contracted endeavors. In 2013, the Israeli
Defense Force would reach out to Naveh and Graicer, and invite them to
come resume SOD education at the General Officer School. Something had
changed, and despite the political fights, egos, and debates over the Second
Lebanon War outcomes, somehow Naveh had accomplished part of his original vision. While it took over a decade and the sacrifice of multiple careers,
reputations, and potentially operational failures for Israel in critical security
affairs, the Israel Defense Force seemed to acknowledge that their organization in 2013 was lacking in something needed in terms of intellectual
growth, transformation, and critical self-examination.

From Design Frustration to Renewed Curiosity: The Israeli Heretic Returns Home

After choosing self-exile from the Israeli Defense Force intellectual community in 2005–2006, Naveh returned once more by 2013 to Tel Aviv.[27]
After years of reflection as well as further experimentation with design education and practice in the American and to a lesser degree other international
militaries, Naveh was asked not just to restore SOD into the IDF, but resume
his SOD education at their highest level of senior officer education and development. Always the brash and abrasive intellectual, Naveh would reflect on
this change of heart with:

> In 2014, one of my arguments [to the Israeli War College] was, you guys
> … take a look at yourselves, look in the mirror. You are shmucks! You

threw me out as the devil eight or nine years ago, and after all that time, you crawl back to me. What does this tell you about yourselves? You are nothing. So maybe you should build up this institutional capacity [for design education].[28]

Naveh would, despite misgivings on past arguments, rescue and restore his latest SOD manifestation for the original audience SOD was developed for. He would be once again at home, within his own military, able to converse in his native tongue, and influence the next generation of IDF general officers.

Naveh, during his time teaching internationally, might have also matured on how he viewed the IDF, the decades of institutional resistance to his manner of conceptualizing military design and innovation, and perhaps the outspoken Brigadier General might have realized some humility in the period of exile and reframing. Naveh reflected on this in a 2019 interview as follows:

We all recall how we fell down in 2005 … we [now] have this huge wall, a defensive wall … for example, think about who are the guys that are really influencing the present [organization you want design for]. Get access to them, mobilize them, recruit them, make them your friend … You have to be creative, manipulate the system, and [at first] start small.[29]

Naveh would move from being a well-connected insider to the disconnected outsider in this shift, which likely stimulated such introspection and interest in changing strategies. Perhaps his first experience exposing design to the IDF came too easily, in part due to his status as a reputable IDF general and accredited intellectual, coupled with the sudden surge of position and resources given to OTRI in the late 1990s. All of that ended in 2005, and Naveh with his core team would approach American design interest with different strategies. This included teaching design to anyone, anywhere, and abandoning the original "Generals only" stipulation. As Naveh attempted a draft SOD design doctrine for TRADOC as early as 2005 through a military contract, he apparently changed his inculcation strategy, if briefly, on design and doctrine too.[30] The author pressed Naveh in an interview on this change of heart between 2005 and 2009 and the shift from SOD 1.0 to the American situated SOD 2.0. He offered:

Look, originally it was all linked to generalship, to the praxis [of design]. I was a bit bitter in 2008 or 2009 … and I was really furious. I was angry. I still think that there are a significant amount of stupid generals are still going around, here in the IDF, but one of the things and I have been admitting this since I returned [to teach in Israel] is I rediscovered the community. One thing I have learned is that first of all, that they are not that stupid. Actually, if you make the right efforts to set the conditions that will enable them to expose the best out of themselves, they and yourself will realize that they are not that stupid. And the second thing is that socially, they operate so well. So smoothly, as a team. This could not be done in the U.S. You put ten Generals in a room, and one is responsible and runs the [inquiry].[31]

Graicer, in the same interview session, expands on the topic of design education for mid-grade officers versus senior leadership.

Going back to your question on Majors and Colonels versus Generals … I think, what we do now [in Israeli SOD 3.0 education at Herzliya] is so different from what we did in OTRI [during the first SOD phase of 1995–2005]. First, OTRI was four and a half months of a design course. More than half of it was frontal presentation, academic teachings, texts … it was brutal.[32]

In the original SOD program provided at OTRI in the late 1990s through 2005, Naveh and Graicer worked with ten other SOD instructors (including eight former Brigadier Generals) to provide a robust, self-paced individual learning period where students had to digest hundreds of pages of dense, intellectual content far removed from familiar military topics.[33]

After that first semi-autonomous phase, the SOD facilitation team moved the students into the hands-on, practical phase with several design inquiries. This original style of SOD 1.0 education would change significantly by 2013 based on their experience and setbacks as well as professional growth. Gracier described the original SOD 1.0 design scenarios as dense, based on imagined topics of security interest but supported by a robust reading pile of difficult concepts and challenging multi-disciplinary theories. "There were scenario briefings and intelligence briefings … Again, we were in love with it so we kept adding more things into it".[34] The theory and philosophy of the original SOD 1.0 program was overwhelming, and directly taught to the students by Naveh and his design experts often using elaborate, esoteric language requiring significant intellectual rigor to grasp. All of this changed in SOD 2.0, with Naveh and his team shifting to a reduction in reading and increased team facilitation within the Army SAMS seminars and at TRADOC-hosted war games.

SOD 2.0 would only be understood and experienced by select students, mostly at Fort Leavenworth, and the majority population would get some variation of SOD 1.0 and SOD 2.0 spanning the 2004–2012 total period of Naveh's focused outreach to American and other international militaries. Based on the numerous SAMS design monographs published by students and faculty in that period, SOD 1.0 would dominate student design comprehension, with no published SAMS monograph (or any student-created document) discussing or depicting SOD 2.0 outside of Graicer's published design articles and conference presentations in 2017–2018.[35] SOD 2.0 seems to be understood readily by Naveh and his core design educators, but there is limited evidence of pronounced SOD 2.0 impact on students in the 2005–2012 period Graicer frames that variant as existing. Regardless, SOD 2.0 would operate as a necessary developmental stage for Naveh and Graicer to conceptualize SOD 3.0, which remains the latest known version in practice through 2022. This also would coincide with their return into the IDF, and implies that once more, Naveh desired a re-branding effort, likely coupled with additional refined thinking on how once more to teach these concepts.

In the northern part of Tel Aviv, Naveh, as of this writing, teaches his latest version of systemic operational design with his long-time teaching partner, Ofra Graicer, to small classes of Israeli General Officers and other senior security leaders across Israel. They do not select the officers for their design education; rather in Graicer's words:

> they are selected by the Chief of General Staff in the military, and his counterparts in the other [Israeli] services. At least in the [Israeli] military, it is considered (informally) an act of faith from the CGS in the plausibility of further advancing in the ranks; and I mentioned [at the Canadian Forces College's design education exercise] last June [2019], more than half of the General Staff are [now] graduates of the course.[36]

As of 2022, Graicer updates this percentage to 80% now.[37] This indicates an improvement over the original SOD student ratios prior to 2006.

The two of them teach in an informal, intellectual setting, dividing the students up into two design teams with several outside mentors contributing their own command experience of application of design into real-world security contexts. Naveh and Graicer attempt to keep a low profile, and in a series of interviews with the author reflected on the last three decades of military design theory and practice. Despite describing this "under-the-radar" new approach, Naveh also admits when the program provided him six Colonels to assist him with design education when he first returned in 2013, he fired them all.[38] Despite the years of reflection and maturity concerning SOD praxis, Naveh had no patience for military practitioners unable or unwilling to attempt to clear the intellectual high bar necessary to apply SOD into application. Or, potentially, Naveh sought to kill off any perceived institutional resistance where those demanding "simplification and standardization" first were dismissed outright. The SOD teacher would in this latest phase, agree to work and teach a select audience of open-minded, curious security professionals.

The previous figure depicting SOD 2.0 in the 2006–2012 period suggests that Naveh and his team would aesthetically reconfigure SOD, complete with new metaphoric devices such as "drift", and focus on how they would guide student design teams through a journey of self-reflection, disruption, and intellectual growth in more effective ways. Again, Naveh infamously would explain dense design concepts through elaborate, almost incomprehensible arrangements of terms, topology, models, and theory in his unpublished, private work.[39] While SOD 1.0 was arranged in one simplified geometric configuration as the graphical conceptualization of the method, SOD 2.0 appears to be an attempt to further articulate the SOD ideas graphically within an even clearer, perhaps simpler configuration. SOD 3.0, termed "disruptive thinking", would replace the geometric shapes in the two previous versions of SOD with a topographic, ever-moving, iterative conceptualization that relies upon the letter "Z" as the SOD design methodological shape. This "Z" shape comes not from Hebrew, but the Greek zeta, inspired by the Phoenician glyph *zayin*, which also would become the same in Hebrew.[40] Unlike a loop or cycle, Naveh's SOD 3.0 features a more deliberately directional dynamic, where designers know their location within a clear symbol instead of somewhere along a circle. This indicates Naveh moved past some limitations he experienced with the SOD 1.0 and SOD 2.0 geometric conceptualizations (Figure 6.2).

Naveh's latest and current (as of 2022) depiction of SOD has once again changed the narrative delivery and graphical conceptualization of SOD, but not the core, underlying design content. On this "z-pattern" used now to explain SOD, Naveh appears to balance the emphasis more elegantly concerning movement, iteration, discourse, and disruption in SOD. He does this now without over-burdening the SOD 3.0 graphic with dense terminology or potentially perplexing geometric relationships. Moving back and forth, iteratively in the "z-pattern" suggests a faster appreciation of the military design process unfolding than earlier versions. Note that in the

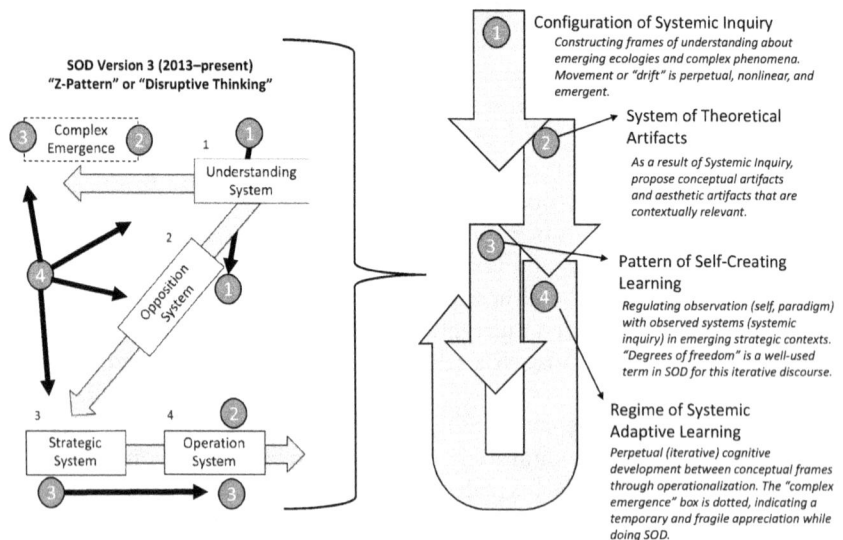

Figure 6.2 Epistemological Framing of SOD 3.0 (2012–2022).
Source: Author's Creation.

upper left of the "z-pattern", the "complex emergence" box is illustrated as dotted, indicating a temporary and fragile appreciation while doing SOD. Emergent processes tend to be marginalized or ignored in classical science, including how the modern military paradigm assimilated a Newtonian stylization based on static, analytical, systematic conceptualizations. Emergence is a phenomenon entirely alien to these perspectives, and thus takes center stage in Naveh's SOD that disrupts Newtonian styled warfare with that of complexity.[41]

This third version is also the least understood or known of Naveh's SOD creations. Aside from Graicer's publications, few outside of Israel know of SOD 3.0, and there is even less published research regarding this latest SOD development. While this latest version uses a new graphic for conceptualization of the design process along with some new terms and techniques, all three versions of SOD retain mostly the same deep philosophical content concerning design for complex military contexts. As Naveh is currently engaging with various military clients on yet some new variation of the "z-pattern" in 2022, one can safely speculate that new developments of a SOD 4.0, 5.0, and beyond may emerge out of Tel Aviv or within Naveh's collaborations abroad. Indeed, Naveh and Gracier continue to transform the design as they design, making any critical study of SOD problematic to those outside the experimentation.

Today, Graicer states, "And then we change the program, it is not the same program. We change the program every year". Naveh follows on with, "sometimes we change it during the course of the year … because opportunities occur". Graicer frames this approach with the term "degrees of freedom" in military design. "You explain to them the degrees of freedom first theoretically, and then something happens which [becomes] an example for you to say, okay, you see what you have done right now? Do you see what you have said?"[42] This mode of flexible design adaptation during the execution of the course represents for Naveh and Graicer a major way the new SOD education in an andragogic sense differs from all earlier versions. Potentially, the freedom offered by the Israeli General Officer school and the low visibility sought by Naveh and Graicer today help foster such an environment that in other professional military environments might not be tolerated. Degrees of freedom suggest deep self-awareness, contextuality, and realizing the uniqueness of where one is "drifting" within an ever-changing, complex system with cunning, resourceful adversaries. War will not stand still, and the notion that rigid, uniform, and universal behaviors are optimal in modern militaries may be obsolete. How far this extends into military decision-making, warfare theorization, and organizational form/function remains to be determined. If anything, SOD remains the radical, heretical option for institutions to unleash in full acceptance that creative destruction may target some things that are institutionally forbidden to challenge.

Today, Naveh and Graicer see their students only one weekend per month in Herzliya, and after each session their students return across Israel to continue leading their organizations in real-world security contexts. In this periodic and iterative mode of design education, students return each month for a few days to discuss design and their own organizational challenges with their peers, and after contemplating novel experiments and alternative perspectives return to their assignments to spend the rest of that month changing their design and experimenting in the real world. This is a unique configuration not readily apparent in other militaries, particularly larger ones such as the United States. U.S. senior officers attend War Colleges in cohorts spanning an entire academic year, while those in distance programs experience semi-autonomous learning that still does not match the Tel Aviv general officer experience. Additionally, Israel's perpetual state of heightened security and ongoing internal and external threats mean that many of the SOD design students get to apply their design education directly into real-world applications between monthly sessions.

Today, Naveh grooms key senior officials across the Israeli forces to become students in his program so that, once they learn and appreciate the design concepts, they subsequently act as a protective shield from others unfamiliar with design that might move to strike the movement. Although Naveh and Graicer have no direct impact upon the selection of students in the Tel Aviv program, they can encourage promising generals to request attendance. As Naveh and Graicer teach design today exclusively to the most senior leaders in the Israeli forces, they can implement this style of self-protection of a minority community of design practice. Naveh also seeks out active and retired Israeli generals that have employed design in their command and uses them in a social network within the IDF as a form of intellectual insulation to help prevent what he sees as a period of misinformation and intolerance of new ideas in the 1995–2005 phase of SOD in Israel.[43] Naveh's new strategy to protect his latest version of systemic operational design is to pursue "building a wall of mentors" to protect his creation.[44] The scarcity of SOD 3.0 content outside of Israel reinforces the secrecy and protection Naveh now has erected to safeguard this third phase of SOD implementation.

In another difference between the original SOD 1.0 design education at OTRI and today's design immersion at Herzliya is the mixed security student audience. SOD 1.0 was exclusively for the Israeli Defense Forces, whereas today Naveh and Graicer host student teams with senior leaders from across the Israeli security and intergovernmental agencies and departments. Bringing non-military security professionals into the classroom requires changes in language, metaphors, and scope. Being able to gather a design team of senior security professionals that are peers but also not a homogeneous group is another distinction. Naveh and Graicer feel that the earlier version channeled into groupthink because the peers came from the same military, and even at times the same unit. Peers today in their Herzliya program are senior but not from the same organization and will challenge the design team when they tunnel in on topics that are vastly more complex than even a military focus area.

Naveh still views the General as the ultimate military professional at the top of their game, with sufficient knowledge and experience to be capable of doing the design he advocates.

> Generals are confident. If you give them a mission, they will do it. They will not be too anxious because, well, they are already Generals ... although, nobody likes to fail. And I mean nobody. Everybody wants to excel ... but Generals usually are cooler.[45]

Graicer takes a nuanced approach to educating Generals where even if they are difficult to educate because they assume their rank assures some completion of their military knowledge, that "generals always think they know best ... and they do know best. Or so we need them to think".[46] Here, Graicer implies that despite the Generals being confident (or even over-confident) of their knowledge and wisdom for war, that confidence should be reinforced by the design educators so that the Generals remainable to make those critical decisions at the top of the centralized military hierarchy where they alone are empowered and totally responsible. Epistemologically, this singular focus on the top of the military hierarchy for most of all design education is unique to the Israeli program.

Today, the Israeli design education is done with small teams of ten General officer students with a rotating responsibility for one of the students to take lead of the design team for the weekend educational session. Their design inquiry topic is always a real security topic, typically drawn from their own organizational challenges within Israel unfolding now. Each design team leader is nominated by Naveh and Graicer about four or five weeks before the class session. At that point, the student leader is tasked to prepare their own design program for their group. There is no formal education in the form of lectures, or any set learning criteria for SOD beyond abstract, andragogic objectives. Naveh explains:

> We don't do any teaching ... no presentations ... we spend ten or twelve hours initially with that leader for an initial briefing, and then he goes and does some thinking, some reading. And then he comes back, and he works out an initial program ... we criticize it, and he goes back ... finally he comes up with a program and he works out [his design approach]. We are not facilitators ... what we do is we actually use the leaders.[47]

These design dialogues between the General student leader and Naveh's educational team take multiple hours and occur in the periods between when they visit together each month for the few days of actual design practice. Naveh does this to enhance the social dynamics of when the student teams are together operating on that design challenge and using that student leader's designed mode of inquiry. Although Naveh and Graicer influence that mode of design inquiry, they do not control it or provide the student leader with a prefabricated design methodology. Once again, Naveh stresses an indirect andragogic mode of facilitation that he declares is not facilitation.[48] In another deviation from earlier SOD educational approaches, Naveh and Graicer do not force students to read huge piles of exotic theory and philosophy or suffer through standardized lectures and intelligence briefings. The learning is self-guided, entirely dependent on the student design teams and their own curiosity and critical thinking. Whether this provides them the intellectual tools and disruptive perspectives or not is an area for future study.

Naveh and Graicer now mentor two design teams of ten to twelve Generals (per team) in separate classrooms, preselected by the Israeli Chief of General Staff as an acknowledgement of these senior leaders' future potential and implicitly an acknowledgement of Naveh and Graicer's impact. Naveh works with one team while Graicer works with the other, and these teams remain separate throughout the design sessions at Herzliya. There are only two design exercise rooms, and the small footprint creates an exclusive and "small-batch" education atmosphere. Within these design student teams, Naveh prescribes breaking them down into three sub-teams of three to four Generals so that they all are engaged and working. He sees larger design teams as impractical because "three will do the work, one will lead, and the rest are zombies and disengaged."[49] In the earlier SOD 1.0 and 2.0 approaches, Naveh and his facilitators worked with larger design teams. This is yet another subtle development where the SOD experience differs, but the core concepts and philosophical framings for military design remain unchanged.

Design intervention by Naveh, Graicer or their non-design mentors associated with the program occurs only in two specific occasions. When the design team gets stuck cognitively, in that they are unable to break through certain intellectual barriers, Graicer explains that "we pull them back". Naveh further clarifies that "we do not tell them what to do. We help them realize what happened to them cognitively, in terms of how they got stuck. This is one mode of intervention, and you have to be very patient. Remember, you are observing, but it is not your creation. It is their deliberation. [You] can jump ahead, see where they are going. When they get stuck, we step in and help them rescue themselves."[50] Their use of the "Z-pattern" metaphoric device appears valuable here, in that such a mode of facilitation using the earlier geometric conceptualizations might be problematic to encourage such nonlinear, emergent design activities intuitively.

The second situation where Naveh will intervene with his latest approach in educating military design teams is when a design team is performing acceptably but they lack self-reflection in the act of doing the design. This is where Naveh states metaphorically

> we begin flogging them ... projecting questions [that] will shatter them. This will lead them to rethink, reframe, and sometimes they are doing brilliantly without being aware of what it is that they have actually done.

So, we stop them and [tell them] this is a miracle, what you have done …
let me explain to you why this is.[51]

This takes significant time and mental energy, which again is difficult when a
military organization wants some designed innovation rapidly and for immediate implementation across the force. Naveh sought this design experience
in earlier versions of SOD, yet now in this third version in Tel Aviv, he
appears to be comfortable in how they can iteratively move students through
un-learning, learning, and re-learning in self-disruptive, intellectually stimulating ways.

   The final design deliverable is presented by the team in Naveh's latest SOD
structure again using significant preparation with the students prior to the
actual event. Naveh does not emphasize the deliverable itself here, but instead
he wants the design teams to explain their cognitive process that produced that
deliverable. Once more, this emphasizes a "thinking about thinking" methodology that enforces what Schön termed "reflective practice" in design.[52] As
opposed to U.S. Army and U.S. Marine design methodology (Chapters 4 and
5) that specifically emphasize a design solution that subsequently is inserted
into the legacy planning methodology, Naveh demands designers to discuss
at an epistemological level why they designed as they did. Designers are asked
to explain their epistemological and ontological choices in their design rather
than describe the methodological outputs alone. Conversations here center
on "why", with less emphasis on "how/what" thinking. This demonstrates an
inversion of where the American Army and Marine Corps' design methodologies in doctrine are as of 2022. For the Americanized versions of military
design, "why" is briefly considered if at all, and rarely toward the institution
itself. External environments are conceptualized so that military plans of
"what/how" can commence quickly, in the indoctrinated manner expected
of the institutions.

   Graicer provides one last development of SOD 3.0 that aligns not with the
facilitators, but with the IDF as a culture. Generations come and go, and the
IDF today is different in Graicer's eyes from the one that first met SOD 1.0
in the late 1990s. In 2019, she sees more open-mindedness of Israeli general
officers compared to the previous generation, perhaps caused by the changing
world we live in.

   Another big change from the OTRI days [1995–2005 SOD period] is the
   world. The world has changed. The Generals that we have today notice
   that it is not working. That this big war that the military is supposed to
   prepare for is not going to happen. I am not saying they do not operate
   under it still; but deep down they know that the system is flawed. So,
   they come [to the design course] more ready to challenge it.[53]

Today's IDF senior leaders appear, at least to the design educators charged
with helping them think critically and creatively, more willing to take

intellectual risks, explore unorthodox ideas, and challenge institutionally cherished beliefs, traditions, and ideas. SOD 3.0 is a maturation of what remains radical, unorthodox, eclectic, and heretical ideas on how to shock the military institution to stimulate new thinking in warfare and about war itself. That Naveh and his core team are once more inside the IDF, with a safer and seemingly more welcome embrace of SOD, indicates that the times are indeed changing.

Variations on Disruption: The Friends, Foes, and Fools of Naveh's Design

Over the decades, Naveh has gained a large following of military professionals with a range of varied intents and positions, as well as a similar legion of SOD contrarians and those that despise his work. Some admire his ideas and see his form of military design as an important and yet incomplete contribution to the profession of war. Others of course dismiss systemic operational design as some postmodern bunk that is if anything a distraction from serious study and application of organized violence. Many have sought to capitalize on the value of design thinking, with a range of applications where their utility seems to depend on how well they understand Naveh's ideas as well as Naveh himself.

Naveh generally discounts most any attempt to perform his mode of military design outside of the unique Israeli context and away from his own ability to mentor and coach as a poor imitation or worse. Even within the Israeli Defense Force where now a larger population of serving senior military generals have received design education from Naveh, only a few admittedly attempt it. Graicer estimates that:

> Of active general officers and their counterparts who participated in the course, I would offer, ten percent will publicly acknowledge their use of SOD, forty percent use it and admit it privately, twenty percent are influenced by it subconsciously, and the rest are either indifferent, or did not get it.[54]

Naveh previously cited similar numbers when teaching design to Australians and Americans in the 2005–2012 period outside of Israel.[55] To them, security design praxis remains a minority within a minority for the novel application of organized violence.

Naveh remains a difficult military designer to work with, as he continues to teach and experiment in his provocative as well as often egotistical style in Tel Aviv today. Of the numerous design theorists, educators, and practitioners over the years that have come and gone from Naveh's inner circle, only Graicer has maintained a strong working relationship as well as a collaboration with Naveh concerning their own academic growth of the SOD style and form. However, as of 2014 Naveh seldom travels outside of Israel, rarely communicates outside of his phone or in person, and until recently has shown no interest in writing or publishing new ideas outside of his small and highly exclusive design program.[56] He regularly dismisses many outside attempts to replicate SOD praxis and harbors some clear biases that his form of design praxis might only be realized within the Israeli Defense Forces, or by Israeli contexts alone. However, upon occasion he also has spoken highly of several non-Israeli military designers and feels that sometimes gifted outsiders can do design praxis sufficiently in his eyes.[57]

As the IDF created a new design education contract for Naveh and Graicer at Tel Aviv since 2013, it could be that Naveh is far too engaged in his current work to have the time or energy to travel or publish publicly. In recent interviews he even seems self-deprecating about how he will not be teaching design or be part of the IDF leadership development in the next five years. Naveh's strong personality, confrontational style of discourse, and his pride have made him many enemies while making it challenging for the few that are willing and able to take up his design movement to freely engage with him. Numerous design offspring and efforts to streamline his multidis-ciplinary and intellectually dense material have frequently oversimplified or entirely misinterpreted military design, including most attempts to render design into military doctrine. Few can understand it thoroughly, with even fewer able to teach it to others. Of those, Graicer is of particular importance concerning where SOD is today, and where it may be tomorrow.

Graicer has over the years contributed to the transformation of systemic operational design through different phases, despite she and Naveh at times in disagreement as well as disunity in how they apply design or for what purpose. Although some differences are technique related or andragogic concerning the style of design education delivery, there are several important distinctions to highlight. She contributed the more recent concept of "degrees of freedom" that SOD has incorporated into its most recent form which Graicer credits as occurring approximately in 2016, which illustrates how fre-quently Naveh and Graicer continue to modify their design methodology.[58]

Graicer describes the concept with:

> in the structuring phase of the inquiry, how you begin, how you motivate yourself, how you position yourself ... to use this concept of "degrees of freedom" to say, "okay – this is what is binding me ... whether it is doctrine", and use it as a point of departure ... talk to me about what represents your mind, talk about it, be aware that it is there, and the opposite [as well] ... and the answer to where the degree of freedom is will usually come from there."[59]

The self-imposed "imprisonment" of one's mind forms the center of this concept.

One's perception as a designer of their own "degrees of freedom" there-fore incorporates reflexive practice in line with the design work of Schön,[60] as well as acknowledging one's preferred paradigm for realizing reality.[61] The "degree of freedom" consideration occurs early at the onset of a design inquiry, with Graicer suggesting that SOD facilitators should not permit a design team to move into another design phase if their degrees of freedom concept has not matured to where they have "opened new cognitive space … if they have not opened this in the beginning, it is just going to get more narrow [in thought] … why wait three days if you didn't give yourself the conditions to begin something new?"[62] Naveh's earlier versions of SOD clearly had different manifestations of this core design concept, however Graicer's original contribution provides the Israeli military students something more readily understood than Naveh's earlier concepts. It represents one of the major contributions to current SOD theory that was not created by Naveh himself yet accepted by him as fundamental to their methodology.

Graicer, as a female teaching design almost exclusively to a male-dominated audience within the IDF, has encountered different institutional challenges than Naveh and over the past nearly two decades of practice has evolved her own unique style of design education that clearly differs from that of Shimon Naveh.[63] This is despite them working side-by-side in adjacent classrooms as well as progressively experimenting and reconfiguring SOD over the years. Although Naveh published and wrote on SOD earlier in the development of their program, Graicer has in the last five years been the public and academic face of IDF design with a series of design articles, online lectures, conference appearances and the international engagement of the growing design community.[64]

In personal correspondence with the author and others, Graicer acknowledges that her gradual design orientation may also be drifting into a new direction where Naveh is potentially not involved or as great an influence in the next decade. Graicer stated that she is in full agreement with the methodology, theory and philosophy concerning this latest form of SOD. However, she differs with Naveh on the application of it as well as perhaps some personal stylistic differences as well as personality in how they educate.[65] Graicer's continued experimentation with SOD as well as her own unique insights and adaptations may branch the current Israeli SOD practice into two different yet arguably authentic and unique design communities of practice within Israel as well as abroad.

Systemic Operational Design is the first military design methodology created, and today in this third version continues to generate much of the original excitement, curiosity, controversy and outrage it did in the early 1990s. Of the core developers of SOD, Naveh would quickly become the figurehead and the father of this movement. His personality, intellect, and military accomplishments made for the right combination of skill, intelligence, and brashness for military design to take on not just the Newtonian stylized, systematic logic for modern planning, but the entire modern military institution for war. Kuhn explained a scientific paradigm shift as one where a new, superior scientific frame would challenge and replace the obsolete one. Social paradigm theorists a decade after Kuhn would modify the scientific paradigm shift to that of multiple social paradigms used by different groups of humans, socially constructing a socialized reality upon the existing complex natural world. While the term "paradigm shift" is often used incorrectly, there are two different shifts that might occur. One is scientific and comprehensive, and the other is sociological and involves gaining awareness and appreciation of worldviews entirely dissimilar to the one we use and assume is universal.

Naveh, in spearheading the military design movement, created SOD to potentially try to do both. Military design, in the original SOD form, seeks to disrupt, deconstruct, and reconceptualize how militaries think, organize, and act in warfare. Military science in the traditional sense is challenged by SOD. Yet SOD seeks to reconfigure how military professionals think about their thinking in war, and realize their own socially constructed limits that prevent them from conceptualizing the unrealized in complex warfare. It is no wonder that Naveh and his SOD proponents went through so many battles, were banished from institutions, and still later would be invited back in. Deep down, modern militaries must know that traditional modes of warfare have a shelf life, and without military design, they might expire before their nations even realize it. When militaries are failing, or are experiencing frustration when their established tools are crumbing in their hands, they appear more open to unorthodox, even alien options. SOD provided this to Israel, then the United States, and later still, into Canada and Australia.

Notes

1 The author was hired upon military retirement by USSOCOM at JSOU to assume design course duties. Naveh's SOD course was terminated by JSOU in 2014, and Mr. Jeb Downing was tasked by the JSOU President to generate a replacement SOD design course; he recruited the author in 2015. The original JSOU intent was to reproduce a SOD-style design course internally, not based on external design contractors. See Jeb Downing, "Draft White Paper: Introduction to Special Operation Forces Design" (Unpublished manuscript, January 15, 2014); Jeb Downing, "JSOU Proposal for Operational Design Instruction to USSOCOM Leadership and Staff (as of 26 Jan 15)" (Unpublished manuscript in author's private collection, January 23, 2015); Jeb Downing, "JSOU Educational Support for Commander Directed Operational Design Instruction for USSOCOM Leadership and Staff (as of 22 Jan 15)" (Unpublished manuscript in author's private collection, January 30, 2015).

2 Municipally, the neighborhood is called Ramat Hasharon, and houses the Military College for the IDF also referred to as their National Defense College.

3 This information is based on author discussions with multiple designers aware of this project as of 2022. This new form of SOD is apparently termed the "K-Method" and advances upon Naveh's "Z-Method" covered in this chapter. At the time of writing, information on the "K-Method" was unavailable.

4 Personal correspondence between the author and Philippe Beaulieu-Brossard, February 2020.

5 Personal correspondence between Graicer and the author, 25 Nov 2022.
6 Shimon Naveh and Ofra Graicer, Naveh audio 14OCT2019 20min12sec Day 1 afternoon.MP3, interview by Ben Zweibelson and Nathan Schwagler, personal interview, October 14, 2019, 11:30.
7 Deleuze and Guattari, A Thousand Plateaus, 376.
8 Cara Wrigley, Genevieve Mosely, and Michael Mosely, "Defining Military Design Thinking: An Extensive, Critical Literature Review," *She Ji: The Journal of Design, Economics, and Innovation* 7, no. 1 (Spring 2021): 104–143.
9 Philippe Beaulieu-Brossard and Philippe Dufort, "Introduction to the Conference: The Rise of Reflective Military Practitioners" (Hybrid Warfare: New Ontologies and Epistemologies in Armed Forces, Canadian Forces College, Toronto, Canada: University of Ottawa and the Canadian Forces College, 2016); Wrigley, Mosely, and Mosely, "Defining Military Design Thinking: An Extensive, Critical Literature Review"; Aaron Jackson, "Design Thinking in Commerce and War: Contrasting Civilian and Military Innovation Methodologies," *Air University Lemay Center for Doctrine Development and Education* LeMay Paper No. 7 (December 7, 2020): 1–78; Ben Zweibelson, "Three Design Concepts Introduced for Strategic and Operational Applications," *National Defense University PRISM* 4, no. 2 (2013): 87–104.
10 Larry Kay, "'A New Postmodern Condition': Why Disinformation Has Become So Effective," *Small Wars Journal*, February 27, 2020, 4.
11 Anthony Giddens, *The Consequences of Modernity* (Stanford, California: Stanford University Press, 1990), 45.
12 Milan Vego, "A Case against Systemic Operational Design," *Joint Forces Quarterly* 53 (quarter 2009): 70–75.
13 Paul Mitchell, "Stumbling into Design: Teaching Operational Warfare for Small Militaries in Senior Professional Military Education" (tri-fold poster, Canadian Forces College, Toronto, Canada, 2015), www.doria.fi/bitstream/han dle/10024/117634/MITCHELL%20Paul_poster_Designing%20Design,%20T eaching%20Strategy%20and%20Operations%20for%20Small%20Militaries. pdf?sequence=2
14 Seth Johnson, ed., *IBM Design Thinking Field Guide Version 3.1* (IBM Corporation, 2015), ibm.biz/idt_fieldguide.
15 Shimon Naveh, Jim Schneider, and Timothy Challans, *The Structure of Operational Revolution: A Prolegomena*, A Product of the Center for the Application of Design (Fort Leavenworth, Kansas: Booz Allen Hamilton, 2009). This is also how Graicer recollects this period, with SOD developments occurring outside of Army formal attempts to simplify the SOD 1.0 content. Correspondence between the author and Graicer, 25 Nov 2022.
16 Butler-Smith, "Operational Art to Systemic Thought: Unity of Military Thought."
17 Graicer, Graicer audio 15OCT2019 17min04sec Day 2 afternoon.MP3, 10:30.
18 Graicer, 13:00.
19 The concept of "drift" has also appeared in strategic management theory. See Henry Mintzberg, "The Design School: Reconsidering the Basic Premises of Strategic Management," *Strategic Management Journal* 11 (1990): 186.
20 As mentioned in the Chapter 4, this early SAMS faculty monograph demonstrates the initial Army interest in SOD, and a desire to promote it across the force. See William Sorrells et al., "Systemic Operational Design: An Introduction"

(US Army School of Advanced Military Studies Monograph, Fort Leavenworth, Kansas, May 26, 2005), ATZL-SWV, OMB No. 0704–0188.

21 Deleuze and Guattari, A Thousand Plateaus.

22 Beaulieu-Brossard, "Encountering Nomads in Israel Defense Forces and Beyond," 2016, 6.

23 Beaulieu-Brossard, "Encountering Nomads in Israel Defense Forces and Beyond," 2020, 14.

24 Graicer, "Self Disruption: Seizing the High Ground of Systemic Operational Design (SOD)," 34.

25 When interviewed by the author and asked this specific question, Naveh would only answer indirectly. He acknowledged both of these reasons in conversations but never settled upon one or the other.

26 Shimon Naveh, "The Australian SOD Expedition: A Report on Operational Learning" (Unpublished manuscript, December 2010).

27 Graicer termed this "self-exile" in that it is unclear if the IDF formally dismissed Naveh and his ideas, or if Naveh chose exile in response to his ideas being blamed unfairly for military failures.

28 Naveh and Graicer, Naveh audio 14OCT2019 2hr28min Day 1 patio.MP3.

29 Naveh and Graicer.

30 Shimon Naveh, *Systemic Operational Design: Designing Campaigns and Operations to Disrupt Rival Systems (Draft Unpublished)*, Version 3.0, unpublished draft (Fort Monroe, Virginia: Concept Development & Experimentation Directorate, Future Warfare Studies Division, US Army Training and Doctrine Command, 2005).

31 Naveh and Graicer.

32 Naveh and Graicer.

33 Personal correspondence with Graicer on February 23, 2020.

34 Naveh and Graicer.

35 Graicer first presented these concepts in Toronto to a small design audience the author participated in, and later produced an unpublished, early draft. That paper later would be published with some editorial changes. See Ofra Graicer, "Self Disruption: Seizing the High Ground of Systemic Operational Design (SOD)," *Journal of Military and Strategic Studies* 17, no. 4 (June 2017): 21–37; Ofra Graicer, "Self Disruption – Beyond the Stable State of SOD," in *Cluster 1* (Hybrid Warfare: New Ontologies and Epistemologies in Armed Forces, Canadian Forces College, Toronto, Canada: University of Ottawa and the Canadian Forces College, 2016).

36 Ofra Graicer to Ben Zweibelson, "Re: Go with This Version Instead (15 FEB Reply)," February 15, 2020. Personal correspondence over e-mail.

37 Personal correspondence between the author and Graicer, 25 Nov 2022.

38 Naveh and Graicer, Naveh audio 14OCT2019 2hr28min Day 1 patio.MP3.

39 Shimon Naveh, "Northern Storm: A Narrative of Reflective Command, Systemic Learning, and Operational Design 2002–2005." This unpublished PowerPoint slideshow of SOD concepts is an example of Naveh's dense, challenging content. He provided copies of this to select students including the author in 2010.

40 Whether intentional or not, in Hebrew, the letter *zayin* is associated with both "weapon" and "penis." The shape of *zayin* in Phoenician symbology is an "I", and in Hebrew a "T", with Greek changing it into the "Z."

41 Jamshid Gharajedaghi, *Systems Thinking: Managing Chaos and Complexity, A Platform for Designing Business Architecture*, Third (New York: Elsevier, 2011), 13, http://pishvaee.com/wp-content/uploads/downloads/2013/07/Jamshid_ Gharajedaghi_Systems_Thinking_Third_EdiBookFi.org_.pdf; Christopher Paparone, *The Sociology of Military Science: Prospects for Postinstitutional Military Design* (New York: Bloomsbury Academic Publishing, 2013), 10–19; Ysanne Carlisle and Elizabeth McMillian, "Innovation in Organizations from a Complex Adaptive Systems Perspective," *Emergence: Complexity & Organization* 8, no. 1 (2006): 2–9; Antoine Bousquet, "Cyberneticizing the American War Machine: Science and Computers in the Cold War," *Cold War History* 8, no. 1 (February 2008): 77–102.
42 Naveh and Graicer, Naveh audio 14OCT2019 2hr28min Day 1 patio.MP3.
43 Personal correspondence with Graicer on February 23, 2020.
44 Naveh and Graicer, Naveh audio 14OCT2019 2hr28min Day 1 patio.MP3.
45 Naveh and Graicer.
46 Why Generals Need to Forget before They Can Become Generals | Ofra Graicer | TEDxTelAvivUniversity (Tel Aviv, Israel: Tel Aviv University, 2015), 2:13, www.youtube.com/watch?v=6pfZM9uSlmg
47 Naveh and Graicer, Naveh audio 14OCT2019 2hr28min Day 1 patio.MP3.
48 Naveh and Graicer, Naveh audio 14OCT2019 2hr28min Day 1 patio.MP3, 1:27:10.
49 Naveh and Graicer, 1:31:00.
50 Naveh and Graicer, Naveh audio 14OCT2019 2hr28min Day 1 patio.MP3.
51 Naveh and Graicer.
52 Donald Schön, "Knowing-in-Action: The New Scholarship Requires a New Epistemology," Change, no. November/December 1995 (1995): 27–34.
53 Naveh and Graicer.
54 Ofra Graicer to Ben Zweibelson, "Re: Go with This Version Instead (17 FEB Reply)," February 17, 2020.
55 Naveh, "The Australian SOD Expedition: A Report on Operational Learning."
56 Graicer to Zweibelson, "Re: Go with This Version Instead (15 FEB Reply)," February 15, 2020.
57 Graicer and the author have placed Naveh in contact with some former students he holds in high regard in this respect since 2019. The author knows of a handful of non-Israeli military designers that Naveh recognizes and appreciates their work.
58 Ofra Graicer, Graicer audio 15OCT2019 11min02sec Day 2 afternoon.MP3, interview by Ben Zweibelson and Nathan Schwagler, personal interview, October 15, 2019, 7:26.
59 Graicer, 6:48.
60 John Gero and Udo Kannengiesser, "An Ontology of Donald Schön's Reflection in Designing" (Key Centre of Design Computing and Cognition, University of Sydney, undated).
61 Gibson Burrell and Gareth Morgan, Sociological Paradigms and Organisational Analysis: Elements of the Sociology of Corporate Life (Portsmouth, New Hampshire: Heinemann, 1979); Dennis Gioia and Evelyn Pitre, "Multiparadigm Perspectives on Theory Building," Academy of Management Review 15, no. 4 (1990): 584–602.
62 Graicer, Graicer audio 15OCT2019 11min02sec Day 2 afternoon.MP3, 9:50.

63 Graicer, Graicer audio 15OCT2019 17min04sec Day 2 afternoon.MP3, 2:10.

64 Graicer, "Self Disruption: Seizing the High Ground of Systemic Operational Design (SOD)"; Graicer, "Beware of the Power of the Dark Side: The Inevitable Coupling of Doctrine and Design"; Graicer, "Between Teaching and Learning: What Lessons Could the Israeli Doctrine Learn from the 2006 Lebanon War?"; Why Generals Need to Forget before They Can Become Generals | Ofra Graicer | TEDxTelAvivUniversity.

65 Personal correspondence with Graicer on February 23, 2020.

7 Designing Further Afield in Canada and Australia

Beyond the Israeli and American military design methodologies presented in previous chapters, several other defense institutions experimented since the mid-2000s with some portion of military, commercial, or hybrid design praxis. These militaries also attempted to integrate it into existing military planning and strategy as well as efforts in doctrine and education. Most do not yet present a formalized mode of military design that expresses a clear and distinguishable theory, or they continue to experiment with a range of design potentials. Some have moved toward structured educational processes for the organization on how to conduct design activities, while others remain uncommitted except in small-batch, limited endeavors. Most feature a dominant population that defends the modern engineering-oriented logic of systematic military thinking and consider any design concepts as a threat requiring elimination or marginalization. However, that tide appears to be changing. In Canada and Australia over the last two decades and more significantly in the last five years, a new generation of military professionals appear dissatisfied with the legacy frames and institutional norms of their armed forces. Many veterans of Iraq, Afghanistan, and other regional security conflicts seem more interested in the disruptive qualities of design than previously, demonstrating the same trend of military reform efforts in Israel and the United States.

Whether Canada and Australia feature fully developed military design methodologies is an area of debate in the military design community. If a military coops operational planning with systems thinking and uses some design terms, does that constitute a military design methodology? Or, if a military university declares itself "agnostic" on any particular design method, does their multi-disciplinary approach become a design methodology itself? In this chapter, they are instead termed "proto-design" efforts, as suggested by Jackson in his own assessment of the Australian Defence Force (ADF). Not to be taken pejoratively, the "proto-design" label distinguishes these efforts with what the Israeli Defense Forces, the American Army, and the American Marine Corps have done through deliberate design of new doctrine, a defined methodology, educational reform, and deliberate design in training.[1] At best, Australia and Canada are invested in experimenting and testing

DOI: 10.4324/9781003387763-8

various design concepts, while also cautiously considering how and what design might best pair with their needs. A proto-design effort suggests that in time, militaries using them will mature those labors into a more established, likely more robust design process that will gain widespread acceptance (or tolerance) across the entire enterprise. Both Canada and Australia feature robust internal debates on whether military design is useful or not, mirroring the same institutional introspection ongoing in Israeli and American forces during design experimentation and testing over the last several decades.

The Canadian and the Australian militaries both have a history of design experimentation to include experience with Naveh's earliest systemic operational design concepts in the first decade of the twenty-first century, along with significant partnership and educational exchanges between the American armed forces and their own. Canada and Australia, particularly in the last decade, have experimented with commercial design applications set to distinct military challenges, and their armed forces have considered ways to form mixed-methodology design hybrids as well. As medium-sized nations, their militaries can change faster than larger nations, but the smaller size and budgets of Australia and Canada also suggest some institutional hesitancy on moving too quickly into a reform or investment that remains ill-defined, confusing, or controversial. The military design movement over the last three decades certainly provides room for both concern and optimism.

Canada's Military Journey Toward Designing in Organized Violence

The Australian and Canadian militaries feature many similar patterns with American Armed Forces with respect to history, geographical context in relation to earlier periods of war, a colonial common parent, culture, ethnic, as well as ideological and economic similarities. To introduce the proto-design movements in both nations, a case will first be established where both military forces experienced a breakdown in their dominant modern military paradigm as well as a potential opening for some postmodern, disruptive considerations. This breakdown in the state-centric application of organized violence as defined in the post-Westphalian modern form (modernity of war) would create opportunity for military design in Canada and Australia just as it would in Israel and America, yet upon different paths and due to cultural, social, and geopolitical differences comprising the Australian and Canadian military identities.

The Canadian understanding of what war is and how Canada ought to execute nests within the strategic alignment of Canada which since 1945 has tilted extensively toward the United States. Canada has both the luxury and misery of being geographically attached to a superpower and the backyard space and climate that has created a historic barrier of natural protection from outside threats. Contemporary Canadian military planning methodologies share and often assimilate United States military doctrinal and theoretical origins.[2] Dalton describes the Canadian Operational Planning Process (OPP) as based on the established North Atlantic Treaty Organization (NATO) planning process, with all of them remaining mechanistic military decision-making processes that are linear in nature and focused on analysis.[3] "[The] theory of planning, a linear, mechanistic, reductionist approach to problem solving, forms the basis of contemporary operational design ... these are the roots of Canadian contemporary operational design".[4] This makes sense in that Canadian military forces routinely act in a supporting role and task organization often to American (since World War II, and to Great Britain prior to that) or Coalition commands, requiring supporting forces to at least be familiar with the theories, models, methods, and military language of the commanding entity.

This served the Canadian society sufficiently through the late twentieth century up into the Post-Cold War period where, like most other Anglo-Saxon forces seeking reinvention and restructuring in the 1990s, the Canadian Army encountered serious issues. The pre-twentieth-century Canadian military history is often summarized by the maxim that Canada was "the Peaceful Kingdom" given its natural geographic advantages, distances, and resources therein.[5] Twentieth-century Canadian military endeavors would dispel that notion to some degree, yet these ideas would remain deep inside the nation's identity.

Canada features several distinct organizational transitions after World War II that would influence not only the dominant military frame for warfare, but also how and why their Armed Forces should function in accordance with Cold War Era Canadian policies and society. First, Canada broke from the Anglosphere in how it organized as well as financed their military instrument of power by combining the previously independent Army, Navy, and Air Force into a single Joint Service in the 1960s. Previously, the three services fought "three separate service campaigns within the larger Allied strategy, and then eyed each other warily during the uncertainties of peacetime demo-bilization".[6] Deemed part of the "Management Era", Canadian policy makers in the 1960s obtained Royal Ascent to unify the separate military services under a single functional command system and rename it "the Canadian Armed Forces" (CAF). This reorganization would have significant cultural, social, and decision-making impacts across the entire force.[7]

The rise of the Cold War, and the transition of the former multi-polar political world into a bi-polar standoff between superpowers, ushered in a new era of conflict preparation and confusion. This would also lead to conditions where the CAF prioritized the analytic optimization of process improvement, risk reduction, and uniformity potentially to the detriment of military curiosity and experimentation. Critics of this consolidation of the three formerly independent services saw the newly unified military structure as "an organization in great confusion, a military profession unsure of its values, its history, or its future, and with the old problems still firmly in place".[8] Despite this unification of the three forces under a combined Canadian defense force organizing logic, the managerial streamlining did little to fix the deeper institutional issues that would arguably culminate in the 1990s.

Despite the ambitious desires of the Canadian military leadership following the defeat of Axis powers in 1945, Canadian policymakers moved to dismantle the large standing forces and return once again to a "small professional body geared toward supporting national mobilization were some future conflict to reoccur".[9] Collapsing the three formerly independent services into a single defense force was essentially a large-scale design experiment in seeking greater efficiencies and budget savings "through the elimination of duplications among the branches of the armed forces ... and an amendment to the National Defence Act to provide for one group of military laws" instead of the previous triplicate structure.[10] Canada's military during the Cold War moved away from the largely white Anglo-Saxon male homogeneity toward a more culturally inclusive military force, largely through social and political efforts to reform the Armed Forces. The French–Canadian population became a prominent feature in Canada's military, reflecting the near quarter of the Canadian population that have French-Canadian origin.[11] In a social tension dating at least back to the 1950s, Canadian officers required bilingual abilities and some officers felt a greater emphasis is placed upon language and cultural ratios in promotion boards over individual accomplishment.

The CAF swung back and forth with various institutional reforms on linguistics-based recruitment initiatives, the weight of performance evaluation associated with bilingual officers, and efforts to make the English and French languages equal in status, rights, and privileges as to their use in the Canadian Forces. Canada's modern military progressed with the rest of the Anglosphere toward greater diversity and heterogeneity, although tension still remains prevalent between ethnic lines of Anglo-Canadians and French-Canadians.[12] The repercussions of ethnic, cultural, lingual, and political-based military actions in both World Wars would have significant long-term impacts in this Cold War period and introduce different tensions that would impact how Canada's Armed Forces organized, thought, and acted in the application of organized violence.[13]

Geographically, Canada in the Cold War faced a threat of Soviet aggression while remaining within the American security blanket by sharing a common and unprotected border. Mitchell remarks on this as follows: "Canada has traditionally lived within the metaphorical equivalent of a gated community, which explains much of this policy diffidence [on the ineffectiveness or irrelevance of Canadian defence policy]".[14] Nowhere within the Cold War could Canada realistically mount any self-defense without immediate support of larger allies. However, in the Post-Cold War Era, a multiplicity of state and non-state threats as well as international security demands no longer permit Canadian isolationism, absolution or deferment toward a majority power in alliance.

One of the defining events in modern Canadian military history remains the Somalia Scandal in 1993, termed the "Somalia Affair" that became a deeply troubling national event. The beating death of a Somali teenager by

Canadian soldiers conducting humanitarian efforts became a sociological trigger point for critically examining a static military culture, and to this day remains a difficult topic within the CAF. The key findings of the Canadian Inquiry Commission indicated that a lack of critical and divergent thinking had created the conditions for the ethical and moral dilapidation within the CAF by the 1990s.[15] The Somalia Affair was but one major event within a series of institutional failings and setbacks that would force institutional self-reflection, usher in reforms, and promote the importance of "understanding how and why the institution thinks and acts". Unlike the Australian Army which in the 1990s remained mired in "the long peace" following withdrawal from Vietnam, the Canadian Army was overstretched, under-educated, and entangled in institutional soul-searching.[16]

In the military generation following the Somalia Affair, Canadian military practitioners debated what institutional changes ought to be applied across the force so that ethical behavior, compliant in CAF values, could become the cornerstone of sound leadership and ethical decision-making in complex defense/security environments for the new post-Cold War security environment and the increasingly dynamic rise of novel non-state actors and threats. St-Denis observes how the dominant institutional form generated groupthink in behavior, language, relationships, and thought:

> Chief among those limits is the necessary restriction of the soldier's freedom to make choices as an individual. Soldiers are required to sublimate in many ways their own individualism for the sake of the group "… military success depends on absolute authority being wielded by the leader, and this same absolute authority works strenuously against the idea of encouraging followers to challenge beliefs and values, and to work things out on their own. It also works against the notion of empowering individuals to seek innovative solutions."[17]

Clermont, citing St-Denis above as well, posits that the Canadian military today continues to be "dominated by behaviorism where human beings are viewed through a mechanistic and functionalist paradigm".[18] In these military organizations, soldiers are conditioned to approach challenges within a reductionist perspective, "and critical thinking is perceived as antithetical to organizations. It is perceived as a challenge and threat to command and leadership. Change and emergence are unwanted".[19] This created clear tensions between Canadian military strategy and planning styles that were of an engineering and Newtonian mindset, and the disruptive and multidisciplinary, free-wheeling improvisational stylings of military design. These tensions would manifest particularly in how the Canadian military establishment would react to new design concepts as they would be introduced into their military education system.

Canada's military paradigm remains wedded to Newtonian stylization, where the CAF OPP methodology is "a mechanistic, analytic approach to problem-solving that seeks to control or impose order on complicated situations characterized by large-scale state on state mechanized warfare".[20] Even though Canadian military doctrine on how to apply the CAF-OPP methodology explicitly warns against misinterpreting the process as linear and quite formulaic, military academic studies of the planning process struggle to illustrate it in anything but a linear sequencing of distinct, discrete activities couched in entirely analytic and rather mechanical language.[21] This mindset of seeking convergent, overly analytical thinking toward warfare would no longer serve the Canadian Forces when they faced irregular and increasingly complex security challenges that were not "solvable" using the previously successful methods derived from the twentieth century's World Wars.[22]

Clermont critiques the CAF of needing "to change the institution's *doxa*, one in which a profound culture of anti-intellectualism is still present, and with it the negative effects it has on professional development".[23] He nests this anti-intellectualism back to at least the Somalia Scandal period of the 1990s, where the Canadian Inquiry Commission determined that the CAF had lacked significant development of intellectual and cognitive skills across their military.[24] The prominence of uniformity, regulatory behaviors, and convergent thinking would promote certain military outcomes in performance while delegating divergent thinking, experimentation, intellectual curiosity, and critical reflection to the backseat for the military institution. This would become a systemic pattern and not something easily overcome for the Canadians.

In the military generation following the Somalia Affair, Canadian military practitioners debated what institutional changes ought to be applied across the force so that ethical behavior, compliant in CAF values, could arguably become the cornerstone of sound leadership and ethical decision-making in complex security environments. Yet in the decades to follow the 1990s transition and disruption, the CAF appeared unable or perhaps unwilling to pivot off a decidedly state-centric and analytic decision-making frame for organized violence. Clermont proposed that the Canadian military (among others) remain organizations "dominated by behaviorism where human beings are viewed through a mechanistic and functionalist paradigm".[25] Canadian soldiers were conditioned to approach challenges within a reductionist perspective, "[where] critical thinking is perceived as antithetical to organizations. It is perceived as a challenge and threat to command and leadership. Change and emergence are unwanted".[26] This would foster opportunities for change within the Canadian military in the 2000s when select officers would be exposed to Naveh and design.

From Modern to Crisis in the Land Down Under: Designing Military Reforms

Australia would too express quite a unique history that would foster different tensions and paradoxes within applying the modern military war frame.

Australia's European lineage, remote geographic location, proximity to emerging non-western competitors and adversaries, as well as the cultural and social qualities that define the Australian warfighter interact within a dynamic and complex network of influences and beliefs so that their military design development takes a distinct journey of trial and error. The Australians first learned of military design through officer exchanges and shared educational programs, with Australian officers and defense academics becoming aware and excited about systemic operational design in the early 2000s.

Naveh interacted with the few Australians in American military classrooms, and later visited Australia at the request of their special operations units. While writing up a design report after work with Australian special operations forces in the 2009 period, he reflected on Australian military spirit. Naveh observed that the cultural littoral between "Anglo-Saxon and South-East Asian traditions" where "the Australian institutional legacy system has always been nurturing an approach of heterodoxy ... by their very nature, Australian officers are more susceptible to critical discourse than their peers in other state military institutions".[27] This anecdotal reflection differs with other Australian published critiques of their military and a sense of "anti-intellectualism" that pervades the force, which will be expanded upon later in this chapter. This also could be bias on account of Naveh operating in a commercial, contracted capacity here with his final report going to the Australian client. Regardless, Australian military culture is nuanced, representing a combination of cultural, ethnic, and political ties to Great Britain as the powerful and original "mother country" for the Anglosphere. Yet Australia's (and New Zealand's) position alone in the Pacific in a rather isolated yet insulated context dates back to their founding as colonies.

World War I and Britain's application of remaining colonial military enablers indirectly created conditions for Australian and New Zealand military mythology and lore that would reinforce Naveh's position of this uniquely Australian war culture distinct from peer Anglo-Saxons. Jackson describes the long-term impact of the Australian and New Zealand Army Corps (ANZAC) landing at Gallipoli in 1915,

> the landings have taken on a legendary status in both countries ... underlying ... for the first time, Australians and New Zealanders performed a noticeable role on the world stage and individual soldiers-or "diggers," as they became known-inspired feelings of honour and pride in their fledgling countries.[28]

This would become powerful symbolism that continues to manifest to this day across the Australian armed forces.

The "Anzac legend" has shaped and influenced generations of Australians (and New Zealanders) since World War I, potentially in politically motivated and culturally exclusive ways. Crotty and Stockings explained in their research into the mythology of Australian military history: "from April 1915 onwards the Anzac legend grew into an inescapable social force tied to the core of national identity … the social values held dear by the collective consciousness of the late nineteenth- and early twentieth-century Australia … were informally but systematically codified into a national legend about 'diggers'".[29] Conformity to these beliefs coupled with fierce independency would make for rather specific military operating preferences as explored below. While individual soldier independence was high, the Australian armed forces overall remained highly dependent upon Great Britain for necessary equipment and advanced weapon systems as well as even British operational and strategic leadership in war through World War II.

At a strategic level, World War II presented to Australians and New Zealanders the first real and immediate existential threat to their nations in their histories after the surprise bombing of Pearl Harbor in 1941.[30] Although Americans associate the sudden Japanese existential threat exclusively with Pearl Harbor, Australians saw the lesser-known Japanese attacks at Darwin, Broome, and the submarine attack on Sydney harbor as far more significant, again demonstrating subtle distinctions between various contemporary western industrialized military organizations despite common cultural and ethnic linkages to the British Empire.[31] These grim lessons served as a sobering realization that Australia was vulnerable to powerful regional enemies able to use advanced technology and resources to strike at what was previously considered geographically isolated and protected territory.

Throughout the Cold War, Australia would move closer to an American foreign policy orbit than that of the United Kingdom, and shift the priority of military modernization and education over to how and what the American military was developing. This shift would manifest in the ADF and contribute to the later conditions of military frustration with the modern war frame as well as the established form and function for what, how, and why to apply organized violence on behalf of their respective societies. The Australians (and the New Zealanders in close orbit), despite being the lone Anglo-Saxon society in a vast Pacific region of the planet with natural geographical barriers and defenses, would need to wrestle with modern societal and technological advancements that demanded military transformation. Australian post-Vietnam defense policy shifted once more, with new doubts developing on US support in complex Pacific issues such as with Indonesia.[32]

The collapse of the looming Soviet threat concluded an important chapter in the Anglosphere concerning the struggle between Capitalism and Communism; however, it ushered in a new and confusing period of uncertainty. "The end of the Cold War had made Australia's strategic situation more uncertain and possibly more dangerous".[33] Australia and New Zealand moved from the Cold War period of an anti-communist Western alliance focused upon deterrence and attrition-based warfare to a 1990s transition period of peacekeeping.[34] Indeed, this occurred throughout the Anglosphere.[35] Jackson describes the new banner of a United Nations peacekeeping force where Australians, New Zealanders, as well as Canadians shifted their focus away from the crumbling "Red Threat" of Moscow toward "coalition building in order to attain a level of international influence that they [could not] attain individually".[36] Layton adds that the 1987 Defence White Paper would bring joint warfare into vogue, pressuring the Australian defense industry to bring advanced, integrated weaponry for a new emphasis on air and naval defense of Australia.[37] The 1990s would be a decade of limited utilization, and a downsizing of Australian forces coinciding with a loss of combat experience and knowledge. While British forces maintained their counterinsurgency experiences with activities in Northern Ireland, things in Australia between their withdrawal from the Vietnam War (1972) and the collapse of the Twin Towers were largely devoid of intensity as well as significant investment by the government.[38]

The effects of 9–11 upon Australia, Canada, and the western alliance of NATO as well as Anglo-Saxon shared national affinity cannot be understated in how things would unfold for this new period of confusion, uncertainty, and desire to consider new ideas concerning warfare and military decision-making in complexity. True to Australia's history, their military deployed just as they did in World War I. They sent small tactical forces overseas to work for a larger, more powerful ally as part of a multi-national, integrated force where the operational and strategic decisions were not made by the Australians concerning the overarching strategic or diplomatic objectives. Indeed, Australia was disappointed with limited US involvement in the 1999 East Timor operation, and while Afghanistan and Iraq were not directly important to Australia, forging a stronger Australia–US relationship was.[39] Tactically, the Australians would excel in Afghanistan and Iraq, yet operationally and strategically at the direction of more powerful allies the partnership would experience systemic failure to achieve any major objectives.

In the same period that the American military grew frustrated in the "War on Terror", both Canadian and Australian forces would also increasingly become open to alternative ideas on how and why to think and act in war. It would challenge Western militaries to conceptualize in novel form their military operations and how to think and act in the application of organized violence.[40] The counter-culture experimentations of Naveh and his minority band of Israeli military designers would gradually spread not only to the American military as explained in the last few chapters, but around the world into the Canadian and Australian militaries as well.

From Tel Aviv to Toronto in a Decade: Canadian Design Development

The CAF took a similar design developmental journey as the Australians in the last nearly two decades, although their path differs in several important ways. Like the Australians, the Canadians do not have their own formal design methodology, nor do they have any formal design doctrine. They do present what they term an "agnostic" mode of design education, and unlike the Australians they have implemented robust military design education at all levels of their professional military education (PME) system. This includes some elementary design exposure to their cadets and junior officers, a well-structured design educational program for their mid-grade officers and also executive design education for their senior officers at the war college level.

Academically, Canadians also would learn of these new military design concepts in contemporary literature, especially by Canadian military officers writing to share their thoughts on design within Canadian journals, along with Canadian officers traveling abroad to study with foreign militaries such as the Americans where design experimentation was ongoing. A growing library of professional articles, monographs, online blogs, lectures, and experiments with military design quickly became international in scope, with some Canadian military personnel publishing on whether Canada might incorporate Naveh's SOD, blend it into existing Canadian operational planning, or perhaps consider some other alternative.[41] Canadian military professionals would in the 2007–2015 period foster much debate on what, if anything, ought to be done with design. A majority of design advocates appear to come from exposure to Naveh or through the U.S. Army SAMS program where design was studied, and many critics tended to align with existing, established Canadian strategic and operational practices.

Although small in numbers, by 2007 sufficient Canadian military professionals, academics, and educators had been exposed to design and began considering whether it had value in Canada. As these Canadian officers returned to their organizations or consumed military design topics in available military journals, blogs, and conferences, they began to trigger an institution-wide curiosity as well as apprehension over design.[42] Lauder's 2009 design article in the Canadian Military Journal and Anderson's subsequent article on design in 2012 in the same journal both addressed Naveh's systemic operational design and are some of the earliest examples of Canadian published discourse on the topic. American design concepts would be introduced into formal U.S. Army doctrine in the 2006–2010 period which coincided with significant Canadian troop deployments to highly dangerous, complex counterinsurgency hot spots.[43]

For Canadian design experimentation, what began largely as one military instructor's personal experimentation with Israeli SOD in the classroom would later mature into a comprehensive, faculty-driven, and institutionally supported effort to inculcate military design thinking across the CAF. Dr. Paul Mitchell, a long-time faculty at the Canadian Forces College, reflected on this as follows:

> Anderson's efforts on the JCSP [Joint Command and Staff Program JCSP] which was called the Command and Staff Course at that point were simple experiments in SOD. It was a single officer's initiative here, rather than a college sanctioned approach. As soon as Anderson was posted elsewhere, the experiment ended.[44]

This personal endeavor by Anderson and his field grade initiative in the Canadian Forces College's JCSP is addressed first, as it utilized extensive elements of Israeli SOD methodology imported from the U.S. Army School of Advanced Military Studies' first experimentation with SOD in 2005–2008.[45]

In the 2007 academic year for JCSP, faculty members launched a three-year unique educational experience exploring whether SOD was a valid alternative to the traditional OPP.[46] Anderson, then a Lieutenant Colonel in the Canadian Air Force, wrote the only available documentation on this design experiment with a 2012 Canadian Military Journal article. Anderson described each of the three academic years from 2007 to 2009 where design was applied through education and exercise application for students. Although the total student population for JCSP is significant, they are separated into about twenty-five students from various services to form a Joint Operational Planning Group (JOPG) to conduct the Canadian Forces College (CFC) exercises.[47] Only one of these JOPGs became a design group, where those students received some systemic operational design education along with some readings associated with Naveh's core concepts of that period.

According to Anderson, the 2007 academic year was quite problematic. The students were unable to complete all the systemic operational design prescribed process, and they additionally decided to perform the seven SOD discourses sequentially instead of how Naveh originally intended them to be conducted.[48] Naveh's SOD is supposed to be fluid and feature aspects of "drift" in the exploration, and designers are expected to move across each of the seven discourses in an emergent, nonlinear experience "of becoming". However, these first-year design students found this too challenging, and it is uncertain if the CFC design faculty were experienced enough in teaching systemic operational design to facilitate such dexterity. The postmodern theory and dense texts for required reading were also frequently critiqued or rejected by the students, leading to subsequent course revisions and concerns that military design might be too abstract or disruptive to be of serious value for the larger force. However, graduates of that academic year would later on advocate design based on Naveh's SOD concepts, to include future Canadian Flag Officers such as Simon Bernard who would contribute to articles on design as well as lead design praxis in Canadian organizations including their Combined Joint Operations Command (CJOC) by 2019.

In 2008, the JCSP faculty reflected on the previous year's frustrations. The original 2007 systemic operational design reading material proved insufficient for explaining SOD as well as how to guide the SOD process. The students lacked necessary theoretical understanding of core systemic operational design concepts such as complexity theory, systems thinking, or postmodern philosophy. There was insufficient time allocated to complete the exercise doing systemic operational design in the manner the Canadian Forces College had originally set out to execute.[49] The 2007 students did express a somewhat positive outlook on systemic operational design potential value, thus the college continued in 2008 while also adjusting several aspects of the education.

The 2008 JCSP student pool shrunk from twenty-five down to a manageable fourteen students, where they enrolled in a specific elective to learn systemic operational design in a more disciplined manner. This smaller, more educated group also departed significantly from the original Israeli crafted design methodology.

The U.S. TRADOC pamphlet *Commander's Appreciation and Campaign Design* from 2008 provided a stronger influence, according to Anderson, upon the Canadian students regarding how it articulated the design process.[50] This CACD pamphlet was characterized as a blend of traditional operational planning with the systems theory and complexity theory of design, and was an intermittent conceptual document between the American Army's *Field Manual 3–24* first design chapter in 2006 and the subsequent design chapter in 2010's *Field Manual 5–0, The Operations Process.*

The lack of Naveh's emphasis on postmodern philosophy potentially made the CACD document a more attractive, albeit oversimplified form of design for education at the college. The Canadian professional military institution drew heavily from U.S. design experimentation at Fort Leavenworth. With the 2008 JCSP students emphasizing the systems theory of the CACD document with some remnants of original SOD theory, Anderson detailed the fruits of their labor with various diagrams and graphics in his 2012 article.[51] These graphics matched many of the concepts and diagrams found in the CACD pamphlet as well as related systems-thinking approaches,[52] while distancing itself from the more abstract conceptualizations and plasticity of Israeli original design theory.[53] While the 2008 JCSP students had advanced the Canadian design experiment and achieved some success with their graphic depictions of complexity in nonlinear and decidedly untraditional campaign design, they struggled with the critical element of the accompanying design narrative as prescribed by SOD methodology.[54]

The third year of the systemic operational design experiment with JCSP occurred in the 2009 academic year where again the faculty divided students further into two groups of seven.[55] According to Anderson, faculty all but abandoned teaching SOD and instead "it was decided to use a systemic approach to operational design and planning that emphasized the contributing theories (complexity, chaos, and systems theory), rather than adhering to the SOD methodology, *per se*".[56] This pattern suggests what both SAMS faculty Ryan and Butler-Smith observed of many American military faculty rejecting the difficult postmodern concepts and opting for military design concepts that reinforced rather than confronted existing belief systems and frames.[57] The postmodern ideas would, in the absence of social paradigm theory and individual frame awareness often foster institutional resistance in that virtually all postmodern work exists outside of what modern militaries deem useful and complimentary knowledge for warfare. The modern war paradigm is one of technical rationalism, engineering language, and systematic logic. Naveh's SOD challenged this paradigm at its philosophical foundations, triggering institutionalized defense of the modern military paradigm to what seemed an outside intruder. Returning to Ryan's observation in his blog, "how do you introduce a new paradigm when you cannot have a conversation at the level of paradigms?"[58] If the Canadian students could not consider what their preferred war paradigm was, how would they be able to incorporate the parts of SOD that deconstructed those very things cherished by the institution?

Interestingly, Anderson suggested that each group might develop their own design methodology, devoid of any adherence to systemic operational design or other military design methodology. "They were free to use whatever means they decided were appropriate for the operational problem and themselves as a group".[59] Of the two design groups, one Canadian team took a technically rationalized, engineering approach with electronic graphics, a blog for the team, and a formal electronic record of their design journey. The other group of Canadian designers used traditional design means with extensive white board work, butcher paper, and other temporary and disposable mediums, in keeping with the fluid, emergent nature of how SOD advocates sought emergent, iterative praxis.

Based on the graphics provided of the 2009 JCSP group, they accomplished making nonlinear campaign design concepts that remained compatible with established OPPs. The design paired with institutionalized planning. The language and organizing logic produced by the designers remained of the same functionalist paradigm, meaning the designers placated the institution by disposing of anything disruptive to the modern war paradigm. However, Anderson closed his article by recommending that the college pursue further study of design, while not necessarily becoming constrained by the underpinning theories upon which SOD was first developed.[60]

While the CFC did implement further design education by 2013, this exercise did not feature multiple design methodologies at first, and continue to largely pacify institutional demands that indoctrinated planning concepts not be challenged. Mitchell, charged with running the design program at the CFC would in 2015–2016 attend a military design educational workshop at the Joint Special Operations University in Tampa, Florida, and also travel to Fort Leavenworth to meet with design faculty at SAMS.[61] Below, Mitchell explained how within the 2015–2017 period the CFC would move to implement a different way of thinking about design methods and education:

> The idea [of shifting to a mixed disciplinary approach to design education] actually emerges much later, about the time you held [the author's] first workshop at JSOU. We were making a silk purse out of a pig's ear [at the CFC] at first, because we didn't have our own approach, and were reliant on multiple contractors, all of whom showed up with different approaches. Only in the second or third year of [design exercise]

"Shifting Sands" by 2016 or 2017 did we recognize there might be pedagogical value in this approach.[62]

Thus, the earlier SOD-heavy design education drawing from postmodern theory and Naveh's concepts would start and end with Anderson's assignment at the college. From 2009 through 2011, there would be very little design education within the Canadian military education system, aside from select individuals potentially providing some design concepts at small scale. Mitchell would restart the design education in 2011 but until 2013 this consisted of design panel discussions with military designers Ryan and Elkus. Ryan would bring deep understanding of SOD along with experience teaching it with Naveh, while Elkus would publish on using design to extend the utility of modern military planning methods.[63] Elkus would shun the postmodern and disruptive aspects of SOD like many others, and promote design in how the U.S. Army would present Army Design Methodology in doctrine. By 2013, Mitchell would invite Ryan to run a student design exercises. Ryan would make significant inroads into fostering a positive military design environment within the Canadian Forces College, yet there would be different tracks for senior officer education and that of the mid-grade officer developmental journey.

While the mid-grade officer design journey from 2007 to 2015 went through many adjustments and reboots, the senior level education for the CAF took a different path. The senior officer level of education had an uninterrupted and somewhat different developmental journey, largely occurring after the first wave of SOD experimentation with JCSP in 2007–2009. In the Spring of 2008 at the Canadian Forces College, faculty combined the existing six-month National Security Studies Programme and the three-month Advanced Military Studies Programme (AMSP) into a single yearlong National Security Programme called NSP for senior officers.[64] This NSP development came from the Canadian Forces seeking to streamline the two senior-most PME programs into a single ten-month program. To accomplish this, CFC faculty sought to update and transform the design course (consisting of thirteen classes) to work in tandem with the other seven courses so that the program was capable of receiving graduate level credit for a master's degree at the Royal Military College.[65] Mitchell, reflecting on the Canadian movement toward design in 2016, attributed the transition as part of a larger effort to reform the senior officer education for the Canadian Forces so that it made "attendance simpler and more effective ... the AMSP had been focused on the operational level [of war], but the NSP is more focused on national security as opposed to operational warfare".[66]

Mitchell placed the Canadian transformation toward operationalized military theory and practice later than contemporary western military forces, occurring after the Americans, NATO, and the United Kingdom introduced operational concepts and art.[67] Part of this had to do with Canada's smaller military size, and the limitations of time and educational resources as well as the traditional PME glide path for officers over their career. According to Mitchell:

> Operational Art began to be introduced into CFC curriculum at the command and staff level in the early 1990s. However, until the mid-2000s, there was a significant bow-wave of senior officers who had completed the intermediate level PME prior to this shift. The AMSP was developed to assist those officers in understanding operational level warfare. By the mid-2000s, this bow-wave had disappeared as a professional cohort, thus eliminating the need to conduct this PME program.[68]

Mitchell and his colleagues at the Canadian Forces College saw this transition occurring by the mid-2000s and the CFC would institute the necessary educational reforms to develop the force.[69] It was right at this transition of Canadian senior leadership education from the earlier pre-operational era methodology toward a more comprehensive one where military design was also introduced into the force in 2008, despite design being distinct and in many ways disruptive of operational constructs. Canadian senior officers by then had mastered the new concepts of operational art and planning, and growing frustration with military setbacks in the American-led "War on Terror" offered a window of opportunity for military design to enter.

Senior officers in the CAF began to experiment with design just as the international design movement did writ large, through the gradual assimilation of small numbers of innovators, researchers, and maverick authors proposing new and disruptive ideas that at first were ignored or taken with a high degree of skepticism. Canada was half a decade behind the Americans in this pattern, with initial design concepts published in 2009[70] and again in 2012.[71] Anderson's article on a three-year design educational experiment in the field grade level schooling gained brief attention at the senior officer educational level, with Mitchell recollecting in 2016 that "I had initially brushed off [systemic operational design] as simply another [Revolution in Military Affairs] buzz word or intellectual fad like [Effects Based Operations]".[72] Mitchell's initial reaction in 2009–2010 to the design movement shares many of the common concerns about Naveh's esoteric, dense constructs, but within two years Mitchell would circle back to the ideas:

> Students were looking for a solution to the epistemological questions we were raising in the DS592 course "Modern Comprehensive Operations." In academic years 2012–2013 I proposed the very first design exercise … conducted in the Spring of 2013. It would be inaccurate to say that the [Professional Military Education] system was "formally considering design." This was Mitchell's science experiment, and I was being handed enough rope to hang myself. The Canadian Armed Forces did not formally acknowledge the value of design in CFC programs until 2016.[73]

Later student research published on design at the Canadian Forces College echoes Mitchell's position where design appeared to be a needed and relevant concept to address military dissatisfaction with existing strategic and operational decision-making methods.[74] Mitchell saw design forced senior leaders to attempt to think beyond the familiarity of tactical military challenges that could be overcome with "algorithmic processes and tactical heuristics".[75] Design seemed to challenge the strict reasoning and scientific rationalism that student tended to favor. Mitchell named this first incarnation "Critical Operational Epistemology" during course development.[76] His intentional use of epistemological reflection is significant in that this nests directly within Naveh's core SOD tenets for military design.[77] Mitchell and his co-developer for this course at the time felt somewhat subversive about the design concepts they were promoting as well as the possibility that if the CAF paid closer attention to what the CFC was teaching, "it probably would have killed the DS592 course immediately".[78] Beaulieu-Brossard, who would later join Mitchell at the college, termed this as Mitchell's "Field of Dreams" strategy. Incrementally, Mitchell was introducing design education on a small scale, and after it gained a steady following, he then would increase the scale and scope further, drawing more Canadian military interest.[79] Mitchell built it, and indeed they would come, as the famous movie line went.

Mitchell and fellow military faculty recognized that a tension existed in the Canadian senior military professional and their desire for relevant and applicable education toward extremely ill-structured and complex twenty-first century military challenges. Mitchell and his faculty recognized that Canadian students desired change and innovation in military affairs, but paradoxically would hesitate to abandon the foundations of how they understood warfare and doctrinal decision-making entirely. Essentially, they wanted the planning process to be repaired, but only with recognized tools, drawing from established and familiar parts, so that the new engine runs just like the old one, but better. Drawing from Jackson's work on the development of military doctrine in the modern era,[80] much of the postmodern content, as well as any complexity theory not easily grasped was ignored. The dense, esoteric yet essential bits of Naveh's original design went missing, and few had the curiosity or intellectual stamina to go hunting for them. Mitchell coined his own term of "missing manuals" to describe this lack of postmodern design and complexity within any available military doctrine or manuals.[81] The Canadians needed a more sophisticated way to get at teaching new ways of thinking about complexity for military applications. They needed a way to bridge the esoteric with the explicit, and the complex with the Newtonian styled.

By 2013, the Canadian design experiment took on a new and profound development. Drawing from Naveh's available yet difficult design materials and the more accessible design translations of other design theorists and educators in various military journals, Canadian faculty made a course correction. They took a geographically focused capstone operational exercise named "Strategic Warrior" and turned the problem-solving into a problem-*framing* design approach.[82] Canadian senior leadership approved of the faculty experimentation, and students moved from making rigid, hierarchical, and mechanistic plans to developing dynamic, sophisticated design alternatives. Mitchell framed the outputs as: "operational concepts that could be taken by a diverse set of agencies on which to base their own detailed plans, rather than crafting the plans themselves".[83] International design experts were brought in to facilitate the education, and a range of design styles and theories were sought to diversify the Canadian approach.

Canadian military design by 2014 introduced this mixed-disciplinary approach where military design methodologies and educators taught seminars alongside commercial design educators from the Toronto area. This resulted in myriad design processes echoing through the CFC hallways, with events in one seminar room often entirely distinct from the ongoings of other classrooms. Mitchell implemented design "hack-a-thons" into student design tutorials, based on his exposure to the concepts at the Rotman School of Management's "Business Design Initiative" at the University of Toronto. With this sudden expansion of formal military education within the highest levels for the CAF, student reactions were mixed. Facilitators brought the foundations of multiple, often dissimilar design communities, including commercial ones that awkwardly sidestepped much of the military traditional planning entirely. During breaks, students expressed interest as well as confusion on why one seminar was learning design one particular way, while the others pursued yet different paths. The year 2015 proved a volatile year at the college where not only would many new ideas be implemented on design, but faculty would *experience a design mutiny* by students rejecting the concepts to defend the legacy system of linear, mechanistic planning.

Mitchell described the open rebellion of the 2015 NSP class toward the design module as essentially a rejection of the new epistemology of the military postmodern design movement. Students hotly debated the value of the module right at the onset of the first design lecture, and by the practical exercise some of the student syndicates flatly refused to do the design process at all, preferring to regress into a traditional military planning methodology for the assignment.[84] Mitchell and his faculty would experience yet another student rebellion a few years later as well, although over time a majority of students exposed to the design concepts in Toronto have trended toward far more positive reception. Johnson, recollecting an American Army example of earlier student rebellion or rejection of design concepts at SAMS, offered an assessment of why design appeared to appeal to younger field grade officers and not as much to Colonels on the edge of potentially making General Officer rank:

> We had tried the first year [2004–2005] using the senior fellows [Colonel ranked students at SAMS studying at a war college level] and what we found was they were not as intellectually curious as the Majors were …

Most of them were posturing for promotion ... the Majors were a lot more enthusiastic about trying something new without seeking any kind of gain out of it ... I think that by the time most guys are [Colonels] they are set in their ways. They are "going to dance with the one that brought them." They are not too interested in trying out "new and different things" that they do not really have an invested stake in.[85]

The Canadians were not the first to reject design in these examples or within factions within their military services. With the American partial rejection of SOD's abstract and openly postmodern elements by 2010, it appears that the Canadian Forces College was following an institutional trend of initial design curiosity followed by rejection by the majority, as well as junior officers being more tolerant of the design praxis and radical concepts compared to senior officers. There is also the matter of context versus content, where at times it is more important to consider who is teaching or facilitating the design concepts rather than argue the validity of the content itself. Although the CFC invited design experts such as former Naveh affiliate and SAMS faculty member Alex Ryan in 2014 to facilitate, Mitchell concluded in private correspondence that a general lack of design expertise on the CFC faculty bench might have contributed to the student rebellions.[86]

Nonetheless, the faculty recognized several of these tensions and continued to develop the design education modules while learning and reflecting upon student reactions. Mitchell later wrote:

Old parts of the course had begun to conflict with the design path it was travelling down, and part of the rebellion clearly reflected not just a knee jerk, anti-intellectual rejection of this new methodology, but also honest professional confusion about what the goals of the course were.[87]

The subsequent year featured a modified design course with far more favorable student reviews. It remains unclear if this is due to superior design education approaches at the college, more experienced faculty, or potentially a changing student perspective as the CAF become more aware and accepting of design concepts.

There is another potential reason for the Canadian design rebellion of a seminar of Colonels in 2015. Naveh in a 2019 interview made some bold and rather derogatory observations concerning that specific rank within any military.

Colonels are the worst people in every military. Colonels are people who actually are frustrated people who believe they should have become Generals, and they have not. They worry about their desk, or their office. Some of them read and write well, so they might get assigned [to a college or academic program], but none of them want to work hard.[88]

On the other hand, there is plenty of evidence to the contrary where rising military stars take greater cognitive risks and demonstrate passion to learn new techniques and incorporate new ideas. It is unfair to attempt to categorize any military group into merely a rank and status context, yet in several militaries there does appear to be a unique pattern of design confrontation at the specific Colonel year-group population. Naveh himself paraphrased the unique, sometimes intellectually arrogant position of senior officers in this phase. "What I know has served me so well; look at me – I'm a general. Why should I change?!! Why should I look for something different, an alternative?"[89]

In October 2016, an academic specializing in postmodern philosophy in International Relations with no prior military experience began engaging with the CAF as well as select international experts in 2015. This engagement was a byproduct of an ambitious postdoctoral research literally following the footprint of Shimon Naveh and his impact from Israel to the US Army, to US Special Operations Command and onto the CAF. Dr. Philippe Beaulieu-Brossard along with his colleague Philippe Dufort organized a workshop of sixteen military design professionals, defense scientists, and design educators including global military leaders.[90] Brossard, with fellow academic Dr. Philippe Dufort, organized the first all-design issue of the *Journal of Military and Strategic Studies* in June 2017, which was a first for the military academic landscape and was a direct result of this early design workshop.[91]

Brossard initially was skeptical of whether the military had the intellectual depth of understanding for using sophisticated post-modern concepts for military applications, and quickly became intrigued with Israeli, American, and Australian endeavors with design. He recalls his feelings of concern and skepticism at the time with the author:

> I was concerned how the military would use such thinkers at the time. I thought this would make things worse for the human condition. After conducting my extensive research including listening to more than 100 officers and working with [Zweibelson], I came to the conclusion … that design can surely bring about a better world if well taught and practiced by officers. I joined the cause instead of being a mere observer … which makes things difficult from an academic perspective regarding so-called objectivity.[92]

By 2017, Beaulieu-Brossard became an assistant professor at the Canadian Forces College and largely responsible for their entire design informed education program for the Joint Command and Staff Programme, while Mitchell retained control of the Colonel's program. Beaulieu-Brossard organized multiple design workshops and conferences across the Canadian military enterprise, including Saint-Jean Sur Richelieu (Montreal) with Simon Bernard, Ottawa and Toronto where he helped import international military design professionals, scientists, educators, and theorists to further stimulate Canadian design interest. These events nurtured and consolidated a robust design community of military professionals. It would lead to the formation of the "Innovation Methodology for Defence Challenges" guild council founded in 2019 by Brossard, Jackson, Graicer, Zweibelson, Dufort, Martin, and several other key design educators, academics, and leaders.

By 2020, Beaulieu-Brossard with Dufort would secure a major grant of CAD 750,000 from the Canadian Department of National Defence's Mobilizing Insights in Defence and Security (MINDS). This grant supports and further develops the Innovation Methodologies for Defence Challenges (IMDC) council and network for Canadian design and innovation capacities. Dufort and Beaulieu-Brossard also manage a nonprofit "Archipelago of Design" program that intends to shape and nurture the future of design thinking in Canada and across NATO members and partners for defense applications. To date, the IMDC guild and their AOD have developed design projects between the CAF and the French Ecole de Guerre in Paris, experimentation with video game design to enable design education, as well as research with the Colombian military on design practice.

Thus, from humble origins in 2007 through 2020, design in Canada's Armed Forces has gone through many phases. The Canadian experiments in design education feature several common themes that reinforce shared experiences across the Israeli, American, and Australian militaries, while also demonstrating a few distinctly Canadian outputs. The CFC pursued a similar dual educational track focusing on mid-level and senior management officer education at their field grade and war-college equivalent schools for select student members. Canadian faculty observed a similar pattern where senior officers in the past generation generally had more resistance to the disruptive nature of design, while younger officers appeared to be more open to the concepts.

Mitchell reflected on this and offered cultural clues to one reason why military professionals appear more open to disruptive concepts dependent upon their position, time in service, and potential for career advancement.

> While many military officers at the rank of Colonel were skeptical of the value of design for military operations, Majors were much more open to it and leaned into the exercises with greater levels of acceptance and enthusiasm than had been encountered in the NSP. We surmise that there is greater room for taking professional risks at more junior levels and thus Majors are more willing to experiment with new approaches than senior managers about to be promoted to flag rank and, thus, "on the edge of greatness".[93]

Although he footnoted an additional thought on this spiraling pattern, Mitchell predicts that these junior officers that currently embrace design may return in future years trapped in similar ways and potentially halt any further progression of design itself.

Instead of implementing a particular design methodology, the Canadian military opts to sample all of them. The CFC deliberately employs a mixed-methods design concept that draws from different military as well as civilian design schools, methods, doctrine, and practice. Canada has yet to adapt a single design construct, and unlike the American and Israeli forces, the Canadians have not placed design into any formal doctrine. Additionally, unlike most other military design schools, the CFC as of 2017 continues to apply postmodernism in their educational approaches[94] despite the controversy and anti-intellectualism across much of the industrialized western military forces toward these disruptive ideas. The CFC design lesson plans for 2018–2019 also featured additional required readings on social paradigm theory, reflexive practice, as well as the incorporation of multiparadigmatic design thinking and postmodern concepts directly into the National Security Programme's design blocks.[95] This pattern continued through 2021 at the time of this writing.

The college went one step further to deliberately set a goal of shaping the CAF *to not produce design doctrine,* which again is unique in the Anglosphere. Although no longer directly affiliated with the design program in 2017, Mitchell continues to maintain strong influence on faculty and remarked in 2016, "institutional challenge will be to project CFC's epistemological agnosticism for design methodology and resist the urge to create a defined design doctrine that concretizes the approach and thus lose the vitality our approach creates for situational appreciation".[96] This is strikingly different from other military forces involved in military design, including even the Israeli forces that since 2015 have created a design manual despite Naveh's and Graicer's objections. Yet in 2022, the CAF, in conjunction with the nonprofit Archipelago of Design and a foreword written by Major General Simon Bernard, published online a forty-five-page design manual entitled *Collaborative Innovative Thinking By Design* intended to function as "a practical guide", that it "is no academic textbook", and Bernard concludes that "our political/civilian/military/academia consortia of the future will include this primer in their toolkit and leverage the proposed mindsets, approaches, methods, and tools developed by the Archipelago of Design".[97] Whether this is a shift in Canadian military thinking on standardizing design or some commercial oriented collaboration instead is yet to be determined.

Educating design remains an ever-shifting aspect of the Canadian military design journey, with all design education and much of the available experimentation occurring within the Canadian Forces College, and through their academic networking across Canadian design industry, military, political, and academic networks. The primary institutional driver in this in 2022 remains the CFC program, and the design educational emphasis championed by Mitchell and now continued by Beaulieu-Brossard.[98] Mitchell, co-authoring with Whale and the author in a 2019 Canadian Military Journal article on the Canadian design movement contributed the following:

In a promising development for 21ˢᵗ Century Canadian Armed Forces education, the CFC now offers design at the Major rank level on their Joint Command and Staff Programme (JCSP) in the Advanced Joint Warfare Studies stream, and "design like" activities have been included for officers taking the Institutional Policy Studies stream since 2015. Therefore, younger cohorts of senior officers have now been exposed to the opportunities and advantages offered by these new operational epistemologies. Furthermore, by 2020 the JCSP will be revamped in such a way as to include design thinking as part of the larger "core" of activities offered to all students taking that program.[99]

Mitchell's account of the most recent educational developments at the Canadian Forces College concerning military design education provides another two epistemological indications concerning where Canada is moving with design theory. First, the Canadians are deliberately attempting to provide military design theory to younger officer populations by shifting exposure into broader and more junior classes. This appears to be in tension with Naveh's personal belief that design must be provided to General Officers, although Naveh also has concurred that his position is limited to the Israeli security context and may not apply outside of the unique circumstances of the Israeli Defense Forces. Canada may be moving in the same direction that the American military, NATO, and other European countries suggest with design education below the senior military leadership.

Second, in the 2009–2019 period of initial design experimentation and institutional setbacks, Mitchell and his design educators attempted to focus on small groups. The fact that the CFC is now embarking on program-wide design educational opportunities demonstrates an institutional willingness to engage in this still controversial, ill-defined, and disruptive mode of thinking about the application of organized violence. Beaulieu-Brossard and Dufort write on these experiences based on a CFC design workshop for students in 2019 that focused on a real-world North American Aerospace Defense Command (NORAD) strategic deterrence project.[100] Later, those same students would again iterate toward a fourth design construct that unlocked previously unimagined and likely institutionally "off-limits" defense ideas that according to Beaulieu-Brossard, Dufort, and the CFC faculty the established military decision-making model devoid of design would be unable to generate. There are several important observations to make on this more recent CFC "agnostic design" educational exercise from 2019.

First, the CFC continued to apply their "agnostic model of design" that is illustrated both methodologically and epistemologically by the author below. The CFC themselves do not illustrate their design "proto-method" in that they use no graphic depictions outside of the various design methods they assemble each year for the students to learn from.[101] For this 2019 iteration, the college gathered a variety of human-centered design methods and sources as well as several Toronto-area human-centered design (HCD) facilitators to work with their student teams. Beaulieu-Brossard introduced various schools of design theory in his own description:

> In 2019, Dr. Beaulieu-Brossard assembled a sequence of three categories of schools of thoughts: a mainstream military, a mainstream civilian, and a post-structural one. CFC design teams began with "Scoping and Framing" inspired by the Australian Joint Military Appreciation Process (JMAP) and facilitated by Dr. Aaron Jackson. Then, they moved to "Creative Destruction" inspired from the Israeli Strategic Operational Design (SOD) and facilitated by Dr. Ofra Graicer. Finally, yet importantly, design teams undertook foresight activities building on [a Government of Canada design] methodology and facilitated by Ms. Dione Scott and her team … this sequence of school of thoughts also enabled inquiring the challenge from radically different perspectives, thus making surprises and the emergence of new knowledge more likely.[102]

Canada, at least within the Toronto military college continues to draw from various HCD design sources as explained in the third chapter, including for the 2019 aforementioned design challenge and Beaulieu-Brossard and Dufort's accompanying article design concepts such as Serrat's "Five Why's" technique first implemented by Toyota. They also pull concepts from Bason and other HCD theorists as well as from complexity theorists such as Ackoff. Whether a conscious decision or not, the CFC were drawing from Naveh's original sourcing of Ranciere and *The Ignorant Schoolmaster* in positioning the students as the central "guides" to their own educational development. The praxis of design education would take on a nomadic, self-guiding form where instructors were not directing and controlling the design learning as Ranciere found in the dominant, traditional "old master" methodology.[103] The "agnostic mode" of design praxis is depicted epistemologically below by the author and frames the 2015–2021 period of Canadian "proto-design" activities largely centered on the Canadian Forces College design praxis.

In Figure 7.1, the graphic on the left is one epistemological framing of how the Canadians currently provide military design thinking and is one of the few that attempts to integrate commercial design methods into military operational planning efforts. However, the first arrow addresses this "agnostic design" epistemological stance where multiple design models must be entertained by the military designer. At the Canadian Forces College, designers then have classroom cross-pollination of design praxis, models, theories, and language as the seminars are directed to interact and share during their design educational journey.

The third arrow is titled "design interplay" and epistemologically takes a multi-paradigmatic position that different paradigms concerning warfare, security, and conflict will also have areas of overlap, tension, and interplay where those two paradigms form novel configurations previously unimagined or unrealized.[104] The fourth arrow illustrates a similarity to Israeli SOD praxis

CFC's Design Epistemology

CFC faculty multidisciplinary awareness by Seminar Lead unknown

Design Method X

- Facilitator skill.
- Facilitator awareness of other designs.
- Awareness of military planning.
- Design X incorporation into military planning method Y or Z.
- Interoperability with other CFC design methods.

Arbitrary design exposure by CFC students enacts majority of their design methodological experiences. *Randomness of* design talent/methods secured by CFC annually compounds this. Many designs used are wittingly or unwittingly exclusive/singular in orientation (*incommensurate* with alternatives).

Quality of seminar design facilitators (systemically) unknown; equal exposure of designs for all students unknown. Internalization of plural design content unknown. Interoperability between seminars unknown.

"Agnostic Design" approach introduces multiple design models for student consideration

"CFC seminar interaction" encourages critical thinking and self-reflection on multiple design models in practice

"Design interplay" demonstrates overlap, tension, and patterns across a variety of designs

"Perpetual Learning" rotating through different design methods should challenge design ontologies, epistemologies

"Emergent Design" the cross-pollination of different designs could spawn novel practice

1

2

3

4

5

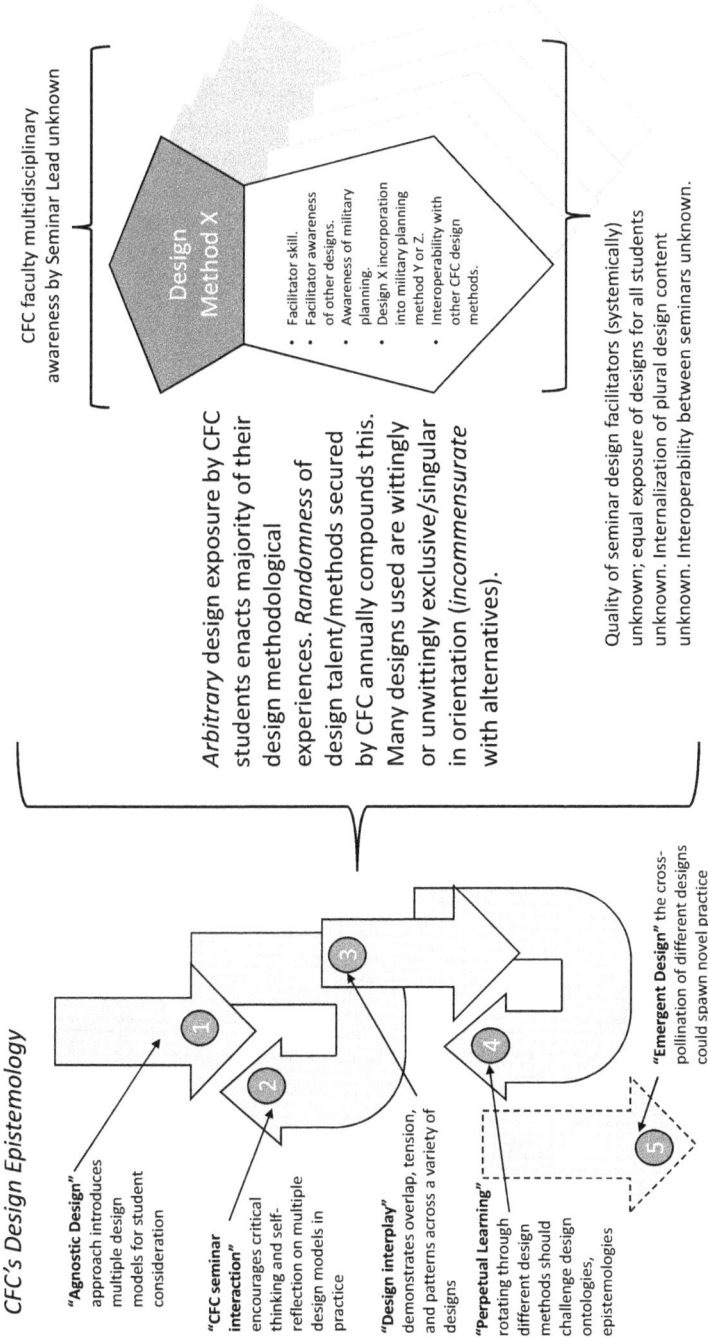

Figure 7.1 Epistemological Framing of Canadian Forces College Design "Agnostic Approach".

Source: Author's Creation.

where designers undergo "perpetual learning" where one "thinks about their thinking" and are as designers framing not just something they see as a military problem, but their own organizations and how they make sense of reality and warfare. Potentially, this leads to the fifth arrow that has a dotted outline and offers a future "emergent design praxis" where designers might generate new hybrids and combinations of military design methodologies. They can realize this only by identifying the various theories, mental models, methods, and language used across a range of different organizations, actors, as well as design praxis so that they might generate something different. This also is why the Canadians are apprehensive about putting any design methodology into their military doctrine, as that would terminate any actions described in that fifth epistemological arrow.

The above illustration is one depiction of how the CFC produces their "agnostic design education" and is not intended to depict a declarative "Canadian form of design methodology" as even the faculty most associated with the current approach consider it a rather non-establishment and potentially less structured form of design for a military. The CAF have no design in doctrine (yet), nor do they adhere to any particular design methodological structure as the American Army, Marines, and Israeli Defense Forces do at the exclusion of other alternatives. Epistemologically, the CFC approach depicted above favors a multidisciplinary appreciation of "design" but does not overtly challenge any distinction between designing for commerce and designing for war as the third chapter explained. This is one potential vulnerability for how the CAF are educated concerning design theory.

The Canadian design interplay does offer at the program-level iterations of divergence (by design seminars) and convergence where the groups of students will participate in presenting their design deliverables to a "strategic sponsor" and a group of design evaluators. Throughout a recent 2019 NORAD themed design challenge, the student teams were able to diverge and experience multiple design methods, yet as the teams converged in to finalize their design deliverables, they collectively broke with the original design "suggested solutions" of the commander at the onset of their exercise. This was highlighted by Beaulieu-Brossard and Dufort below and demonstrates how at the epistemological level, the CAF appear better positioned to violate a rather rigid tenet of the modern military decision-making methodologies of which the CAF follow in their own doctrine and in operational planning applications. Only in their design activities could the student teams challenge, deconstruct, and even contradict or refute the commander's initial design guidance through their design journey and subsequently provide that design narrative to the leader in the final presentation without facing rebuke.

> Beyond the Canadian Forces College's design education philosophy, generating robust challenge-solution pairs became even more possible with the NORAD team welcoming it … even if the approach generated by the students eventually diverged from the approach … advocated by

the NORAD commander. In other words, the openness of the NORAD team allowed CFC students to ignore the solutions already advocated by the commander.[105]

With some design educational highlights in this proto-design approach, there also are some problematic areas. Although in recent years the CFC have introduced greater reading and research requirements as prerequisites to foster this multidisciplinary framing, the exclusive engagement with a single design method and facilitator for their exercise period still acts as a vulnerability in those students spending two weeks doing SOD with their small-group instructor only get a peripheral view of alternative design methods. As the CFC moves year-to-year with budgetary restrictions and availability of various design facilitators is a yearly challenge, the variety and quality of meta-design content is randomized each academic year and potentially non-repeating.[106] This challenge appeared to be due to the availability of skilled design facilitators who are often contractors, as well as CFC contracting issues that are outside of the actual academic content.

The educational experience of each of these facilitators is also irregular, with some able to provide detailed design educational experiences complete with student guides, lesson plans, established exercises that are organized along a series of days to culminate in specific design learning objectives. Other facilitators arrive with little to no content and arrange a series of somewhat unrelated design activities and techniques or follow a design method that is arguably far less connected to any established military decision-making process as the process duplicates their commercial client engagements.[107] However, under-performing design facilitators are not invited back to the CFC, and those that perform exceptionally tend to become the template for what the CFC faculty desire from all facilitators the subsequent year.

One more significant point should be addressed on the ratio of design education at the Canadian Forces College to traditional, mechanistic military planning methods and doctrine. Like nearly all PME programs, the CFC has a large mid-grade officer throughput where the JCSP annually educates the bulk of Canadian mid-career field grade officers either in residence, distance or in some combination therein. The NSP program reflects a much smaller population of senior officers at the Colonel rank which itself is representative of the shrinking top of the military centralized hierarchal pyramid of rank, status, and population. Primeau, a Canadian Officer and instructor in the JCSP for design as well as other courses, reflected that the educational deliverables of the JCSP remain wedded to CAF structures that expect JCSP graduates to possess joint planning skills, "and this can only be achieved through extensive joint linear planning experience/training (JOPP)." Primeau claims that the general JCSP curriculum in 2020 for majors "had 134 hours for JOPP [traditional linear planning methodology] for 23 hours of design".[108] He went on to suggest the following on why this is and what it may be doing to the current Canadian forces:

I suggest(ed) that this understanding is either dated direction that warrants review, or an assumption that made it into programing for ease and convenience. In any case, this trains our majors "soon to be Lieutenant Colonels and Commanding Officers" to successfully fight previous traditional (modern) wars, not to address the complex challenges they actually do face or are more likely to face in the five to ten years after graduation in our small, centralized, culturally-aware, institutionally-focused military, poised for action in complex operating environments at home and abroad.[109]

In addition to the institutionally mandated emphasis on modern planning methods, the CFC design faculty are internally limited on design experience, although since 2017 they have intentionally hired specific design experts in full-time faculty roles to address this concern, such as Beaulieu-Brossard and Rebecca Jensen.[110] These experts provide a mix of design backgrounds from human-centered commercial design methods as well as distinctly postmodern ones repurposed into military design contexts through the inspiration of Naveh. Recent design initiatives in the CFC as of this writing indicate that a philosophical emphasis on "challenging the challenge" to better appreciate and reframe complex defense challenges may become more mainstream in the CAF provided that the CFC continues to emphasize postmodern design, reflexive practice, and multidisciplinary design methodology for a growing population of mid-grade and senior-level Canadian leaders and staff officers.

As a way forward for the CFC, their newly founded "Department of Innovation Studies" established in 2019 and the Ontario College of Art and Design's Professor Michell Mastroeni will combine efforts in seeking a new design methodological approach. According to Beaulieu-Brossard, this new project is intended to rely on ethnography and consider tacit, reflexive practices in how the CAF construct and employ knowledge. This combined research team intend to develop an entirely new design model for potential implementation at the CFC and ideally across the CAF. Whether Beaulieu-Brossard and Mastroeni and their network of design facilitators are able to accomplish this will be seen in the next several years.

The Desire for Thinking Differently in War for the Australian Armed Forces:

In the Australian Armed Forces, the experimentation with design has been fraught with sporadic and often fragmented periods of usage coupled with a broader institutional skepticism as well as potentially some anti-intellectual expressions from within Australian military culture. This is not indicative of the ADF in general, rather that the institutional resistance toward somewhat dense theoretical foundations of Israeli SOD, the tensions between complexity theory and the linear, mechanistic planning methodologies of traditional modern militaries, and a somewhat reluctant military curiosity beyond technological or tactical advancements within military contexts appears to have delayed any substantial Australian design initiatives until just recently.

The Australian military features many overlaps with American, Canadian, and British Armed Forces with respect to similar forcing functions in significant security challenges such as the "War on Terror" and the disenchantment with initial counterinsurgency doctrine at the onset of the post 9–11 era, frustration with unorthodox and emergent adversaries that did not manifest in the familiar Westphalian nation-state form, and increasingly complex military contexts that became quagmires for forces deployed there and well prepared with formal military doctrine, education, and the best practices of earlier military periods. As Naveh and his band of SOD heretics departed the Israeli Defense Forces in 2005–2006 to explore American military interests in this exotic new design theory, Australians would encounter Naveh and SOD through exposure at the U.S. Army School of Advanced Military Studies through international student attendance as well as through various academic and professional engagements between the Anglosphere and partnered militaries.

Somewhat later in the international military design movement when Leavenworth had cast Naveh and his radical SOD ideas out of the American Army to replace it with the rather formulaic "Army Design Methodology" in 2010, Naveh (paired with select Australian systems theorists and a contracted team) visited Australia and conducted at least one deliberate design educational exercise with Australian Special Operations Forces.[111] Despite Naveh's narrative of his unpublished report of the success of this education and his own personal observations on Australian military culture and capability for doing design as he envisions it, there is no evidence that this isolated SOD exercise continued Australian interest or extended beyond this single session with a small, elite unit within the larger Australian Armed Forces. While Naveh engaged directly with the Australian military in 2010, the earlier period of 2005–2009 features less direct connections of Naveh's design ideas to the continent "down under."

In other examples of isolated and fragmented proto-design developments, the Australians attempted to incorporate elements of complexity theory into existing Australian operational planning doctrine and practice, with differing degrees of failure, assimilation, and disinterest by the force. Since 2019, individual units across the Australian Services started experimenting with certain civilian design applications predominantly from HCD contemporary and commercial organizations. These efforts appear centered on technological and tactical applications for new military products, user experiences, and other commercial designs applied toward warfare without addressing the design within warfare itself.[112] However, this trend may influence the Australian military to revisit operational planning, strategy, and their institutional framing for how and why they engage in warfare, defense, and security affairs through future design adaptations.

Through 2022, while there is a recent surge in military design experimentation within the past few years, the current state of design affairs of the Australian Armed Forces remains in a "proto-design" situation devoid of any formalized or declared design methodology, lacking any design theory in existing military doctrine, and little to no formal military design education at any level in the Australian professional development program.[113] The COVID lockdowns in Australia which effectively shuttered its borders made 2020–2021 particularly slow for any military development in design or other endeavors. Despite that interruption, things seem on the rise in 2022 with enhanced design interest across their forces, particularly in their military education programs and training centers. However, this summary of the Australian design movement through 2021 commences at the earliest detections of design theory in their organization. This rewinds the clock back to the early 2000s, precisely when Australian and other partners entered with American forces into multiple counterterrorism and counter-insurgent operations. Here, a flurry of military academics, operational planners, doctrine writers, educators, and some visionary military leaders began to experiment with complexity theory, tinker with the traditional military decision-making methodology, and for a smaller population still, engage with Naveh and his design circle.

Australian armed forces gained exposure to Naveh's SOD early in the initial exodus from Israel. As Naveh began lecturing and showcasing military design in American military schools, the international exposure quickly followed. American partner nations such as Australia regularly exchanged officers for academic development; nearly every military professional development school in the U.S. Department of Defense has Australian students, and in many cases, visiting Australian faculty. Indeed, this is one of the earliest ways in which Australia gained awareness of military design.

Alex Ryan, an Australian defense academic with extensive background in blending complexity theory into Australian doctrine and planning practices, would meet Naveh in Fort Leavenworth and begin working with him as well as studying and exporting SOD through student engagement and academic outreach. Ryan first operated as a scientist in a defense research lab in Edinburgh, South Australia in 2006.[114] Working with another academic interested in complexity theory applications to warfare, Ryan and Anne-Marie Grisogono aided the Australian Army's incorporation of systems thinking into Army doctrine that would become termed "Adaptive Campaigning".[115] This methodology was not military design per say, but it overlapped with shared theoretical concepts primarily within complexity theory and systems theory. Adaptive Campaigning would introduce complexity theory and systems thinking concepts in a manner that challenged or contradicted the institution's Newtonian styled deliberate planning logic.

Adaptive Campaigning represented the Australian development of the first path where militaries moved beyond the proposed conceptual limitations of "Effects Based Operations" and uni-minded systems made popular in the

1990s and into tighter embrace of complexity theory and nonlinear, emergent systems. Adaptive Campaigning also drew from U.S. Air Force Colonel John Boyd's academic work on applying systems theory and complexity theory toward military decision-making methodologies.[116] The Australian "adaptive campaigning" cycle would later echo much of Boyd's observe–orient–decide–act (OODA) loop, oriented toward using these models to improve the operational design of complex military campaigns and strategies.[117] This again refers to operational design and how militaries since the early 1980s associated the construction and orchestration of military campaigns through a strategic design that would in turn cause all operational and tactical planning activities. Operational design sought to make the modern war paradigm work, warts and all. Adaptive Campaigning would be a uniquely Australian effort to modify how operational design for campaign planning was conducted due to earlier ADF frustration with the existing NATO and American inspired methods and doctrines.

Ryan recalled that in 2006 during an Australian workshop on systems thinking, he first met with Brigadier General (retired) Huba Wass de Czege, founder of the SAMS program at Fort Leavenworth, Kansas and a keynote speaker at the workshop.[118] Wass de Czege informed Ryan of the experimental design Naveh was providing at SAMS, and secured an invitation for Ryan to visit the school and see military design firsthand. Upon walking into a small group led by Naveh, Ryan was fascinated by the unusual nature of the exercise. "There was no evidence of division of labour according to the Staff System. There was no sequential process being followed. There were no PowerPoint Slides being produced. Most surprising of all, there were no doctrinal manuals in sight", Ryan later wrote on his first exposure.[119] Naveh's bravado and ego aside, he frequently would captivate observers with the depth of his ideas, and the radically different way that SOD pursued novel warfare opportunities in a manner the traditional military planning process could never reach. Ryan was hooked, and wanted to learn more.

Ryan returned to Australia and undertook a journey of self-discovery on SOD, Naveh, and design applied for military considerations. As Naveh rarely wrote much on SOD in the 2000–2007 period, Ryan reviewed publicly available student monographs on SOD as well as the SAMS student reading list and courseware created by Naveh with SAMS faculty.[120] After Ryan was invited back to lecture further on complexity theory, SAMS offered him a faculty position in 2008. Ryan would be the first foreign civilian to work on the SAMS faculty, and this placed him in direct contact with Naveh and his small team of private contractors teaching Israeli design to the elite Army student planners.

In the three years Ryan taught at SAMS (2008–2010), the program graduated over six hundred field grade officers educated or exposed to some military design education. SAMS faculty are paired with students for monograph research, with student topics largely determining what faculty are best suited to mentor them through a year of extensive research and writing.

Ryan, as a design expert, frequently supervised SAMS students focusing on design in their monograph topic of interest, including a few Australian students.[121] Ryan would quickly become a SOD insider, working extensively with Naveh and his few trusted design facilitators. As he mastered SOD, Ryan took lead roles such as heading the design team to produce the *Art of Design: Student Text* student guide for SAMS in 2008, and also was part of the SAMS writing team creating draft design doctrinal chapters for TRADOC consideration. Ryan would co-author design articles with Colonel Banach, the eleventh director for SAMS, and mentored select students on monographs focused on design.[122]

Ryan was not the only Australian actively critiquing how the ADF were behind the times on operational and strategic innovation for complex warfare. Lieutenant Colonel Trent Scott, writing in 2011 as a Commander of 3rd Battalion, The Royal Australian Regiment, criticized the Australian Defense Forces with being outstanding tactical innovators and adaptors while largely failing at the broader operational and strategic developments. Scott, a graduate of the U.S. Marine Corps' School of Advanced Warfighting,[123] would challenge the outdated manner in how Australia continued to view designing in warfare as the orchestration of campaigns. He remarked: "The ADF's doctrinal approach to 'design', like its doctrinal approach to problem solving, is mechanistic, reductionist and inadequate for an increasingly complex battlespace and array of missions".[124] Scott would conclude that ADF leadership needed to recognize their campaign planning doctrine "is based on flawed assumptions and processes which are losing relevance in an increasingly complex and uncertain operating environment".[125] Rectifying this situation required the ADF to invest intellectual capital into "incorporating creative and critical thinking ... incorporating operational design into planning doctrine", and generate numerous SOD-themed recommendations that Scott would spell out in his conclusion.[126]

What is interesting on how Scott researched and presented his arguments is that in the 2009–2010 period it was written, the Australians had already implemented their Adaptive Campaigning doctrinal changes, and Scott essentially argues that these minor improvements to an otherwise archaic, engineering-oriented, mechanistic mode of warfare were already irrelevant. Scott would introduce military design in his monograph in the last chapter, placing General Mattis' vision for U.S. Joint Forces and design thinking as central to his arguments on fixing the ADF's operational art. As covered in the fifth and sixth chapters, Mattis would see Naveh early in the American experimentation with SOD, and then attempt to bring it into the U.S. Marine Corps. Scott, an Australian attending the Marine's version of the SAMS program at the same time, likely picked up the Marine adaptation of SOD instead of what was occurring at Fort Leavenworth and the Army Design Methodology transformation. This reinforces the intellectual impact that the American military has upon partners and allies, and how Naveh's SOD experimentation at Fort Leavenworth would produce significant waves that would extend not just across the U.S. Department of Defense, but into allied militaries such as Canada, Great Britain, and Australia.

Scott's 2011 monograph is also prescient, in that the author anticipated that the Australians might adapt some sort of design, but quickly eliminate the flux and improvisation that Naveh's SOD would advocate in lieu of any indoctrination into a design manual. This Australian officer knew his military culture and feared what might occur. He warned:

> that any design methodology will inevitably become an institutionalized dogma, dependent on checklists and templates, and will go the same way in application as previous fads such as [Effects Based Operations]. The potential for this to occur is great given the ADF's predilection towards training vice education.[127]

Scott would not be alone in being concerned with how overly analytical, reductionist, and linear the ADF operational planning was.

Colonel Brett Andersen participated on the "Training Needs Analysis for Adaptive Campaigning – Future Land Operating Concept" initiative from 2010 to 2011. This effort by the ADF would create a second wave of "Adaptive Campaigning" initiatives and doctrinal revisions seeking to usher in more design initiatives. The results of the effort were mixed, with these last revisions to Australian operational design and planning representing the final formal effort to redesign Australian warfighting above the tactical, technologically oriented level. In 2020, Anderson reflected on this as follows:

> I've had a look at the latest [Australian] doctrine for joint planning (2018 and 2019) and not much has changed in my opinion … [it] has not changed in 20 years. From my perspective, not many people actually understand the process or how it actually functions. In this regard, I agree with Trent [Scott] in that it is reductive in nature – and it actually says it in ADDP 5.0 that "planning is also analytical and reductionist, in the sense that it is bounded by the laws of physics and provides a means for addressing complex problems in manageable ways (the science)." In fact, I would offer that we are usually bounded by two other key factors – guidance from the government and guidance from senior military leaders.[128]

Since 2011, the Australian military experienced nearly a decade of design stagnation, with little evidence of any further thought or experimentation beyond two ADF doctrinal revision efforts, and the various individual actions of military innovators and educators such as Alex Ryan in America. This is not to say that design died down under, as a small population of design educators and innovators did attempt to continue the momentum. This includes Aaron P. Jackson, who started as a public servant (civilian) joint doctrine writer in the Australian Department of Defence in 2010, and who is also an Army Reserve officer. Jackson would integrate into an international military design community that would take shape in the 2014–2020 period, particularly centered between Canadian and American military academics.[129]

First focusing on military culture and doctrine, Jackson would learn of design while researching the philosophy of military doctrine (ontology and epistemology). Throughout the 2010–2020 period, Jackson would be one of the louder military academic voices, writing journal articles, monographs, and engaging with the international design community in workshops and conferences. In 2019, Jackson provided a robust historical study of the entire Australian design movement in an introductory chapter for the first all-design edition of the Joint Studies Paper Series. Later, he would co-author a 2021 review of Australian military design thinking, and would provide a rich social media presence on military design in Australia.[130]

For Jackson, the Australian development of "adaptive campaigning" in 2006 was a dramatic departure for the ADF from what had been previously a rigid, linear, and a Newtonian styled sort of endeavor that worked as a recipe for warfighters to follow. The adaptive campaigning concept was not a design methodology, nor was it strictly an extension of previous military operational planning. It introduced a new generation of ADF personnel to some of John Boyd's OODA Loop model along with thinking systemically for complex, dynamic systems.[131] Adaptive campaigning brought in "problem framing" instead of the traditional "problem-solution" formulations of earlier military doctrines. Jackson credits that change to the influential theories of Rein and Schön, which again would be prominently featured in Naveh's SOD focus for experiential learning. The OODA cycle by Boyd would clearly make the strongest mark in how the Australians sought to conceptualize their adaptive campaigning, in that both appear quite similar in composition. Below, the full version of the OODA loop is illustrated on the left, yet only a slimmed down, oversimplified version would make it into Australian doctrine. It is assumed that doctrine writers likely referenced the robust version but substituted a simpler OODA for brevity.[132]

Boyd's cyclic concept of self-reflection, decision-making, and awareness of the rival's thought process came from his background as a fighter pilot during the mid-twentieth century, and subsequently enhanced by his deep study of complexity theory, general systems theory, and a host of other disciplines that were rarely encountered in any military school or doctrinal publication. Naveh, like Boyd, would do the same in how SOD would be developed, with Naveh drawing from his career as an Israeli paratrooper. The Australian "adaptive campaigning" cycle would later draw inspiration from Boyd's OODA loop, and more subtly, from Naveh's SOD concepts. Yet despite these influences, the Australians from 2006 to 2009 crafted a hybrid that was part design, part operational planning. That concept would fade over the following decade as the ADF collectively halted their debates on design and innovation. The Adaptive Campaigning methodology would be captured in two major ADF doctrinal publications in 2006 and 2009, and both would conceptualize the process through a cycle depicted below next to a depiction of Boyd's OODA loop (Figure 7.2).

Graphically, similarities are clear in the representation of Australian Adaptive Action and John Boyd's earlier work that would become established in American Joint planning along with Air Force and U.S. Marine Corps operational art. Jackson critiques the seemingly ADF predilection with sampling from other military experiments by observing: "[The ADF has] predominantly engaged with design thinking by applying externally-developed methodologies in preference to developing its own".[133] The Australian military would assimilate Boydian concepts literally, and potentially several Naveh SOD concepts less obviously.[134] Naveh's concepts required a clandestine touch in that the Australian institution, as Scott, Jackson, Palazzo, and others had argued, there was an anti-intellectual sort of rejection of anything that disrupted well-held beliefs on warfare, or were not otherwise easy to grasp with minimal perusing.[135] Naveh's contributions to the ADF at this time are limited, and potentially overestimated. Ryan and Grisogono's early 2000s work with complex adaptive systems would directly, clearly influence ADF adaptive campaigning development, as would Australians familiar with Boyd's earlier systems thinking.[136] Naveh may have influenced various ADF operators in the 2004–2010 period, but the bulk of authors creating doctrine were unfamiliar with him or rejected SOD concepts in favor of strictly systems thinking and complexity theory.

According to Jackson, a drafted first version of the ADF Publication (ADDP) 5.0 – *Joint Planning* was circulated in 2013 internally, and it drew

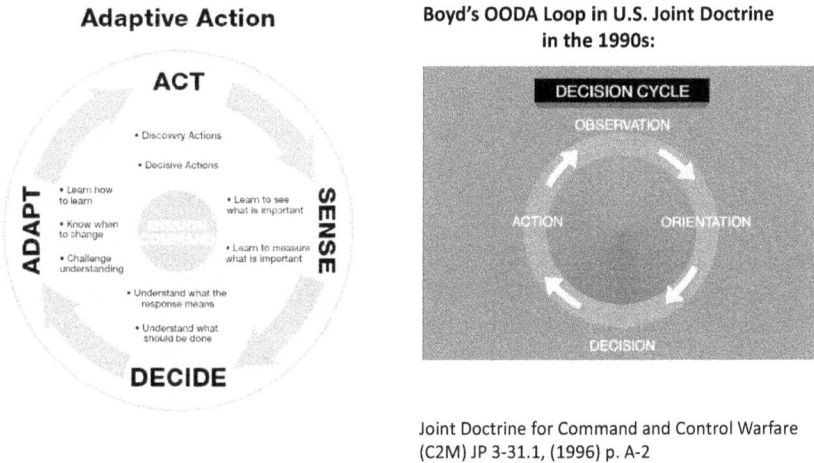

Joint Doctrine for Command and Control Warfare (C2M) JP 3-31.1, (1996) p. A-2

Figure 7.2 Australian Adaptive Campaigning Cycle and Boydian Influences.
Source: Two Non-Copyright Military Doctrinal Publications.

extensively from Naveh's SOD concepts. The author was a US Army War College graduate and yet another example of how impactful Naveh's ideas were to the international students attending at that time. The draft version of ADDP 5.0 "contained many elements of SOD – including the much-criticised dense and obscure philosophical language. As a result, this draft was rejected by Joint Doctrine Centre (JDC) staff as too esoteric for implementation by the [Australian Defence Force]".[137] Once again, the difficult prose and postmodern concepts in SOD would prove difficult to implement into the military institution.[138] Earlier Adaptive Campaign content lacked any direct Naveh influence, yet this later 2013 period would demonstrate clear SOD influence, even if only in draft and rejected formats.

SOD was designed to disrupt and challenge the very paradigms that maintained the modern military frame for making sense of complex reality and the application of organized violence. A British military student who attended SAMS and studied under Naveh also attests to the difficulty of learning such unorthodox, esoteric concepts. According to Edward Hayward, then a major at the school: "these terms do not come easily, as with any technical terminology ... as with learning any new language, translation can only be attempted after the original language has been mastered ... there are no shortcuts to this process".[139] Naveh's ideas would remain outside of any Australian design doctrine or formal education during the decade of Australian design stagnation, with lone advocates taking risk at being too vocal or critical to the ADF establishment. Individual heretics aside, Naveh's inculcation of SOD into the ADF until recently is limited, scattered, with negligible details publicly available.

By 2010 the impact of Naveh's systemic operational design and any trace of postmodern theory largely terminated within the ADF, at least from a formal position of doctrine, training, and education. From 2010 forward, only select components of complexity theory and systems thinking made their way into Australian operational design and campaign planning, with much of that confined to advanced military schooling and individual self-study. Broadly, the ADF continued as it did in the 1990s, excelling tactically while ceding operational and strategic thought to whatever coalition or leading military partner nation had developed. Tactical and technological design continued during this period, yet designing new tools for war did not necessarily translate to much beyond improving the tactical effects and local, immediate impacts against technologically inferior opponents. As discussed in the second chapter, this technological overmatch is quickly dissolved by cunning opponents regardless of resource shortages.

Military design would continue to be applied in clandestine ways in Australia. The second edition[140] of the Australian "Joint Military Appreciation Process," published in 2015 and written primarily by Jackson, with several other Australian military officers contributing, would incorporate similar operational planning concepts while subtly referring to Naveh's SOD concepts. Particularly in complexity, nonlinearity, framing, and emergence, this second edition would encourage SOD-like disruption of the legacy system under-pinned by a Newtonian stylization of modern warfare.[141] Despite Australia not actually producing a design methodology, the doctrinal updates to how they framed operational and joint planning in complex, dynamic security contexts continued to blend influences from design thinking, systems theory, complexity theory, as well as more subtle attempts to disrupt the traditional modern military frame.

Despite the new design terminology, Australian operational planning doctrine maintained the same traditional mechanistic military paradigm of the modern era for war, complete with the required identification of a desired end state, objectives, and decisive points. Australian planning doc-trine stipulates the use of center of gravity analysis, the establishment of lines of operation as well as most all other elements of mechanistic, reductionist military planning.[142] However, Adaptive Campaigning did introduce some fundamentals of sensemaking, reflective practice, and disruptive thinking into how the ADF operationalized military campaign design. These additions are difficult to spot in the methodological constructs depicted in their doctrine without additional enhancement and study. According to Jackson, "adaptive campaigning was not considered with the second edition of JMAP was being written in 2014. We used American sources discussing military design thinking instead."[143] The below graphic depicts one way of considering the epistemological framework of how Australian "proto-design" functions as a methodology, using the Adaptive Campaign model for the core methodology despite the apparent confusion within the ADF doctrine on what precisely this proto-design sought in form and function.

Figure 7.3 demonstrates the author's epistemological framing of what could be fairly termed Australian "proto-design" based on their earlier Adaptive Campaigning methodology as well as how the ADF incorporated some design concepts between 2006 and 2015. The "proto-design" may be now shifting into a deliberate effort to create something uniquely Australian for their ADF in 2022. Jackson, in observing the grassroots movement gain speed remarks: "In the last few years the application of design thinking methodologies outside of activities directly related to operations planning has increased in frequency across [Australian] Defence".[144] Should the ADF formally implement some Australian military design methodology complete with the necessary theoretical and educational depth, their training centers, military education programs, and even their doctrine may experience a sudden development similar to how the American Army and Marine Corps did in over the last decade of design implementation. They also may experi-ence more of the institutional power struggles, battle for ownership, and even the division between design purists and pragmatic planners.

Regardless of proto-design or full-blown military design status, where the Australian military is currently and what they have been doing over the last two decades requires the same philosophical framing as other treatments offered in this book. Below, an epistemological mapping by the author

An Epistemological Frame

"Mission Command" is central to entire methodology:
- Centralized hierarchy
- Control

Framing the system requires self-reflection and feedback loops for learning

Adaptation: complexity requires a "probe-sense respond" construct that permits failure in order to learn

Emergence: the learning cycle is perpetual, requiring reflective practice to disrupt static behaviors

Institutional Values: ADF uses "matters", "important" for an institutionally centric position to decide

1
2
3
4
5

Adaptive Campaigning

Adaptive Action

SENSE
- Learn to see what is important
- Learn to measure what is important

ACT
- Discovery Actions
- Decisive Actions

ADAPT
- Learn how to learn
- Know when to change
- Challenge understanding

DECIDE
- Understand what the response means
- Understand what should be done

1
2
3
4

MISSION COMMAND

This involves institutionalized beliefs, values, and ADF core identity (wittingly, unwittingly) and might fall within "decision" systemically.

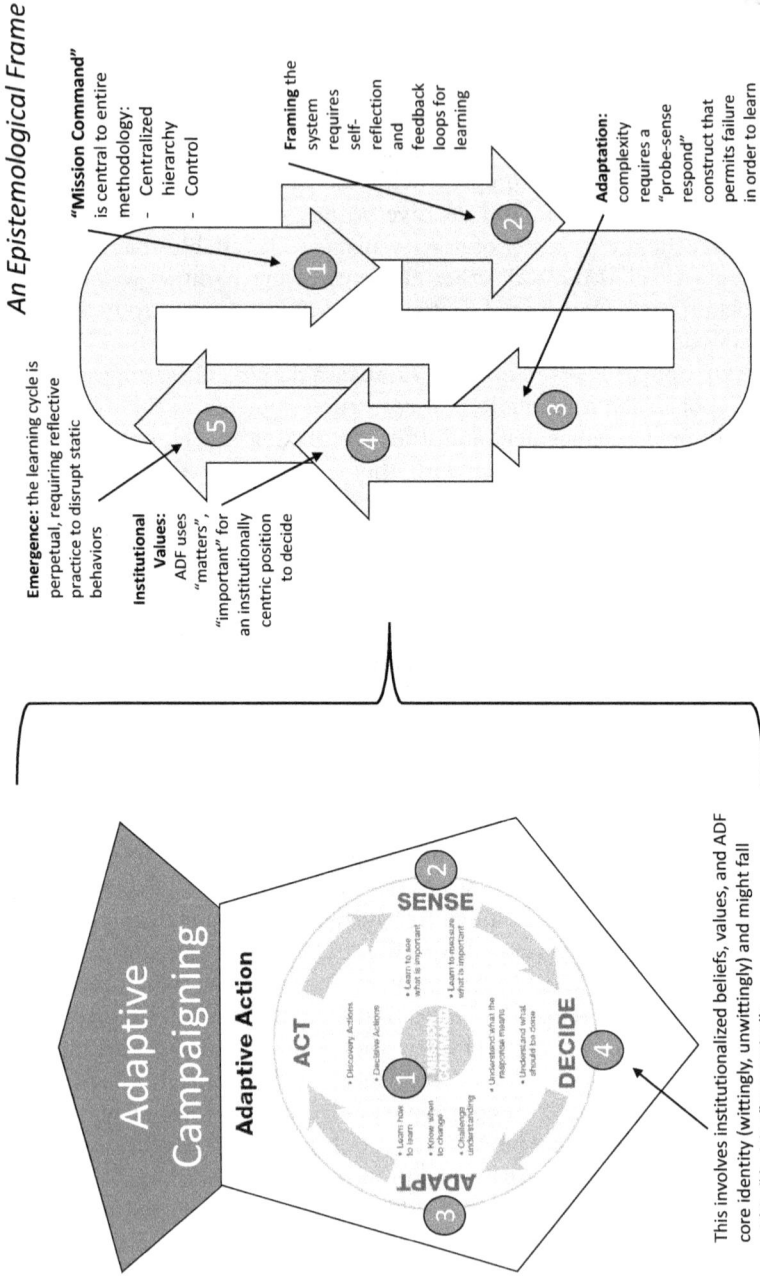

Figure 7.3 Epistemological Framing of Australian "Proto-Design" Through 2022.
Source: Author's Creation.

represents one of many ways to conceptualize what design is pursued by the ADF, how they have gone about doing it, and *why* they appear to pursue design activities within their military organization, culture, and belief system. Of all the military organizations researched, the Australians represent the most dynamic group in 2022, given the recent end of pandemic quarantining for their nation, and a significant increase in design experimentation over the last twelve months. While the analysis below represents the dominant military design developments and "proto-methods" of the last two decades, this may change soon, and potentially into something quite unlike what the ADF have allowed to go stagnant over the last decade. The ADF, like their Canadian counterparts, appear to be on the verge of a significant, grassroots-styled design revolution in 2022–2023.

With respect to Adaptive Campaigning as depicted in their military doctrine in the Figure 7.3, we now move to an epistemological framing. The Australian military positions "mission command" as the central cognitive model around which their entire methodology and decision-making praxis occurs. Mission Command is a prominent philosophy of leadership and decision-making within modern militaries, and also a style that pairs trust, decentralization, and an emphasis on the maneuvering subordinate unit making rapid, contextual decisions that higher elements may be unaware of, or unable to conceptualize during the original orders issuing.[145] It differs from heavy-handed, centralized hierarchical decision-making that continues to remain popular in militaries such as Russia, China, and forces that operated under Warsaw Pact era training practices, but militaries including the United States may have misinterpreted original German origins of the concept,[146] and potentially some of those flawed concepts in American doctrine have extended into the ADF.[147]

The first arrow above illustrates this epistemological choice to prioritize decentralized decision-making as central to their military processes in complex warfare, corresponding to their Adaptive Campaigning graphic. "System framing" occurs as the next sequence where Australians may draw indirectly from original SOD and Naveh's emphasis of systemic conceptualization over the traditional systematic logic; this also could only be reference to systems thinking and a stronger Ryan/Grisogono influence in the early 2000s. As both drew from complexity theory and systems thinking, the epistemological framing remains valid. Australian planners are expected to consider feedback loops and other systems thinking fundamentals that also suggest some sociological options, if the organization is aware of social paradigm theory. Otherwise, ADF practitioners might fall into similar cognitive traps as American soldiers using ADM to force a singular paradigmatic view. This is where the "operational environment" is detached, like a sterile laboratory that strategists are independent from while looking down and in. Unfortunately, Australian planners may take the institutionally complicit path here and ignore sociological differences, projecting their own worldview upon all others through a technically rationalist, incommensurate war paradigm exercise.

Next, Adaptive Campaigning planners move to "adaptation" which again is a nod to complexity theory, and may also extend back into original SOD depending on how far the ADF will tolerate change in the complex defense challenge. Adaptation done in merely technological, tactical, or within overarching ADF planning doctrinal rules and beliefs will turn their designs into what American military members are conditioned to do. Design becomes subordinate to planning, and adaptation merely echoes what available options might exist that do not challenge anything cherished by the Australian military institution.

Through adaptation as a deliberate activity depicted above, there is a suggestion that Adaptive Campaigning should have experimentation, iteration, or some manner of prototyping concepts so that change can occur, and there is a reference to the experiential learning Naveh emphasized in SOD with "learn how to learn" specified in their methodology.[148] This also could come from direct experiential learning theorists as well, yet Naveh's emphasis on this would be made apparent to Ryan when he learned of SOD, and ultimately come to ADF consideration formally after 2014.[149] The phrase "know when to change" accompanies "challenge understanding", while under "sense" the ADF directs planners to "learn to *measure* what matters" [emphasis added]. Measurement may be quantitative, or qualitative, and the design logic could function through deductive, inductive, or abductive reasoning. SOD disrupted deductive and also certain inductive reasoning methods so that abductive thinking could gain prominence in ways that modern militaries generally do not consider. Australian doctrine does not get to this level of theoretical clarity, nor does the doctrine offer useful primary source content or explanation beyond the simple methodological statements. Epistemologically, the Australians could move toward what original SOD proponents advocate for in these sequences, but institutional preferences and biases likely would prevent anything radical from challenging their understanding of overarching planning processes, core doctrinal beliefs, or their preferred war paradigm that mirrors the rest of the Anglosphere.

The fourth arrow depicted above in the epistemological mapping is an interesting and uniquely Australian focus toward institutional values. The ADF do seem to prioritize understanding in a contextual mode that suggests some of Schön's reflective practice theory, but this also could inadvertently correspond to the concepts of Boyd's OODA loop which clearly inspired the ADF graphical conceptualization of their method. Proponents of Boyd can embrace sociological paradigm concepts, yet many may be drawn to quantitative interpretations of systems thinking that remain entrenched in a single overarching paradigm where scientific beliefs occur.[150] Institutional awareness suggests that the ADF could, using this proto–design methodology, challenge deeply held institutional beliefs and present alternatives for experimentation. Without greater detail or theoretical concepts offered in the doctrine, we are left with leaving these options up to the individual units attempting to operationalize "adaptive action". Without any education in these unorthodox military disciplines and topics, it is more likely that the ADF moves toward how the American military tends to position design subordinate and obedient to institutional planning endeavors.

"Learn to see what is important" is stated in the methodology where "sense"-making occurs in their cycle. In HCD methodologies covered in the first chapter, this is equivalent to the "empathy" stage. Importance suggests a contextual quality where the ADF might be in a system with multiple actors, adversaries, and stakeholders and there are different positions on what is important to whom. Systemic thinking is better suited for these complex defense challenges over any reductionist, systematic, formulaic mode found in the deliberate planning methodologies shared by NATO forces and the Anglosphere.[151] Adaptive campaigning as a proto-design suggests that planners could, with sufficient education and experience, function in the orientation that Naveh's original design suggested, although there is scant evidence in ADF military design history of anything like this except in isolated, often unpublished cases.[152]

The fifth arrow in the above illustration circles back into the entire process indicating that emergence is a paramount quality of complex, dynamic systems. This ever-changing complexity is not explainable in the Newtonian stylings of modern military planning methods, and while this OODA loop-inspired cycle of perpetual learning suggests that Australian planners might challenge deep institutional beliefs and processes, according to deep insiders such as Jackson studying ADF design activities, he finds no evidence that they have done this to date.[153] Like the American army and marines, Adaptive campaigners might be unable to escape institutional demands to not disrupt established doctrine and deeply held beliefs on warfare, and war itself. Hypothetically, the Australian planners could remain in perpetual loops of reinforcing the current legacy system instead of offering disruptive innovation, or articulating the unrealized to an organization unwittingly tied to a singular war paradigm. Whether this occurs or not requires additional research, to include further analysis of whether the disciplines combined by Naveh in SOD are even present in ADF professional education and training.

Jackson provided in his introduction to a 2019 collection of military design articles published in the Joint Studies Paper Series a detailed summary of multiple design workshops, exercises, and design methodologies that occurred in Australia between 2016 and 2019. Yet

> while these presentations indicate a growing awareness of and interest in design thinking within the ADF ... none of these presentations were able to go beyond an introductory-level explanation of design thinking and a few of its constituent methods ... [To] date, the ADF has not developed a design thinking education course that goes beyond cursory

awareness-raising. Nor has it conducted a detailed, comparative evaluation of different design thinking methodologies.[154]

Jackson concluded with the acknowledgement that there is no ADF organizational approach to developing its own military design theory, methodology, as well as internal faculty expertise and capacity to inculcate across the ADF. Instead, the only current options remain the direct import of other military design methods whether through international education, doctrine, or direct contracted assistance. The ADF also employ civilian design methods, largely for force projection and technological applications, with little or no adaptation done for military contexts.[155]

In Jackson's conclusion to his introduction chapter in the 2019 *Joint Studies Papers Series* volume, he effectively framed an emergent Australian military demand for design thinking. This reinforces the overarching position of this research that militaries in the postmodern context for war require the self-disruptive, multi-paradigmatic and transformative design constructs of reflective practitioners that think and act in the application of organized violence. The range of military design theory and diversity of disciplines in recent articles published by the Australians also indicates a potential further shift away from the single-paradigm emphasis of legacy military decision-making logics of the modern era.

> It is worth noting that these contributions span a range of disciplines and paradigms … none of these contributions is exclusively single-disciplinary … It is fitting, therefore, to conclude this introductory paper with a reminder that one of the strengths of design thinking is that it is ill-defined and multidisciplinary. What is important is to ensure that whatever design methodology one employs, that methodology is conceptually robust and rigorously implemented.[156]

Conclusion

This chapter presented two "proto-design" movements from Canada and Australia, although as of 2022 in both countries, military designers are rapidly accelerating in practice, theory development, potential military doctrinal publication options, and also clear interest in formal design education across their forces. While much of this momentum seems fixated upon technological, explicit, and war artifact designs, any increase in design interest suggests institutional frustration with legacy practices and methods.[157] Design interest in Australia continues to develop, and whether applied toward warfighting tools or toward how militaries understand complex security challenges and warfighting itself, the design movement provides necessary conceptual tools and techniques to aid militaries in transformation. Canadian military design is more advanced and organized within their military forces compared to

Australia, yet Canadian design remains largely within advanced education programs and is not yet found in doctrine.

Doctrinal publication of a design methodology is not a requirement for design to advance, and indeed it often signals a calcification of design thinking or institutional assimilation of design into legacy planning processes such as in the American army and marines. What remains to be seen in Canada and Australia is a broadening of design practice, theory development, and a deepening community of military designers that advocate for military design practice to be integrated formally across military services. Planning already is well established in both forces, where planning teams might be observed operating for virtually every mission, activity, or objective their armed forces encounter. The same cannot be said of design, and while design is not required for all planning endeavors, the Canadian and Australian design advocates do agree that when ill-structured defense challenges are presented, their military institutions may or may not consider design before rushing into planning. This remains the high hurdle to clear in both forces, and over the next decade, how each nation's military seeks to address this institutional challenge will provide insight into where Canadian and Australian military design transforms toward next.

Notes

1 Jackson used "proto-design" in an unpublished early draft version of this article, but later switched to "early design" during peer review. See: Aaron Jackson, "Introduction: What Is Design Thinking and How Is It of Use to the Australian Defence Force," *Unpublished Draft*, 2019, 11; Aaron Jackson, "Introduction: What Is Design Thinking and How Is It of Use to the Australian Defence Force," *Australian Journal of Defence and Strategic Studies*, Joint Studies Paper Series, no. 3 (December 2019): 1–22.

2 Craig Dalton, "Systemic Operational Design: Epistemological Bumpf or the Way Ahead for Operational Design?" (monograph, US Army School of Advanced Military Studies, Fort Leavenworth, Kansas, March 2006), 7.

3 Dalton, 15; S. L. Smith, "Canadian Armed Forces Decision Making" (Canadian Forces College, 2016 2015), 1, www.cfc.forces.gc.ca/259/290/301/305/smiths.pdf

4 Dalton, "Systemic Operational Design: Epistemological Bumpf or the Way Ahead for Operational Design?," 12.

5 Some Canadian military historians discount any military actions prior to 1917 as either British or French military history in Canada, and that as a nation Canada had no military identity until a Canadian actually led a military Corps in World War I. In this context, the author correlates any military activities involving Canadians or on Canadian soil as "Canadian military history" in a broader context here.

6 Hugues Canuel, "Canadian Civil-Military Relations in the Early 'Command Era,' 1945–1955," *Canadian Military History* 24, no. 2 (Summer/Autumn 2015): 111.

7 Canuel, 2.

8 Douglas Bland, *The Administration of Defence Policy in Canada, 1947–1985* (Kingston, Ontario: Ronald P. Frye & Company, 1987); Canuel, "Canadian Civil-Military Relations in the Early 'Command Era,' 1945–1955," 2.

9 Canuel, "Canadian Civil-Military Relations in the Early 'Command Era,' 1945–1955," 7–8.

10 Canuel, 12.

11 Aaron Jackson, "Doctrine, Strategy and Military Culture: Military-Strategic Doctrine Development in Australia, Canada, and New Zealand, 1987–2007" (Ministry of National Defence, Australia, 2013), 16.

12 Jackson, 17.

13 Jatinder Mann, "'To the Last Man and the Last Shilling' and 'Ready, Aye Ready': Australian and Canadian Conscription Debates during the First World War," *Australian Journal of Politics and History* 61, no. 2 (November 2, 2015): 199.

14 Paul Mitchell, "Educating Strategic Leaders: The Foundations of Strategic Level PME at Canadian Forces College (Final Editorial Draft)," in *The Education of an Air Force*, ed. Randall Wakelam, David Varey, and Emanuel Sica (Lexington, KY: University Press of Kentucky Press, 2020), 4.

15 Francis Clermont, "Design: An Ethical and Moral Project Conscious Intention for the Cybernetician," *Journal of Military and Strategic Studies* 17, no. 4 (June 2017): 56.

16 Aaron Jackson, "Moving Beyond Manoeuvre: A Conceptual Coming-of-Age for the Australian and Canadian Armies," *Australian Defence Force Journal*, no. 177 (2008): 87–89.

17 Thomas St-Denis, "Transformational Leadership: Not for the Warrior," *Canadian Military Journal* 5, no. 4 (Winter – 2005 2004): 85.

18 Clermont, "Design: An Ethical and Moral Project Conscious Intention for the Cybernetician," 58.

19 Clermont, 58.

20 Dalton, "Systemic Operational Design: Epistemological Bumpf or the Way Ahead for Operational Design?," 46.

21 L. Bruyn, T. Lamoureux, and B. Vokac, "Function Flow Analysis of the Land Force Operations Planning Process," *Defense Research and Development Canada Toronto*, no. DRDC No. CR 2004–065 (2005): 12.

22 Bruyn, Lamoureux, and Vokac, 1; Smith, "Canadian Armed Forces Decision Making," 5; Jackson, "Moving Beyond Manoeuvre: A Conceptual Coming-of-Age for the Australian and Canadian Armies," 88.

23 Clermont, "Design: An Ethical and Moral Project Conscious Intention for the Cybernetician," 59.

24 Clermont, 56.

25 Clermont, 58.

26 Clermont, 58.

27 Shimon Naveh, "The Australian SOD Expedition: A Report on Operational Learning" (Unpublished manuscript, December 2010), 4.

28 Jackson, "Doctrine, Strategy and Military Culture: Military-Strategic Doctrine Development in Australia, Canada, and New Zealand, 1987–2007," 5.

29 Martin Crotty and Craig Stockings, "The Minefield of Australian Military History," *Australian Journal of Politics and History*, 2014, 580–582.

30 Daniel Leach, "The Other Allies: Military Security, National Allegiance, and the Enlistment of 'Friendly Aliens' in the Australian Armed Forces, 1939–45," *War & Society* 32, no. 1 (March 2013): 39.

31 David McCraw, "Change and Continuity in Strategic Culture: The Cases of Australia and New Zealand," *Australian Journal of International Affairs* 65, no. 2 (April 2011): 170.

32 Peter Layton, "The Third Age of Australian Defence White Papers Continues," *The Strategist*, March 7, 2016, www.aspistrategist.org.au/the-third-age-of-austral ian-defence-white-papers-continues/

33 McCraw, "Change and Continuity in Strategic Culture: The Cases of Australia and New Zealand," 172.

34 Jackson, "Doctrine, Strategy and Military Culture: Military-Strategic Doctrine Development in Australia, Canada, and New Zealand, 1987–2007," 7.

35 Markus Mäder, *In Pursuit of Conceptual Excellence: The Evolution of British Military-Strategic Doctrine in the Post-Cold War Era. 1989–2002* (Bern, Germany: Peter Lang AG, 2004), 45–49.

36 Jackson, "Doctrine, Strategy and Military Culture: Military-Strategic Doctrine Development in Australia, Canada, and New Zealand, 1987–2007," 9.

37 Layton, "The Third Age of Australian Defence White Papers Continues."

38 Australian security concerns would include events such as the fall of the Berlin Wall, and even small deployments of observers to Somalia in 1992, but these matters were less strategically pressing to Australia as compared to the United States, Great Britain, and Canada.

39 Layton, "The Third Age of Australian Defence White Papers Continues."

40 Jackson, "Moving Beyond Manoeuvre: A Conceptual Coming-of-Age for the Australian and Canadian Armies," 85.

41 John Anderson, "From Systemic Operational Design (SOD) to a Systemic Approach to Design and Planning: A Canadian Experience," *Canadian Military Journal* 12, no. 3 (2012).

42 Anderson; Matthew Lauder, "Systemic Operational Design: Freeing Operational Planning From the Shackles of Linearity," *Canadian Military Journal*, Operational Planning, 9, no. 4 (2009): 41–49.

43 Jackson, "Moving Beyond Manoeuvre: A Conceptual Coming-of-Age for the Australian and Canadian Armies," 88.

44 Paul Mitchell to Ben Zweibelson, "Your Chapter (Personal Correspondence)," August 6, 2020.

45 Ben Zweibelson, Kevin Whale, and Paul Mitchell, "Rounding the Edges of the Maple Leaf: Emergent Design and Systems Thinking in the Canadian Armed Forces," *Canadian Military Journal* 19, no. 4 (Autumn 2019): 25–33.

46 Anderson, "From Systemic Operational Design (SOD) to a Systemic Approach to Design and Planning: A Canadian Experience," 38.

47 Anderson, 39.

48 Anderson, 40; Ofra Graicer, *Two Steps Ahead: From Deep Operations to Special Operations – Wingate the General*, Special Edition (Dayan Base, Tel Aviv, Israel: Israeli Defense Forces, 2015), 33–38.

49 Anderson, "From Systemic Operational Design (SOD) to a Systemic Approach to Design and Planning: A Canadian Experience," 40.

50 Anderson, 41; Department of the Army, *Commander's Appreciation and Campaign Design*, version 1.0, vol. TRADOC Pamphlet 525-5-500 (Fort Monroe, Virginia: Training and Doctrine Command, 2008).

51 Anderson, "From Systemic Operational Design (SOD) to a Systemic Approach to Design and Planning: A Canadian Experience," 41–44.

52 Russell Ackoff, "Towards a System of Systems Concepts," *Management Science* 17, no. 11 (July 1971): 661–671; Department of the Army, *Commander's Appreciation and Campaign Design*; Robert Dixon, "Systems Thinking for Integrated Operations: Introducing a Systemic Approach to Operational Art for Disaster Relief" (monograph, US Army School of Advanced Military Studies, Fort Leavenworth, Kansas, March 2006); Jamshid Gharajedaghi and Russell Ackoff, "Mechanisms, Organisms, and Social Systems," in *New Thinking in Organizational Behaviour*, by Haridimos Tsoukas (Oxford, United Kingdom: Butterworth-Heinemann Ltd, 1994), 25–49.

53 Shimon Naveh, "Between the Striated and the Smooth: Urban Enclaves and Fractal Maneuver."

54 Anderson, "From Systemic Operational Design (SOD) to a Systemic Approach to Design and Planning: A Canadian Experience," 42.

55 Anderson, 42.

56 Anderson, 42.

57 Alex Ryan, "A Personal Reflection on Introducing Design to the U.S. Army," *The Medium* (blog), November 4, 2016, https://medium.com/the-overlap/a-personal-reflection-on-introducing-design-to-the-u-s-army-3f8bd76adcb2

58 Ryan.

59 Anderson, "From Systemic Operational Design (SOD) to a Systemic Approach to Design and Planning: A Canadian Experience," 42.

60 Anderson, 43.

61 The author, as a design educator and contractor for U.S. Special Operations Command, organized the JSOU event and invited Dr. Mitchell as well as numerous other design theorists and educators.

62 Mitchell to Zweibelson, "Your Chapter (Personal Correspondence)," August 6, 2020.

63 Adam Elkus and Crispin Burke, "Operational Design: Promise and Problems," *Small Wars Journal*, 2010, 1–21.

64 Paul Mitchell, "Stumbling into Design: Action Experiments in Professional Military Education at Canadian Forces College," *Journal of Military and Strategic Studies* 17, no. 4 (June 2017): 84.

65 Mitchell, 84.

66 Mitchell to Zweibelson, "Your Chapter (Personal Correspondence)," August 6, 2020.

67 Mitchell, "Stumbling into Design: Action Experiments in Professional Military Education at Canadian Forces College," 87.

68 Mitchell to Zweibelson, "Your Chapter (Personal Correspondence)," August 6, 2020.

69 Paul Mitchell, "Stumbling into Design: Teaching Operational Warfare for Small Militaries in Senior Professional Military Education" (tri-fold poster, Canadian Forces College, Toronto, Canada, 2015), 87, www.doria.fi/bitstream/han dle/10024/117634/MITCHELL%20Paul_poster_Designing%20Design,%20T eaching%20Strategy%20and%20Operations%20for%20Small%20Militaries. pdf?sequence=2

70 Lauder, "Systemic Operational Design: Freeing Operational Planning From the Shackles of Linearity."

71 Anderson, "From Systemic Operational Design (SOD) to a Systemic Approach to Design and Planning: A Canadian Experience."

72 Mitchell, "Stumbling into Design: Teaching Operational Warfare for Small Militaries in Senior Professional Military Education," 89.

73 Mitchell to Zweibelson, "Your Chapter (Personal Correspondence)," August 6, 2020.

74 R.J. Moore, "Operations Assessment in the Contemporary Operating Environment: How Design Methodology Can Contribute to More Effective Operations Assessment" (Master's Degree Thesis supporting paper, Toronto, Canada, Canadian Forces College, 2015), www.cfc.forces. gc.ca/259/290/317/305/moore.pdf

75 Mitchell, "Stumbling into Design: Teaching Operational Warfare for Small Militaries in Senior Professional Military Education," 88.

76 Mitchell, 88.

77 Shimon Naveh, *Systemic Operational Design: Designing Campaigns and Operations to Disrupt Rival Systems (Draft Unpublished)*, Version 3.0, unpublished draft (Fort Monroe, Virginia: Concept Development & Experimentation Directorate, Future Warfare Studies Division, US Army Training and Doctrine Command, 2005).

78 Mitchell to Zweibelson, "Your Chapter (Personal Correspondence)," August 6, 2020.

79 Beaulieu-Brossard to Zweibelson, personal correspondence, August 24, 2020.

80 Aaron Jackson, *The Roots of Military Doctrine: Change and Continuity in Understanding the Practice of Warfare* (Fort Leavenworth, Kansas: Combat Studies Institute Press, 2013).

81 Mitchell, "Stumbling into Design: Teaching Operational Warfare for Small Militaries in Senior Professional Military Education," 92.

82 Mitchell, 94.

83 Mitchell, 94.

84 Mitchell, 97.

85 Bob Johnson, Interview with Bob Johnson on U.S. Army Design 30 SEP 2021, interview by Ben Zweibelson, mp3 Audio File, September 30, 2021, 27:28–29:28.

86 Paul Mitchell and Ben Zweibelson, "RE: Question on One of Our Past Email Exchanges," January 14, 2020. Personal correspondence between the author and Mitchell.

87 Mitchell, "Stumbling into Design: Teaching Operational Warfare for Small Militaries in Senior Professional Military Education."

88 Shimon Naveh and Ofra Graicer, Naveh audio 14OCT2019 2hr28min Day 1 patio.MP3, interview by Ben Zweibelson and Nathan Schwagler, personal interview, October 14, 2019.

89 Shimon Naveh and Ofra Graicer, Naveh audio 15OCT2019 18min02sec Day 2 morning.MP3, interview by Ben Zweibelson and Nathan Schwagler, personal interview, October 15, 2019, 15:15.

90 Philippe Beaulieu-Brossard, "Encountering Nomads in Israel Defense Forces and Beyond" (unpublished draft provided by author, University of Ottawa, Canada, 2016), 1.

91 Philippe Beaulieu-Brossard and Philippe Dufort, "Conclusion: Researching the Reflexive Turn in Military Affairs and Strategic Studies," *Journal of Military and Strategic Studies* 17, no. 4 (June 2017): 273–289.

92 Beaulieu-Brossard, personal correspondence with the author, August 24, 2020.

93 Mitchell, "Stumbling into Design: Teaching Operational Warfare for Small Militaries in Senior Professional Military Education."

94 "Canadian Forces College 2013–2014 Joint Command and Staff Programme DS/CF 548 Lesson Plan: Advanced Joint Warfighting Studies" (unpublished – Canadian Forces College internal document, March 2014).

95 "Canadian Forces College 2018–2019 National Security Programme DS/CF 592 –Modern Comprehensive Operations and the Complexity of Contemporary Conflicts" (CFC internal document for education module, Toronto, Canada, 2018), 24; "Canadian Forces College 2018–19 National Security Programme CF575 – Strategic Thinking and Formulating National Strategies" (CFC internal educational document for courseware, Toronto, Canada, 2018), 4–6.

96 Mitchell, "Stumbling into Design: Teaching Operational Warfare for Small Militaries in Senior Professional Military Education," 101.

97 Archipelago of Design, *Collaborative Innovative Thinking by Design: For the Canadian Armed Forces* (Toronto, Canada: Archipelago of Design, 2022), 1–3, https://aodnetwork.ca/wp-content/uploads/2022/10/Primer_10-27-22_downs ave.pdf

98 In an interesting overlap, Beaulieu-Brossard is the lead design faculty for the CFC and the Co-President of the non-profit Archipelago of Design entity. AOD created the design primer in conjunction with the CFC, and features a strong investment of Canadian military, governmental, industry, and academic design contributors that work within CFC exercises, serve as faculty, and also operate within the AOD enterprise.

99 Zweibelson, Whale, and Mitchell, "Rounding the Edges of the Maple Leaf: Emergent Design and Systems Thinking in the Canadian Armed Forces," 12.

100 Philippe Beaulieu-Brossard and Philippe Dufort, "Enhancing Challenge Formulation in Defence Organisations: Towards Reflexive Methods" (Draft unpublished provided to author, July 2020).

101 By 2020, Dr. Mitchell would begin using the author's graphic on the following page in his own design presentations to explain how the CFC approached "agnostic design" while crediting this research.

102 Beaulieu-Brossard and Dufort, "Enhancing Challenge Formulation in Defence Organisations: Towards Reflexive Methods," 14–15.

103 Jacques Ranciere, *The Ignorant Schoolmaster: Five Lessons in Intellectual Emancipation*, trans. Kristin Ross (Stanford, California: Stanford University Press, 1991), 5–20.

104 Majken Schultz and Mary Jo Hatch, "Living with Multiple Paradigms: The Case of Paradigm Interplay in Organizational Culture Studies," *Academy of Management Review* 21, no. 2 (1996): 529–557; Ben Zweibelson, "The Multidisciplinary Design Movement: A Frame for Realizing Industry, Security, and Academia Interplay," *Small Wars Journal*, January 2019, https://smallwars journal.com/jrnl/art/multidisciplinary-design-movement-frame-realizing-industry-security-and-academia-interplay

105 Beaulieu-Brossard and Dufort, "Enhancing Challenge Formulation in Defence Organisations: Towards Reflexive Methods," 15.

106 The author participated in design education at the CFC from 2017 through 2020.

107 This may have changed since 2018–2019. Beaulieu-Brossard states that each design facilitator must now submit an agenda and preparatory material a week in advance to the CFC faculty. Beaulieu-Brossard, personal correspondence with the author, August 24, 2020.

108 Mathieu Primeau to Ben Zweibelson, "Updated Draft Chapter with Fixes to Canada Portions (Personal Correspondence)," August 14, 2020.

109 Primeau to Zweibelson.

110 Beaulieu-Brossard has been with the CFC since 2017, while Jensen was hired in 2020. The CFC has retained Donna Dupont as the design exercise director for the Field Grade officers in 2020 but is not full-time faculty at this time. Beaulieu-Brossard, personal correspondence with Zweibelson, August 24, 2020.

111 Naveh, "The Australian SOD Expedition: A Report on Operational Learning."

112 Cara Wrigley, Murray Simons, and Aaron Jackson, "Australian Military Design Thinking: A 2021 Review," *The Archipelago of Design Website*, February 15, 2022, https://aodnetwork.ca/australian-military-design-thinking-a-2021-review/

113 Select military units and schools are introducing military design education such as special operations forces in Sydney, but this remains a case-by-case, individualized approach across the force.

114 Ryan, "A Personal Reflection on Introducing Design to the U.S. Army," November 4, 2016.

115 Anne-Marie Grisogono, "The State of the Art and the State of the Practice: Implications of Complex Adaptive Systems Theory for C2," *Land Operations Division, Defence Science and Technology Organisation, South Australia*, June 2006, 1–19; Ryan, "A Personal Reflection on Introducing Design to the U.S. Army," November 4, 2016.

116 John Boyd, "Destruction and Creation" (Unpublished manuscript, 1976), www.goalsys.com/books/documents/DESTRUCTION_AND-CREATION.pdf; Frans Osinga, *Science, Strategy and War: The Strategic Theory of John Boyd* (Routledge, 2006).

117 Jackson, "Moving Beyond Manoeuvre: A Conceptual Coming-of-Age for the Australian and Canadian Armies," 91.

118 Ryan, "A Personal Reflection on Introducing Design to the U.S. Army," November 4, 2016.

119 Ryan.

120 Ryan.

121 Michael Bassingthwaighte, "Adaptive Campaigning Applied: Australian Army Operations in Iraq and Afghanistan," *U.S. Army School of Advanced Military Studies Monograph*, May 19, 2011.

122 Stefan Banach and Alex Ryan, "The Art of Design: A Design Methodology," *Military Review* 89, no. 2 (April 2009): 105–115; Ryan, "A Personal Reflection on Introducing Design to the U.S. Army," November 4, 2016.

123 The Marines run a smaller equivalent of the SAMS program that also studied design. See Chapter 5.

124 Trent Scott, "The Lost Operational Art: Invigorating Campaigning into the Australian Defence Force," Study Paper, Land Warfare Studies Centre Study Papers (Canberra, Australia: Land Warfare Studies Centre, February 2011), 100.

125 Scott, 114.

126 Scott, 114–116.

127 Scott, 102.

128 Brett Andersen to Ben Zweibelson, "RE: FW: SDW (Personal Correspondence)," August 12, 2020.

129 Philippe Beaulieu-Brossard and Philippe Dufort, "The Archipelago of Design: Researching Reflexive Military Practices," The Archipelago of Design: Researching Reflexive Military Practices, 2017, www.militaryepist emology.com; Philippe Beaulieu-Brossard and Philippe Dufort, "Introduction to the Conference: The Rise of Reflective Military Practitioners" (Hybrid Warfare: New Ontologies and Epistemologies in Armed Forces, Canadian Forces College, Toronto, Canada: University of Ottawa and the Canadian Forces College, 2016).

130 Jackson, "Introduction: What Is Design Thinking and How Is It of Use to the Australian Defence Force," December 2019; Aaron Jackson, "Innovative within the Paradigm: The Evolution of the Australian Defence Force's Joint Operational Art," *Security Challenges* 13, no. 1 (June 2017): 59–79; Wrigley, Simons, and Jackson, "Australian Military Design Thinking: A 2021 Review"; Aaron Jackson, "Design Thinking in Commerce and War: Contrasting Civilian and Military Innovation Methodologies," *Air University Lemay Center for Doctrine Development and Education* LeMay Paper No. 7 (December 7, 2020): 1–78.

131 Australian Army doctrine in the 1990s did introduce some of Boyd's ideas, so senior personnel still serving in the ADF had some awareness of these ideas. Jackson's work re-introduced it to the next generation.

132 Boyd had even more elaborate diagrams for his OODA loop that can be found in Osinga's research. However, the Osinga examples are unlikely to have been studied in detail, particularly due to the 2006 publication date and a lack of awareness of the full OODA concept. See: Osinga, *Science, Strategy and War: The Strategic Theory of John Boyd*.

133 Aaron Jackson, "Introduction: What Is Design Thinking and How Is It of Use to the Australian Defence Force?," *Polemus: The Australian Journal of Defence Studies*, December 2019, 13.

134 Naveh's influence upon ADF design is problematic. Naveh, Graicer and other SAMS faculty in interviews suggest a direct, causal linkage with ADF students and the elite SOF engagement under a contracted requirement. However, in the ADF interviews of academics and military leaders for this research, there is scant evidence of any direct influence, and potentially a categorical rejection of SOD"s intellectually challenging content as well. Naveh's contracted work with the ADF features aspects of self-promotion that further introduce bias.

135 Palazzo, a senior researcher at the Land Warfare Studies Centre, would accuse the ADF of failing to foster any real debate on the future character of warfare. See: Albert Palazzo, "The Future of War Debate in Australia: Why Has There Not Been One? Has the Need for One Now Arrived?," Land Warfare Studies Centre Working Papers (Canberra, Australia: Land Warfare Studies Centre, August 2012).

136 Alex Ryan and Anne-Marie Grisogono, "Hybrid Complex Adaptive Engineered Systems: A Case Study in Defence," *Defence Science and Technology Organisation*, 2004, 1–8; Grisogono, "The State of the Art and the State of the Practice: Implications of Complex Adaptive Systems Theory for C2"; Alex Ryan, "The Foundation for an Adaptive Approach," *Australian Army Journal for the Profession of Arms* 6, no. 3 (2009): 69.

137 Jackson, "Introduction: What Is Design Thinking and How Is It of Use to the Australian Defence Force?" 13.

138 Aaron Jackson, "A Tale of Two Designs: Developing the Australian Defence Force's Latest Iteration of Its Joint Operations Planning Doctrine," *Journal of Military and Strategic Studies* 17, no. 4 (June 2017): 174–193.

139 Edward Hayward, "Planning Beyond Tactics: Towards a Military Application of Philosophy of Design in the Formulation of Strategy" (U.S. Army School of Advanced Military Studies, May 22, 2008), 4.

140 Of note, the first edition would have no references to design at all. A third edition is in publication as of 2022 and whether any design content will remain is unknown.

141 Jackson, "Introduction: What Is Design Thinking and How Is It of Use to the Australian Defence Force," December 2019, 13.

142 Jackson, "Moving Beyond Manoeuvre: A Conceptual Coming-of-Age for the Australian and Canadian Armies"; Jackson, "Innovative within the Paradigm: The Evolution of the Australian Defence Force's Joint Operational Art"; Aaron Jackson, "Center of Gravity Analysis 'Down Under': The Australian Defence Force's New Approach," *Joint Forces Quarterly* 83 (October 2016).

143 Private correspondence between Jackson and the author, November 20, 2022.

144 Jackson, "Introduction: What Is Design Thinking and How Is It of Use to the Australian Defence Force?," 15.

145 Anthony King, "Mission Command 2.0: From an Individualist to a Collectivist Model," *Parameters* 47, no. 1 (Spring 2017): 7–19; Harris Stephenson, "Mission Command-Able? Assessing Organizational Change and Military Culture Compatibility in the U.S. Army," *Journal of Military and Strategic Studies* 17, no. 1 (2016): 104–143; Christopher Lamb, "The Micromanagement Myth and Mission Command: Making the Case for Oversight of Military Operations," *National Defense University Press*, Strategic Perspectives, No. 33, August 2020, 1–59.

146 Ricardo Herrera, "History, Mission Command, and the Auftragstaktik Infatuation," *Military Review*, August 2022, 53–66; David Devine, "The Trouble with Mission Command: Army Leadership and Leader Assumptions," *Military Review*, October 2021, 36–42.

147 Russell Glenn, "Mission Command in the Australian Army: A Contrast in Denial," *Parameters* 47, no. 1 (Spring 2017): 21–30.

148 Naveh, "The Australian SOD Expedition: A Report on Operational Learning," 6.

149 Alex Ryan, "A Personal Reflection on Introducing Design to the U.S. Army," in *Cluster 2* (Hybrid Warfare: New Ontologies and Epistemologies in Armed Forces, Canadian Forces College, Toronto, Canada: University of Ottawa and the Canadian Forces College, 2016).

150 Gibson Burrell and Gareth Morgan, *Sociological Paradigms and Organisational Analysis: Elements of the Sociology of Corporate Life* (Portsmouth, New Hampshire: Heinemann, 1979); Dennis Gioia and Evelyn Pitre, "Multiparadigm Perspectives on Theory Building," *Academy of Management Review* 15, no. 4 (1990): 584–602.

151 Jamshid Gharajedaghi, *Systems Thinking: Managing Chaos and Complexity, A Platform for Designing Business Architecture*, Third (New York: Elsevier, 2011), 107, http://pishvaee.com/wp-content/uploads/downloads/2013/07/Jamshid_Gharajedaghi_Systems_Thinking_Third_EdiBookFi.org_.pdf; Haridimos Tsoukas, *Complex Knowledge: Studies in Organizational Epistemology* (New York: Oxford University Press, 2005), 326; Antoine Bousquet and

Simon Curtis, "Beyond Models and Metaphors: Complexity Theory, Systems Thinking and International Relations," *Cambridge Review of International Affairs* 24, no. 1 (2011): 43–62.

152   Naveh, "The Australian SOD Expedition: A Report on Operational Learning."

153   Private correspondence between Jackson and the author, November 20, 2022.

154   Jackson, "Introduction: What Is Design Thinking and How Is It of Use to the Australian Defence Force?," 15–16.

155   See Chapter 1 on how commercial design methods are routinely inserted into military applications without any serious evaluation of the commercial concepts. The ADF in 2021–2022 do use commercial design in a variety of tactical, technological, and educational efforts that lack any rigorous military adaptation, usually with design teams treating the military as another equivalent client.

156   Jackson, "Introduction: What Is Design Thinking and How Is It of Use to the Australian Defence Force?," 18.

157   Wrigley, Simons, and Jackson, "Australian Military Design Thinking: A 2021 Review."

Conclusion

The Destruction of Old Monsters by New Ones: A Design Insurgency Continues

The maxim "science advances one funeral at a time" is alleged to come from the Nobel-Prize winning theoretical physicist, Max Planck. This isn't precisely what Planck said, but the idea is wholly appropriate to the focus of this conclusion chapter. Institutions gradually transform over time not just by the realization and acceptance of new ideas, but by the elimination of inflexible vanguards of the old ways of thinking one way or another. Kuhn quotes the original statement by Planck in his groundbreaking book on scientific paradigms as follows: "A new scientific truth does not triumph by convincing its opponents and making them see the light, but rather because its opponents eventually die, and a new generation grows up that is familiar with it".[1] This summarizes the current situation in modern military institutions worldwide and the military design community of practice. Design innovators and experimenters continue to generate critical, creative, and highly disruptive war designs that challenge the institution, while pragmatic or unwitting single-paradigm, traditionalists defend against design incursions through battles over education, training, doctrine, philosophy, and military identity.

The next few decades should present considerable change in how modern militaries understand and act in complex warfare, not just through technological change and the wickedly novel security developments that will unavoidably make future wars unlike earlier ones, but also through design thinking. Militaries will continue to deconstruct, challenge, and improvise experimentations well outside the institutional boundaries, ushering forward radically different ways to frame a future war paradigm unlike current ones. This will not come easily, and much of it likely will occur after the retirement, removal, or changing of the guard in various institutions, as well as funerals.

Many traditionalists hostile to military design will never agree to the ideas, the philosophy, or the possibility that long-held beliefs on war might be limited, or even obsolete. They will continue to preach some orthodoxy of an earlier military framework for decision-making in warfare that may no longer be found on any future battlefields or practiced by any future adversaries. At the same time, emergence and change are unending, and perhaps

DOI: 10.4324/9781003387763-9

the only true constant in complexity is that change is continuous. There will never be some set or static decision-making praxis that employs some ultimate combination of war theories, mental models, and decision-making methodology for security affairs that requires no new language, nor any new challenge or critiques. The certitude of a scientific frame for modern warfare comes with the ontological and epistemological stances of those that crafted a natural science ordering of war through mimicry of scientific methods and phenomenon outside of war. If such a frame was useful, it manifested in the seventeenth through late nineteenth or perhaps early twentieth centuries only, and now is continuously extended well the past expiration date. Perhaps this is human nature and relates once more to our inability to break away from the application of organized violence as an expected solution to political, societal, and organizational problems. Yet it is also natural for humans to creatively design anew, and war as a human generated construct cannot be exempt from such change.

A Shift toward Multidisciplinary Design

This conclusion presents some predictions on normatively how military design ought to develop. First, military design, as an international movement, should progress toward increased multidisciplinary or heterogeneous, divergent combinations of designs, theories, and fields. Over time, it should retreat from single-domain, single-service, single-design-school/discipline/field inspirations except in clear contexts of using a type of design toward an obvious application. Design of a new weapon should be done with an industrial, tool-oriented sort of design, while designing new military software should be done in software design methods. What to do with such new tools, and how they might be applied in future complex security contexts likely requires military designs that will progressively advance beyond any of the ones covered in this book. Services will likely continue to desperately cling to service-specific design concepts for largely self-relevant matters or beliefs. Over time, this strategy may fail, encouraging services to drop such pretenses and consider systemically how designing in complex warfare need not remain in a single paradigm, or cater to a single service.

Peter Rowe, while writing from an architectural design perspective, observed: "In the give and take of problem-solving situations in the real world … we discover there is now such thing as *the* design process in the restricted sense of an ideal step-by-step technique".[2] Rowe, writing in the 1980s at the surge of commercial "design schools and methods" golden era, was critiquing civilian design. However, his comments certainly apply to military communities in the 2020s with myriad service-oriented, doctrinally branded design methods. There is an American Army "way to design", and another that instructs the U.S. Navy, while the U.S. Marines formed yet another. USSOCOM created yet another including one misguided attempt to "brand a SOF design way", and today additional militaries, agencies, and adjacent security entities seem poised to create their own design methods.[3] Each features a heavy dose of doctrine, and usually stipulates where design functions as a subordinate component within broader institutionalized planning. Military designers continue in many organizations to be instructed step-by-step, how and when to do design, in order to produce institutionally palatable deliveries to those waiting to perform unchallenged planning methods. This pattern may continue for the short term, but over sufficient time and reflection, intellectually rigorous designers should overcome these barriers.

Future single service designs along with their doctrines may need to become interoperable across joint and coalition communities, particularly if the United States and allies no longer are able to retain peer-level deterrence against adversaries such as China. If Chinese military advancements in the 2020s and 2030s strip America of any single-nation deterrence capabilities, the United States Department of Defense will have no other option than reconceptualizing some integrated deterrence with strong allies and partners across multiple domains, and through forging new strategic and operational designs. Such conceptual efforts may be spoiled if the U.S. Army, for example, demands all other nations comply with Army Design Methodology. Greater concerns abound for whether such a multinational, multi-domain, and multidisciplinary challenge ought to feature designs crafted in any particular doctrine. Instead of wondering which military design might become "king of the hill" among peers, should a *systemic* design manifest that is valuable in different contexts to most all involved? This hypothetical does not yet exist, but future possibilities seem inclined toward heterogeneous over homogeneous configurations.

Future wars will rarely appear simple, or mirror historically sound analysis of past patterns of conflict. For every "Gulf War" situation where a dictator employs his forces "by the book" and is soundly defeated, myriad others will seek to avoid such contexts. Rarely will the world throw at us something simple to tackle with straight military planning and an "off the shelf" campaign plan that we are rehearsed and prepared for. Rarer still will any potential adversary decide to attack and fight us in a manner that plays to our clear strengths in technology, resources, education, and established doctrine. There are decades of rational actor (game theory) research indicating that few will oblige to take risks that cost them dearly, despite the problems with all actors being treated as rational. Life and death battles are, regardless of the ideological, cultural, societal, or institutional belief systems, still quite literally life and death battles where the stakes cannot be higher. Multidisciplinary design unlocks previously unrealized, unimagined opportunities, but only to those willing to wittingly violate institutional norms.

By multidisciplinary, this means a combination of often unusual or even antagonistic disciplines, fields, competing theories, and essentially "strange bedfellows" of design. While complexity theory and systems thinking are two clear disciplines integrated in nuanced, often mangled ways by modern militaries, the treatment of postmodern ideas in modern militaries present one prominent example of how the modern military paradigm resists serious self-examination and critique. Naveh's SOD has the highest utilization of such ideas, and demonstrates the strongest institutional pushback. Postmodernism is not necessarily some magical key to unlocking complex warfare strategies, but the near hysterical reaction to designers using elements of it is worth emphasizing if future multidisciplinary designs for war are expected to develop further. Design purists, and artists in general are drawn to the very things an institution says one cannot do like moths to the flame. That all design attempts by militaries over the last three decades appear to try to remove the unwanted postmodern content from SOD is an important pattern. That Naveh's SOD continues to be practiced by a thriving international design community is yet another, and whether those that follow military design in doctrine (devoid of such content) have yet to be fairly evaluated against non-doctrinal, more intellectually diverse design is another important point of consideration.[4]

Hayward, as a British Army student in the SAMS program during Naveh's tenure, would deploy to combat and informally experiment with SOD. He reflects on how SOD's interdisciplinary strengths, including postmodernism, would give him entirely novel insights:

> I never felt we were exposed to Checkland, Senge et al and systems thinking – in the broad sense and with the varied models. I think they help staff – as they are expansive, and it is reductionist like military tools [are]; They help lay out relationships/underlying intent/logic and rules (behaviour in short) … For me, to discover this at the end of the [SAMS] course in private study was a breakthrough and hugely disappointing. So, I think the military want the tools and not the essences/insight behind it. [For me], Naveh explaining fractal manoeuver was for me the breakthrough. I finally saw the form, function and logic … The brilliant part was not the operational aspects of swarmed [Command and Control] that forced the terrorists out into the street – but the total reversal of the landscape (Christopher Alexander) … I don't think this level of praxis and reorganisation through action is possible without the deeper philosophical foundations. To me – this was the essence of Derrida and "reading a text back against itself", but it was pan domain and in the real world. I think postmodern theory was part of the reason [SOD] was rejected, and why there is no clear design model (unlike Sanford 5, Double D, or 6–9 steps; HCD et al). I also believe it lies at its heart and cannot be removed.[5]

Graicer in 2021 echoes Hayward and sees the military design movement as requiring a multitude of theories, disciplines, fields, concepts, and flexible methods. Militaries preoccupied with service-specific, branded design doctrines and an oversimplified methodology seem to put themselves at a disadvantage. Militaries are today expected to think, strategize, plan, and win in the most complex, dynamic, and uncertain of security contexts. Failure occurs such as with the collapse of Afghanistan during the American withdrawal in 2021, yet militaries struggle with what ultimately is to blame in these strategic calamities. Often, failure is used not to call for institutional reflection, reform, or critical self-examination of deep, unquestioned tenets. Instead, failure becomes the guiding light to safeguarding institutional belief systems, identity, and renewed efforts to prepare for the next war in ways that pair not with national security needs, but with institutionally preferred acts of self-interest.[6] Militaries unwittingly tied to a Newtonian stylization of modern warfare cannot seem to let go of these conceptual tools, whereas the multidisciplinary design of SOD intentionally includes postmodernism as a powerful tool to disrupt these institutional practices. Graicer expands on this as follows:

> Once you begin to uncover the roots of failure, you may need to go even deeper and wider than originally expected. You cannot change your understanding of emergence without accepting self-regulation, negation of hierarchy and limits of scientific fact/truth. For that you need systems/ chaos/ complexity language; And readings like Maturana and Varela's *Tree of Knowledge*. You cannot change the relationship between strategic – operational – tactical command spheres without changing knowledge/ power meta-structures, discourse analysis and meta-narratives. For that you'll read Lyotard, Foucault, Derrida, Delanda, etc. You cannot begin to comprehend a true alternative to Western thought and action without reading *A Thousand Plateaus*. [To support] that alternative, nomad, rhizomic option you'll find in anthropological texts and post-colonial studies. You cannot conceive [of] alternative modes of manoeuvre without new impressions of space, speed, and time: Paul Virilio is an essential read. You cannot challenge your state-person's take on policy, national interests and strategy without meta–cognition, mental models, cultural intelligence, potential, (de)stability and destruction/ creation. For that you'll need the likes of: Schon's *Beyond the Stable State*, Buckminster Fuller's *Operating Manual for Spaceship Earth*, Richard Heuer's *Psychology of Intelligence Analysis*, Fernand Braudel's *History of Civilizations*, Francois Jullien's *A Treaty on Efficacy*, and John Boyd. The list goes on and on. Since war and culture/society are the Janus face of each other, globally shifting systems of values, logic and form entailed new language to surpass a dissolving modern, industrial, symmetric era. The question should therefore not be whether to use them, but how to communicate their need prior to introducing a new meta-theory and language for effective utilization.[7]

The next wars will call for new war tools, new systems that generate new military experiences, and continuously new design needs. At the same time, security forces must take these new war tools, systems, and defense abilities rendered by commercial design activities and weave *yet another activity* of security design where one organizes violence toward societal, institutional, and political desires. New technological and user-oriented war tools require a second-order design that seeks synthesis of that which is well outside of commercial design frames. At this nexus of human war aims, societal war experiences, and the application of new war tools, there also must be a military design to realize the unimagined, the unexplored, and the unanticipated in war itself. This includes ways to deter the horrors of war itself, in that possibly the best war design yet discovered is one that prevents future wars from manifesting.

Traditions Disrupted: Designing War Anew, to Realize New Ways of Warfare

All traditions are challenged and eventually replaced, regardless of how long such beliefs remain in use. Military design in informal earlier manifestations is how military innovation has always occurred, just as less formal planning methods would be replaced in modernity with a pseudo-scientific, mechanistic, sequential, and rationalized process for warfare. Today, militaries publish volumes of instructions on exactly how to think, act, and impose measured order upon the chaos of warfare. Some of this formal decision-making must remain central to future military needs, yet other aspects will be disrupted, challenged, and destroyed. Planning will occur in the absence of design, while designing will occur in the absence of clear, predicted victories. Failure is still a powerful stimulant for reform and experimentation, and historically, militaries tend to safeguard traditions until they catastrophically fail in warfare.

Perhaps a better way to frame the question is how much disruption will a military institution tolerate? Failure comes in many forms, and while the Afghanistan collapse would be an embarrassment to the American Department of Defense, politically the Biden Administration did accomplish much of their short-term goals in terminating American involvement with the country. In 2022, the same Department of Defense continues with all other missions, and aside from various introspective questions, debates, and obvious political fallout, the American military machine continues in a way unlike serious military defeats such as the destruction of the Iraqi Army in 2003, the Fall of Saigon in 1975, and the destruction of the Army of the Republic of Vietnam (ARVN), or the final defeat of the Russian monarchy's White Armed Forces by the Bolsheviks in 1923. Those armies (and their nations) underwent tremendous transformation (if not elimination) in defeat.

The real question is, how much in terms of radical displacement and transformation through alternative theories, mental models, and alternative methods in warfare? This indeed will be a philosophical argument in the end, and one where the linkages of theories to models and methodological construction drawing from those models interact with reality in warfare. We need to question whether contemporary military decision-making still requires the models and metaphoric devices reliant upon classical mechanics, models drawn from biology, geology, Newtonian physics, mathematics, and systematic logics entirely. Is failure in war linked to one primary issue, or a host of them? If the single or interrelated issue includes how a military makes sense of war including the arrangement of ideas and actions within the conduct of warfare, can that military "pull itself up by its own bootstraps" philosophically? If the dominant war paradigm is incomplete, irrelevant, or counterproductive, how might a military reform this?

Military design provides this opportunity to transcend localized, convergent, and limited adjustments to a system that requires *systemic* overhaul. Technological innovation only gets a military so far, and the extensive history of technologically advanced militaries being soundly defeated by low-tech, under-resourced peers is long and distinguished. Acts of innovation, regardless of if taken from warfare or other contexts, occur in the defiance of future expectations, making such designs the definition of surprise, uncertainty, and emergent, complex future systems. Doing what worked yesterday today because one assumes tomorrow will mostly be stable only works when systems manifest simple or complicated patterns. These of course exist, and generally most societies have strong rules and processes to get through daily events that subscribe to such stability and order.[8] It is when system behavior is dynamic or chaotic that such expectations fail spectacularly. The best plans evaluated by the institution assuming a stable tomorrow are the same ones that are entirely wrong to permit their formulation today, yet only through military design might an organization lacking a crystal ball have the chance to thwart these patterns.

Military design, in order to be taken seriously in such complex, dynamic affairs, should be granted no intellectual limits for ideation, critical examination, and creative experimentation. This is not to say that designers can do as they wish in war. Instead, just as military planners and strategists will huddle together to attempt to think differently while using the institutionally prescribed tools to construct recognized strategies and war plans, designers must be granted the same. Designers ought not have doctrine that is like planning doctrine (thus design becomes subordinate to planning), if there is design doctrine at all. Instead, designers should be given an environment that allows deep, intellectual debate, introspection, growth, and experimentation that often may seem fantastic or absurd. There cannot be anything within a military institution "off limits", and virtually everything should be open for critique, reform, modification, or destruction. What this means is that ultimately, military design should be a specialized, carefully nurtured, and also quite dangerous activity for the organization to manage. It cannot be managed like a planning effort, nor should new designs reinforce existing patterns of behavior when the barriers to systemic advantage are those institutional expectations themselves.

Growing Military Designers: Investing in Innovation with Deadly Seriousness

Today, the modern military institution creates the deep experts that they require in a wide range of disciplines, fields, and professions that support how one can tackle complex security challenges. The modern military needs doctors, lawyers, fighter pilots, as well as special forces operators skilled in quite particular abilities that the larger infantry and armor forces simply do not need or possess. Select recruits enter these programs and follow a clear pipeline of development, education, experience, and selection to advance through merit to higher levels of responsibility and influence. Militaries that need senior Flag Officers cannot bring them in from the outside, as the institution itself generates them over decades of service. Virtually every specialized skill is created within the military institution, meaning that professionalization requires a particular career path, resources, recruitment, and growth incentives. There is nothing like this today for designers.

There is no clear profession for military designers, as there is not even yet any broad agreement on what military design entails except in service-specific, often limited circumstances already covered. Civilian designers fare better with clear schools, programs, and general career paths for them to pursue. Most top commercial designers may not have exactly the same skills or backgrounds, but they tend to demonstrate some patterns of overall professionalization across the commercial design fields. That said, in newer areas such as computer software and technology, designers often do not follow subscribed paths. Militaries demonstrate this with the uncertainty over new areas of warfare, such as with quantum, space, cyberspace, and artificial intelligence. The best computer hackers rarely meet traditional military recruitment goals or even security clearance requirements.

Design continues to be a bit of a black box in that the combination of artistry, creativity, imagination, and subjective aspects that create exceptional designers does not render into convenient or quantitative ways for pedagogic (and andragogic) modes for education and training. Even high-level companies that rely on design thinking often recruit top designers not based on some developmental path or record of degrees and qualifications, but often on their reputation, performance, and body of work. Today's militaries tend to do the same, in that there is no real military design formal education other than however each service, branch, or military as detailed in past chapters decided to go about inculcating design.

Should militaries treat designers the way they do military strategists, and create entire advanced career paths and specialized education for them while managing where they are assigned and what they do for security forces? Might militaries forge entire careers for designers, treating them as a rare yet essential intellectual modifier for the entire force's survival and relevance? Indeed, such a shift would generate military philosophers of design within the force. Or should design be spread across the entire force, through basic design education at the lowest levels of military training and development. This would create a broad base for the institution to saturate design ideas, so that over time, those select few with unique design skills might rise to the top for special assignment and development. Yet another option is for militaries to outsource design requirements, assuming that commercial designers are not confused with those with particular military abilities. Just as the Department of Defense and ministries around the world employ specialized think tanks, military design could become a select ability developed in these organizations and applied by militaries for the deep design requirements identified. Any option could be valuable, and currently with the informal, ambiguous setting for military design today, militaries struggle to separate the real design practitioners from those seeking profit and influence devoid of such abilities.

If We Destroy to Create What Is Needed Next, Do We Become Monsters?

Ultimately, design facilitates change. This is not the change that occurs when one systematically sets up a problem-solution framework and directs action to change the future state so that we reach that envisioned future just as we planned it. Instead, security design addresses most every other sort of change that occurs outside and beyond that logic. This is where surprise, frustration, disappointment, confusion, and even fear exist, in that when our best-laid plans unfold just as designed, we often never experience any of those things. Victory is sweet, but only when we achieve it.

Institutions as well as collections of humans in groups, organizations, or teams tend to fear change, again unless it is more of the same. Tomorrow needs to be understood in how we reflect on yesterday, so that we can extend what was valuable then into tomorrow's uncertainty to maintain order and control. There is an emphasis on gaining efficiency with the standardized tools one has already mastered, and an institutional preference to protect these "tools" whether cognitive or actual artifacts encoded with cultural symbolization if only to retain what worked yesterday to sharpen it for tomorrow's expected use.[9] Militaries want to prepare for the next war by becoming better at winning the last war. This is an often-repeated maxim, but tragically it endures regardless of how often our decision-making methodologies we cultivate and teach continue to fail in execution. We are quick to blame the operators, and ignore the theories, models, methods, and language we demand operators to apply.

The modern war frame seeks linear causality in complex systems that have no such mechanism. Modern military doctrine and most strategic to tactical methodologies capitalize on engineer-based and reductionist methodologies that espouse a Newtonian physics reality where classical mechanics metaphors correlate to theories of warfare. This frame seeks to reinforce past lessons to validate experience beyond the unique contexts from which it became relevant *to create and maintain universal laws and tenets for war.* This belief is that over time, as one gains greater depth of experience and more information as well as precision with the established tools, there will be an incremental and progressive gain as each day brings further mastery and validation that yesterday's ideas will work even better tomorrow once the organization collects, describes, and improves upon the current framework. Modern warfare has a purpose therein, and the science of war generates valuable knowledge for the institution bent on progress, while anything lacking scientific discipline and verification through repeated confirmation is a lesser desired or even undesired form of knowledge. Such ideas cannot support any purposeful war logic, from a modern military frame at least. U.S. Secretary of Defense Robert McNamara perhaps said it best when dismissing a qualitative assessment by social scientists on Vietnam War strategic analysis of the war: "Where is your data? Give me something I can put into the computer. Don't give me your poetry".[10]

While this systematic, quantitative military mindset frequently works well at technical and tactical levels, within simple and complicated systems, it fails our militaries in complex and adaptive system frames. Just as the U.S. forces were technologically and organizationally able to completely overmatch their North Vietnamese and Viet Cong adversaries in pure tactical combat metrics, Gorka sees parallels in this thinking with respect to the U.S. continued overmatch against radical Islamic terrorists in the twenty-first century:

> we are peerless in our capacity to apply kinetic force on target … But counting Reaper hits against jihadi high-value targets is just as bad a metric of victory today as counting Viet Cong body bags was during the Vietnam War.[11]

Yet nearly fifty years after the last bombing run on Vietnamese adversaries, there was an institutional belief that kinetic strikes against Taliban forces coupled with technologically crafting a miniature version of U.S. forces in the Afghan Security Forces would suffice. Those expectations came crashing down not in the presumed months or years, but in weeks in 2022.

Indeed, this becomes the great yoke that designers must perpetually carry for security organizations asking for innovation, reflective practice, and institutional change. They need to fix what is broken or no longer working, do this without threatening anything institutionally upheld, and also so that such beliefs are extended into tomorrow. Naveh stated: "Design is the way of creating an alternative world. Tacticians are not supposed to create an alternative world. They are operating in the existing world with the existing instruments and the existing tools".[12] The desire to fight in a stable reality where the accumulation of data permits incremental advantage in stronger analysis over time is necessary for the epistemological belief that technological advancements can cede advantage in war.

This may promote institutional defenders of modernity and warfare to see information technology as "the panacea which will permanently lift the friction and fog of war which has plagued all past militaries"[13] so that divergent thinking and acts of destructive innovation are substituted by the streamlined execution of predetermined, automated responses through universal models enhanced through human–machine teams. As readers discovered in the third chapter as well as later in the American and other militaries, Naveh's SOD praxis breaks things that need to be broken, but in doing so raises all sorts of institutional rejection and resentment. Security design in execution is therefore often shunned as a detractor for how "metadata" and technological rationalism should finally clarify, perhaps purify warfare into a routine protocol controlled and managed by technologically advanced systems of man and machine. War science must trump war artistry; validation implies future control and predictability while uniqueness in war promotes ontological insecurity *in knowing for certain* how the next war might express its logic. We want a "techno-war" that marches to our beat, not a wicked, messy, and ever-changing challenge that somehow turns all of our advantages into disadvantages.

Security design seeks to produce military reflective practitioners that can gaze into the mirror and see that it is not a design monster on the laboratory table being sewn together, but something that generates fear out of what the monster will destroy to produce the necessary transformation. This is where those around it fear uncertainty most of all, but project that upon any design creation as monstrous and requiring a mob with torches and pitchforks. So, we can appreciate that while teleological stances on warfare and technology as well as sophisticated military organizations might offer the promise of stability, increased control, prediction, and certainty – we need to instead become far more realistic about how complex reality will learn and adapt far faster. Instead of running away from uncertainty in war, we need to change direction.

Senior military leaders and the next generation of military practitioners must *become profoundly comfortable with being uncomfortable* … this is the new normal that security design praxis embraces. New "monsters" replace older models; the real monsters are not those on the laboratory table created by design praxis, but those enduring and invisible ones that force security organizations to keep trying to repeat outdated or irrelevant theories, mental models on warfare, and failed decision-making methodologies that orient toward winning past wars with the expectation that this somehow improves our prospects on tomorrow's war. Potentially, military professionals with implicit design abilities may be rare and require special identification, selection, and utilization. Lefebvre mused on the difficulty in spreading design thinking across the Marine Corps as follows:

> There are not a lot of people able to do [design]. Making everyone a design expert may not be possible. But it is important to identify those people who have an intuitive, [General] Mattis-like ability with a history of reading and synthesizing, who are capable of using this process for these really complex problems.[14]

They are able to realize the tools they favor, drop them and experiment with new tools iteratively and in sophisticated ways.

The ability to drop tools is in design potentially more significant for change than the ability to use those legacy tools more efficiently, particularly if that efficiency is tied to a behavior or action that is only relevant retrospectively. If yesterday's complexity is no longer a threat for tomorrow, can we let go of constructs that may now be ossified or ritualized to never leave our grasp? Military design praxis, when executed in how Naveh and his fellow heretics found it most potent, sought military practitioners to gain the ability to

> learn to "unuse" weapons as much as one learns to use them, as if the power and cultivation of the affect were the true goal of the assemblage, the weapon being only a provisory means. Learning to undo things, and undo oneself, is proper to the war machine.[15]

The postmodern qualities expressed here would carry forward yet also create significant resistance in the spread of security design internationally.

Naveh's SOD praxis went through many changes over the last twenty-five years through experimentation, attempts at indoctrination, international applications, different methodological reorganizations, and plenty of discourse concerning war itself. Readers saw how Naveh, and his fellow practitioners and educators, would experiment, develop, expand, and inspire many different versions and variations of this new and different way of thought and action in the application of organized violence. However, many of the core lessons that Naveh stressed would not necessarily be adapted by his own Israeli Defense Forces at least directly into their current design doctrine, nor would many of them translate well into foreign militaries such as the U.S. Army, U.S. Marines, and elsewhere. Naveh is the first significant security design theorist in this shift from the twentieth century highly mechanized and industrial phase of modern war, and in this regard as the pathfinder for numerous design offspring around the world. Naveh will continue to be regarded as provocative, disruptive, and controversial. He is the first postmodern disruptor of war theory, the security heretic willing to deconstruct cherished belief systems to create vital cognitive space to reform them anew. Whether his core concepts remain, modify, or collapse under critical reaction are still a story in development.

In Closing: Long Live the Design Insurgency!

There are many different communities of security design praxis today and many are unaware of others, indifferent, or perhaps antagonistic in that they prefer that in security contexts we need only refer to their style of design to be successful in war. There are thriving schools that teach design, with many discovering security design praxis or integrating in different ways into how security organizations think, innovate, and transform. One does not have to attend a particular school or follow one set community's praxis to be a security designer, although often within a bounded community there will be differentiation between "one of us" and "outsider" like most any other group or discipline. There must be some universal qualities that most all formal and informal security designers might agree upon. It is these that we can conclude this chapter on. Designers must think reflectively and often at abstract levels that require some sort of philosophical acknowledgement; we must think about our thinking to learn about our learning.

To do this in any structured or unstructured way requires a relentless pursuit of an unattainable design ideal – the pursuit of "why". This can indeed frustrate and antagonize those that travel this path, leading to the counterweight argument of "we need to stop admiring the problem and get on with solutions!" Yet when we do this and revert right back to our preferred military conceptual models that rely upon select theories that employ the same warfare methodologies for decision-making, we tend to precisely solve the problems we feel we want to and ignore the ones we really need to try to tackle. On this illusive pursuit of the "why" in design inquiry and reflective practice, Army Special Forces Lieutenant Colonel Grant Martin shares a recollection on his own design team's failings in Afghanistan:

> Somehow the Design group had to be able to question underlying assumptions and that questioning had to be able to permeate out to the rest of the command. Underlying assumptions like questioning the motivations of those you are working with, why they are doing what they are doing, and why they aren't doing what you want them to. Assumptions like why we are there and what we are driving at. Assumptions like what "success" will look like, what our people will support, and what our politicians will accept. And assumptions about what drives people or groups of people to do what they do. We can't accept doctrine or popular psychology as dogma. We can't be attracted to the conventional wisdom of the day. We have to constantly question "why" we think something is the way it is.[16]

Military security organizations must transform to compete in future conflicts and complex security scenarios that likely are unlike what we have seen before, and incompatible with many of our legacy modes of logic for warfare. Not everything must change, and not every wall in the building need be knocked down. New designs will call for increased flexibility of the future organizations that attempt to provide security, safety, and extend societal and political desires on future states of reality into some realization of those goals through military design praxis. To do this, we must learn, *and not be afraid to learn about learning*. To accomplish this means that not just military decision-making methodologies and subsequent doctrine and practices need refinement, alteration, or elimination, but we must design new language, new metaphoric devices underneath of those new words so that new theories can express new mental models on security affairs as well as grow entirely new methodologies for how to link new ideas to new actions in warfare. It indeed calls for an intellectual revolution in military studies that will force institutions to reevaluate certain uncomfortable positions on what is valued in war, and what is not. Gal Hirsch provides an important distinction in why this will be a challenging endeavor for learned military organizations:

> One of the popular conclusions [on the IDF failings in the Second Lebanon War] was that the new terminology of operational art [Systemic Operational Design] was at fault. I was blamed, as before, for using an exalted language which, it was claimed, was not fit for the IDF. There were those who had waited for this chance to attack the field they had always despised and get rid of OTRI methodologies. I continue to insist that in a modern army there must always be the combination of "*safra v'sayfa*" -book and sword. High language, learning, research, and writing are virtues, not faults.[17]

Through the long reach of military history, those with bold new ideas would, if organization and luck were at their side in the moments when only new thinking could create the opportunities the organization desperately needed, design in warfare the necessary innovation that gave them just the advantage required at the hour of desire. If we only can perpetually repeat ritualized and institutionally sanctioned ideas and practices toward the conduct of applying organized violence in security affairs, we will achieve maximum certainty without understanding why reality responds only with more uncertainty in war. If we want to create a great future opportunity, we must acknowledge that the path to that development must involve quite a bit of destruction. It is understood that this statement itself appears brutal, even coldblooded. The meaning of "destruction" must be redesigned as well, something modern militaries tend to misinterpret due to the technical rationalized, systematic frames underpinned by pseudo-scientific beliefs.

This destruction has nothing to do with physical violence or bloodshed, and everything to do with our own ideas on warfare, organization, military decision-making, our belief systems, and those of others around us that are also swirling in this dynamic, every-changing, complex reality. Technology will not save us, nor will uniformity, increased control, analytical optimization, or more efficiency gains. The security demands of tomorrow's conflicts and challenges will be entirely of how we design them in comparison to whether rivals and adversaries might design differently and beyond our imagination. Creative destruction is designed, not planned. *Homo sapiens* are the ultimate destructive creature of this world, and war remains the paramount human creation that organizes violence for entirely conceptual desires and goals. Previously, modern militaries designed future wars in pseudo-scientific, technologically rationalized strategies and planning activities. Military design is a formalization of disrupting this, so that novel experimentation occurs to transform future militaries so they are relevant not in how past wars occurred, but the unrealized, unimagined ones of tomorrow.

Notes

1 Thomas Kuhn, *The Structure of Scientific Revolutions*, 3rd ed. (Chicago: University of Chicago Press, 1996), 151. Kuhn cites Max Planck, *Scientific Autobiography and Other Papers,* trans. F. Gaynor (New York, 1949): 33–34.
2 Peter Rowe, *Design Thinking,* Seventh (Massachusetts: MIT Press, 1998), 2.
3 US Special Operations Command Headquarters, "United States Special Operations Command White Paper: Design Thinking for the SOF Enterprise" (USSOCOM Headquarters, January 29, 2016).
4 One case study suggests this experimentation, but was conducted informally and only on one occasion. Martin experimented with some Special Forces students using SOD, others using ADM, and a control group using just traditional planning activities in the Green Beret final qualification exercise in North Carolina. See: Grant Martin, "Deniers of 'The Truth': Why an Agnostic Approach to Warfare Is Key," *Military Review* 95, no. 1 (February 2015): 42–51.
5 Edward Hayward to Ben Zweibelson, "Re: Design for Defense Book PDF Manuscript," September 21, 2021.
6 Carl Builder, *The Masks of War: American Military Styles in Strategy and Analysis* (Baltimore: John Hopkins University Press, 1989).
7 Ofra Graicer to Ben Zweibelson, "Re: Design for Defense Book PDF Manuscript," September 21, 2021.
8 Lorraine Daston, *Rules: A Short History of What We Live By* (Princeton, New Jersey: Princeton University Press, 2022).
9 Russell Ackoff, *Redesigning the Future* (New York: John Wiley & Sons, Inc, 1974), 30–32.
10 Douglas Lovelace, *Terrorism: Commentary on Security Documents Volume 141: Hybrid Warfare and the Gray Zone* (United Kingdom: Oxford University Press, 2016), 120.
11 Sebastian Gorka, "Adapting to Today's Battlefield: The Islamic State and Irregular War as the 'New Normal,'" in *Beyond Convergence: World Without Order,* ed. Hilary Matfess and Michael Miklaucic (Washington, D.C.: Center for Complex Operations; Institute for National Security Studies, 2016), 355.
12 Naveh and Graicer, Naveh audio 15OCT2019 34min22sec Day 2 morning. MP3, 26:21.
13 Bousquet, 929.
14 Paul Lefebrve, Interview with Major General (retired) Paul Lefebvre on Marine Design Theory 17 AUG 2021, mp3 Audio File, August 17, 2021, 01:00:35 to 01:01:04.
15 Deleuze and Guattari, A Thousand Plateaus, 400.
16 Grant Martin, "A Tale of Two Design Efforts [And Why They Both Failed In Afghanistan]," *Small Wars Journal,* July 7, 2011, 14.
17 Gal Hirsch, *Defensive Shield: An Israeli Special Forces Commander on the Front Line of Counterterrorism, the Inspirational Story of Brigadier General Gal Hirsch* (Jerusalem: Gefen Publishing House, Ltd, 2016), 394.

Index

For Product Safety Concerns and Information please contact our EU
representative GPSR@taylorandfrancis.com
Taylor & Francis Verlag GmbH, Kaufingerstraße 24, 80331 München, Germany